Molecular Plant Breeding

Molecular Plant Breeding

Molecular Plant Breeding

Edited by Bridget Harrington

SYRAWOOD
PUBLISHING HOUSE

New York

Published by Syrawood Publishing House,
750 Third Avenue, 9th Floor,
New York, NY 10017, USA
www.syrawoodpublishinghouse.com

Molecular Plant Breeding
Edited by Bridget Harrington

International Standard Book Number: 978-1-64740-066-8 (Hardback)

Cataloging-in-publication Data

Molecular plant breeding / edited by Bridget Harrington.
     p. cm.
Includes bibliographical references and index.
ISBN 978-1-64740-066-8
1. Plant molecular genetics. 2. Plant breeding. 3. Plants--Molecular aspects. I. Harrington, Bridget.
QK981.4 .M65 2022
572.865 2--dc23

TABLE OF CONTENTS

TABLE OF CONTENTS

Permissions

List of Contributors

Index

PREFACE

Plant breeding is the practice of growing plants with improved traits for achieving nutritional and economic advantage, and food security. Genes are the determiners of the qualitative and quantitative traits that a plant exhibits. While desirable characteristics can be achieved through selective propagation or crossing, modern plant breeding applies the techniques of molecular biology to select and insert the traits required. This application of molecular biology or biotechnology for plant breeding is known as molecular plant breeding. The important areas of molecular breeding are genomic selection and marker assisted selection, genetic engineering, QTL mapping or gene discovery and genetic transformation. Molecular plant breeding is an emerging field of science that has undergone rapid development over the past few decades. The various studies that are constantly contributing towards advancing technologies and evolution of this area of study are examined in detail in this book. It will help new researchers by foregrounding their knowledge in this emerging field.

The researches compiled throughout the book are authentic and of high quality, combining several disciplines and from very diverse regions from around the world. Drawing on the contributions of many researchers from diverse countries, the book's objective is to provide the readers with the latest achievements in the area of research. This book will surely be a source of knowledge to all interested and researching the field.

In the end, I would like to express my deep sense of gratitude to all the authors for meeting the set deadlines in completing and submitting their research chapters. I would also like to thank the publisher for the support offered to us throughout the course of the book. Finally, I extend my sincere thanks to my family for being a constant source of inspiration and encouragement.

Editor

PREFACE

Plant breeding is the practice of growing plants with improved traits for achieving nutritional and economic advantage, and food security. Genes are the determiners of the qualitative and quantitative traits that a plant exhibits. While desirable characteristics can be achieved through selective propagation or crossing, modern plant breeding applies the techniques of molecular biology to select and insert the traits required. This application of molecular biology or biotechnology for plant breeding is known as molecular plant breeding. The important areas of molecular breeding are genomic selection and marker assisted selection, genetic engineering, QTL mapping or gene discovery and genetic transformation. Molecular plant breeding is an emerging field of science that has undergone rapid development over the past few decades. The various studies that are constantly contributing towards advancing technologies and evolution of this area of study are examined in detail in this book. It will help new researchers by foregrounding their knowledge in this emerging field.

The researches compiled throughout the book are authentic and of high quality, combining several disciplines and from very diverse regions from around the world. Drawing on the contributions of many researchers from diverse countries, the book's objective is to provide the readers with the latest achievements in the area of research. This book will surely be a source of knowledge to all interested and researching the field.

In the end, I would like to express my deep sense of gratitude to all the authors for meeting the set deadlines in completing and submitting their research chapters. I would also like to thank the publisher for the support offered to us throughout the course of the book. Finally, I extend my sincere thanks to my family for being a constant source of inspiration and encouragement.

Editor

Genetic transformation of Tomato with three pathogenesis-related protein genes for increased resistance to Fusarium oxysporum f.sp. lycopersici

B. Dolatabadi[1,2*], Gh. Ranjbar[1], M. Tohidfar[3], A. Dehestani[2]

1. Department of Agronomy and Plant Breeding, Sari University of Agricultural Sciences and Natural Resources, Sari, Iran.
2. Genetics and Agricultural Biotechnology Institute of Tabarestan, Sari University of Agricultural Sciences and Natural Resources, Sari, Iran.
3. Agricultural Biotechnology Research Institute of Iran, Karaj, Iran.
   *Corresponding Author, Email: bandd_7159@yahoo.com

Abstract

Fusarium wilt caused by Fusarium oxysporum f.sp. Lycopersici is one of the major obstacles to the production of tomato which causes huge losses in tomato products worldwide. In order to increase the tolerance to this disease, a triple structure containing PR1, chitinase and glucanase genes controlled by 35S promoter was transferred to tomato. Eight days after planting on pre-culture medium, explants were inoculated by Agrobacterium tumefaciens strain LBA4404 containing the aforementioned plasmid. When the regenerated shoots grew to 2-3 cm, they were cut and transferred to rooting medium.The plantlets were then transferred to pots filled with a soil mixture of peat moss and perlite for further acclimatization. The putative transgenic plant lines were analyzed by multiplex PCR and the transcription of the transgenes was confirmed by RT-PCR method using the specific primers. The estimated value for the frequency of the simultaneous transfer of chitinase, glucanase and PR1 genes to tomato was 2.7%. Protein extracts of transgenic plants expressing chitinase, glucanase and PR1 genes inhibited in vitro hyphal growth of F. oxysporum f.sp. lycopersici. Compared with non-transgenic control plants, despite some alterations in chlorophyll content no other morphological changes were observed in transgenic plants. The total content of chlorophyll "a" and "b" in transgenic plants were 31.8 and 36.2 % higher than that of control plants, respectively, which may be attributed to metabolic changes due to simultaneous expression of three transgenes.
Key words: Chitinase, Fusarium oxysporum, Glucanase, PR1, Transgenic Tomato.

Introduction

Today increasing food production commensurate with population growth is one of the main objectives all over the world Reducing the yield loss caused by plant diseases have been focused in recent years and various technologies were implemented to achieve this goal (Giovanni *et al*, 2004). Tomato (*Lycopersicon esculentum*) is one of the world's most important crops due to the high value of its fruits both for fresh market consumption and numerous types of processed products (Giovanni *et al*, 2004). Pathogenic microorganisms such as viruses, bacteria, and specially fungi cause severe losses and drastic decrease in the annual fruit production (Barone & Frusciante, 2007). The *Fusarium* wilt disease causes loss of tomato crops worldwide and first described in 1895 by G.E. Massi (Jones & Woltz, 1981). This fungal disease has been reported from all geographical areas (Tanyolac & Akkale, 2010).

There are 3 common procedures for controlling fungal diseases: 1) agricultural operations, 2) utilization of chemical compounds and 3) using resistant varieties (Barone & Frusciante, 2007). *Fusarium oxysporum* f.sp. Lycopersici is a soil-borne pathogen, it remains in contaminated soils for several years. Therefore, controlling *Fusarium* wilt in the first two months of planting is very difficult and expensive. Meanwhile, generating resistant cultivars can play a significant role in increasing tomato production (Jones & Woltz, 1981). At the present time, one of the most prevalent strategies is producing transgenic plants which are resistant to fungous diseases. A group of plant-coded proteins induced by different stress stimuli, named "pathogenesis-related proteins" (PRs) are believed to have an important role in plant defense against pathogenes (Edreva, 2005). These proteins are commonly induced in resistant plants, expressing a hypersensitive necrotic response (HR) to pathogens of viral, fungal and bacterial origins (Van Loon, 1985). Toxicity of PRs can be generally accounted for their hydrolytic, proteinase-inhibitory and membrane-permeabilizing ability. Thus, hydrolytic enzymes(β-1,3-glucanases, chitinases and proteinases) can be valuable tools in weakening and decomposing of fungal cell walls (van Loon *et al*, 2006).

Chitinases are found in a wide range of organisms including bacteria, fungi and organic plants which play various physiological roles (Felse & Panda, 1999). Chitinase produced in microorganisms is reported as the main bio-control agent for different kinds of fungal diseases in plants (Freeman *et al*, 2004). Furthermore, β-1,3- glucanase which was well examined at physiological and molecular levels, plays an extensive role in defense reactions (Simmons *et al*, 1992). β-1,3-gluconase together with chitinase are expressed in response to pathogenic pollutions, wounds, ethylene treatment and chemical tensions (Li *et al*, 2001). The enzyme β-1,3- glucanase is able to decompose the available glucan on cellular wall of fungus and decrease the damages. On the other hand, this enzyme is able to exacerbate the activity of chitinase enzyme. Therefore, when these two enzymes are available in a transgenic

plant, the plant will be able to represent a better and permeated resistance in a wide spectrum of pathogenic fungi. In cereals, for example, it will display high resistance to diseases, including yellow rust, brown rust and powdery mildew (Selitrennikoff, 2001). Several studies on transgenic tobacco which contains chitinase and glucanase genes controlled by CaMV35S revealed that the growth of *Rhizoctonia solani* was reduced which demonstrated the fact that the genes were applied separately (Jach *et al*, 1995). The genes encoding bean chitinase and tobacco β-1,3 glucanase were introduced into the tomato line A53 (*Lycopersicon esculentum* cv.A53) via an Agrobacterium mediated transformation system. Transformants were obtained and confirmed by PCR and Southern blot analysis. The transgene copy numbers ranged between 1 and 8 copies. The foreign genes expression in the obtained transgenic plants showed resistance to *Fusarium* wilt disease (Bo *et al*, 2003). Regarding the importance of the proposed issue, this research was considered to produce transgenic tomato resistant against *Fusarium* disease with simultaneous transfer of three resistant genes including PR1, chitinase and β-1,3- glucanase.

Materials and Methods

Plant materials and growth conditions

Seeds of commercial tomato cultivars (Sheffellat) were provided by Agricultural and Natural Resources Research Center of Mazandaran. For sterilization, seeds were first immersed in 70% ethanol for 30 seconds and then rinsed by distilled water and incubated in 1% solution of sodium hypochlorite for 15 min. Finally, it was rinsed three times (each time for 3-5 min) by sterile distilled water and disinfected seeds were cultured on Ms-medium (Murashige & Skoog, 1962) including 30 gl^{-1} sucrose and 8 gl^{-1}agar for germination. Afterwards, they were incubated at 25°C and 16/8 light/dark photoperiod. The cotyledons were separated as explants from 8-day plantlets and were cultured on pre- treatment medium containing MS basal medium, 0.1 mgl^{-1} naphthalene acetic acid (NAA) and 0.1 mgl^{-1} 6-benzylaminopurine (BAP) and were incubated at 25°C under dark conditions for 72 hours.

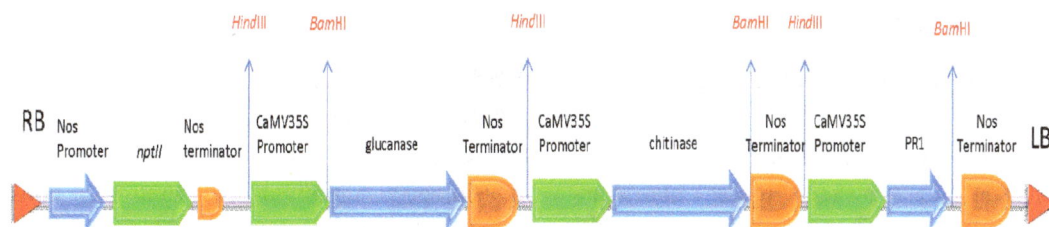

Figure 1. Schematic profile of plasmidic vector PBI121 Chi Glu PR1(+)

Gene Construct

Strain LBA4404 of *Agrobacterium tumefaciens* containing plasmid PBI121 ChiGluPRl (+) (Raufi *et al*, 2012) was used in this study. This plasmid contains three genes of chitinase, glucanase and PR1 with separate promoter (*CaMV*35s) and terminator (Nos). The selectable marker gene was neomycin phosphotransferase with Nos promoter and terminator (Figure1).

Preparation of bacterial suspension and inoculation of explants

Agrobacterium containing neo-compound plasmid was grown in Luria-Bertani (LB) medium containing 50mgl⁻¹kanamycin, 50 mgl⁻¹ rifampicin and 200 µM acetosyringone at 28°C with shaking (220 rpm). The explants which were previously incubated on pretreatment medium, were transferred to bacterial suspension and were shaken for 30 min at 28 °C. The explants were then blotted on sterile filter papers and were placed on a medium containing basal MS salts, 0.1 mgl⁻¹ NAA, 0.1 mgl⁻¹ BAP and 200 µM acetosyringone at 25°C and darkness for 48 hours.

Regeneration and selection of transgenic plants

After co-cultivation, infected slices were rinsed by MS medium and distilled water, which had an appropriate concentration of cefotaxime (500 mg/l), for Agrobacterium removal. They were then transferred to basal MS with 0.5 mgl⁻¹Indoleacetic acid (IAA), 0.5 mgl⁻¹ zeatin-riboside, 300 mgl⁻¹ cefotaxime and 25 mgl⁻¹ kanamycin. When the adventitious shoots grew to about 2-3 cm, they were transferred to rooting medium (basal MS with 200 mgl⁻¹ cefotaxime and 25 mgl⁻¹ kanamycin).

Molecular confirmation of probable transgenic plantlets with multiplex PCR

Genomic DNA was extracted from leaves using Dellaporta method (Dellaporta *et al*, 1983). To confirm the simultaneous integration of genes including chitinase, glucanase and PR1 in genomes of putative transgenic plants, multiplex polymerase chain reaction (multiplex PCR) was performed in 25 µl containing 1x PCR buffer, 0.2 mM dNTPs, 1.5 mM MgCl₂, 0.32 µM specific forward and reverse primers of both chitinase and glucanase genes, 0.4 µM specific forward and reverse primers of PR1 gene, 40ng of genomic DNA and 1 unit of Taq DNA polymerase (Sina Gene). Amplification consisted of 3 min at 94°C (initial denaturation), 35 cycles of 1 min at 94°C (denaturation), 1 min at 58°C (annealing), and 3 min at 72°C (extension) followed by 5 min at 72°C (Final extension). PCR products were separated in 1% agarose gel. For reliable screening, bacterial contamination was also checked by amplification of virG gene in PCR using virG gene specific primers. The virG containing plantlets were considered as false positive transgenic selection. The sequence of primers used in this reaction is as follows (Raufi*et al*, 2012):
R (chi) 5' GCCATAACCGACTCC AAGCA3'
F (chi) 5' GAGTGGTGTGGATGCTGTTG 3'
R (Glu) 5' TCTCCGACACCACCACCTTC 3'
F (Glu) 5' CA GGTCCAAGGGCATCAA CG 3'

R (PR1) 5' TTAGTATGGACTTTCGCC 3'
F (PR1) 5' GTCATGGGATTTGTTCTC 3'

Analysis of gene expression through RT-PCR

Total RNA was isolated from leaves of transgenic and non- transgenic tomato plants using Trizolr agent. Then the first strand cDNA was generated using the oligo (dT) by the "first strand cDNA synthesis kit" (Fermentas). PCR amplification was achieved using the first strand cDNA as template. This reaction was performed in a 25 µl containing 1x PCR buffer, 2 µl cDNA, 0.2 mM dNTPs, 1.5 mM MgCl$_2$, 1 unit Taq DNA polymerase and 0.4 µM each of primers. PCR was carried out as follows: an initial denaturation at 94°C for 3 min followed by 35 cycles of denaturation at 94°C for 1 min, annealing at 60°C for chitinase and glucanase genes and at 57°C for PR1 gene for 1 min, extension at 72°C for 1 min and a final extension at 72°C for 5 min. The PCR products were separated by electrophoresis on 1% (w/v) agarose gel.

Disk diffusion Bioassay

The effect of protein extracts of transgenic plants on the growth of *Fusarium oxysporum* f. sp. Lycopersici was studied on PDA media. The leaf material (700 mg) was grounded to a fine powder in liquid nitrogen using a mortar and pestle. About 750µl of extraction buffer (acetate sodium 100 mM, 2-mercaptoethanol 8 mM and phenylmethylsulfonyl fluoride 1mM, PH 6.5) was added to the leave powder. The extracts were then shaken for 1 h at 4°C and sub-sequently centrifuged at 13000 g for 15 min at 4°C. Protein concentrations were estimated using Bradford method (Bradford, 1976). A piece of agar including the fungal isolates was placed at the center of each of the PDA petri dishes. Petri dishes were then kept at 24°C and the six paper discs were placed in such a way that they surrounded agar segment symmetrically and the samples (containing proteins) were added to discs. They were again kept in incubator at 24°C. In order to compensate the reduction in enzymatic activity of samples during mainte-nance, a protein sample was again added to discs 18 hours later.

Studying morphological changes in transgenic plants

Transgenic plants were evaluated for probable physiological alterations compared with control plants. For this purpose, the chlorophyll levels of transgenic and control plants were measured as follows: About 0.5g of plant leaf was well crushed with 10 ml acetone. The solution was then filtered using filter paper and the volume reached to 25 ml. The absorption level was then measured at wavelengths of 662 and 645 respectively by spectrophotometer and the values of chlorophylls "a" and "b" were calculated using the following formula:

Chlorophyll a = 11.75 × A662 - 2.35 × A645
Chlorophyll b = 18.61 × A645 - 3.96 × A662

Results and Discussion

In current research, 82 among the 960 explants were regenerated into rootless green stems. When the plantlets grew to 2-3 cm, they were transferred to rooting medium (basic MS) and 40 of the 82 plantlets generated roots. They were transferred to the pots filled with a soil mixture of peat moss and perlite (Figure 2). Figure 3 shows the PCR analysis of the putative transgenic plants in the presence of chitinase, glucanase and PR1 genes. Two of the 25 plants were positive for all of the three genes (lanes 2 and 3 in Figure 3), however one plant was positive for chitinase and glucanase gene and negative for PR1 gene (lane 4 in Figure 3). Plasmid PBI 121 ChiGluPRl (+) was used as positive control and water was used as negative control (lane 6 and 1, Figure 3, respectively). A total of 25 putative transgenic plants were generated, out of which 24 plants contained chi, glu and pr1 genes and only one of the plants contained chi and glu genes (as determined by PCR). Chang *et al* (2002) reported the achievement of double transfer of chitinase and glucanase genes into pea genome at the level of 1.6%. In this study the rate of simultaneous transfer of chitinase, glucanase and PR1 genes was

Figure 2. a) Potential transgenic plantlet which produced root in regeneration medium. b) Potential transgenic plantlet after transferring on soil.

estimated as 2.7%. Nookarajuand Agrawal (2012) transferred chitinase and β-1, 3- glucanase genes from wheat to grape genome using agrobacterium to increase resistance to *Plasmoparaviticola* and observed that the transgenic plants demonstrated high levels of resistance to the pathogen.

When the resistance to a specific disease is conditioned by a single gene, resistance breaking events would frequently occur. Using a plasmid construct containing multiple resistant genes encoding for antifungal proteins under the control of individual and strong promoters would be an appropriate approach for producing fungal resistant transgenic plants (Mohsenpour *et al*, 2008).

The results indicated that one of the obtained transgenic plants contained chitinase and glucanase genes while PR1 gene was not present in this line. T-DNA transformation into plant cell is initiated at the right border and terminated at the left border (Mohseni Azar *et al*, 2012). The right border sequence promotes T-DNA transfer and integration (Gheysan *et al*, 1998). Therefore, gene sequences adjacent to the right border are more likely to be integrated into the host genome. The breakage probably happens more frequently in DNA regions away from the sequences of right border. Chen *et al* (1997) showed that 44 % of transgenic wheat lines carried incomplete T-DNA segments. Most of these breakages occur at the left border (Wu *et al*, 2006). Among 260 transgenic barley plants, only 3 percent had complete T-DNA (Bartlett *et al*, 2008). The deletion of some parts of T-DNA can interfere with the performance of the gene. In a study on transgenic plants, it was determined that 37.5 percent of transgenic plants have broken T-DNA segments (Hensel *et al*, 2012).

To verify the absence of agrobacterium in putative transgenic plants, polymerase chain reaction was performed using specific primers for vir G gene. Among the tested plants, three plants exhibited the 390 bp band corresponding to virG gene, demonstrating that the agrobacterium cells are present on plant tissues. Further analysis making were necessary to confirm whether they are real transgenics or false-positive results due to agrobacterium contamination. The absence of this band in other transgenic plants indicates that they are real transgenics with transgene integration. The results indicated that cefotaxime application did not remove the agrobacterium cells and the resulted adventitious shoots were somehow infected with these cells.

Several authors reported similar results when regenerating transgenic shoots after agrobacterium infection (Pena *et al*, 2010).

It was previously reported that the presence of bacterial colonies resistant to kanamycin in some tissues, especially at the cut zone of explants, reduces the antibiotic toxicity and allows the reproduction of non-transgenic cells in the selection medium (Dominguez *et al*, 2004).

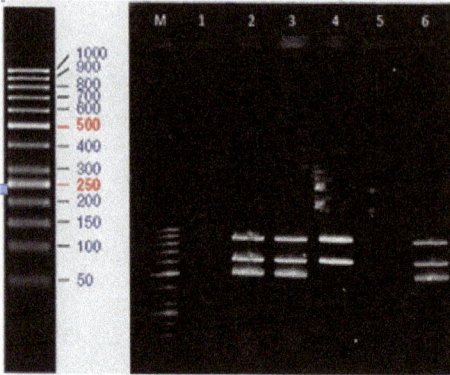

Figure 3. Polymerase chain reaction using the specific primers of genes including PR1, chitinase and glucanase. M: 50 bp DNA ladder, 1: negative control (water), 2 & 3: Transgenic plant containing all three genes, 4: transgenic plant containing two genes of glucanase and chitinase, 5: control plant , 6: positive control (Plasmid PBI121 ChiGluPR1 (+)).

Gene transcription analysis through RT-PCR

The presence of 872, 629 and 510 bp fragments indicates transcription of chitinase, glucanase and PR1 genes, respectively. The lanes 5, 9 and 15 are positive controls corresponding to chitinase, glucanase and PR1 genes, respectively. The lanes 1, 10 and 11 are negative controls. The lanes 4, 8 and 14 correspond to non-transgenic plants. The lanes 1 and 3 correspond to chitinase gene, lanes 6 and 7 are related to glucanase gene. The lanes 12 and 13 are related to PR1 gene (Figure 4).

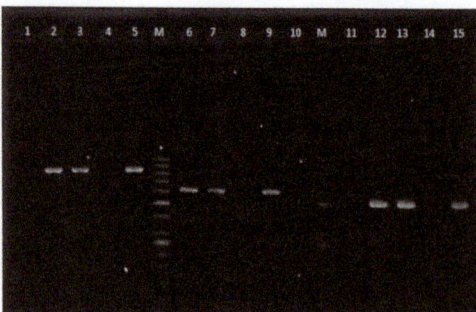

Figure 4: RT-PCR test for transgenic tomato lines with pBI 121 PR1 chi Glu. M: 50 bp DNA ladder , 5, 9 , 15 positive control related to genes including chitinase , glucanase and PR1 , 1 , 10 , and 11 negative control , 4 , 8 , 14 control plant , 2 and 3 related to chitinase gene, 6 and 7 related to glucanase gene, 12 and 13 related to PR1 gene.

Evaluation of antifungal activity

Inhibitory activity of recombinant chitinase, glucanase and PR1 proteins in the protein extracts of transgenic plants was evaluated on *Fusarium oxysporum* f. sp. lycopersici.

For this purpose, inhibitory effect of the protein extracts on the growth of fungal hyphae was assessed using PDA plates. Results showed that protein extracts containing recombinant proteins of the transferred genes, inhibited the growth of fungal hyphae (5 and 6, Figure 5).

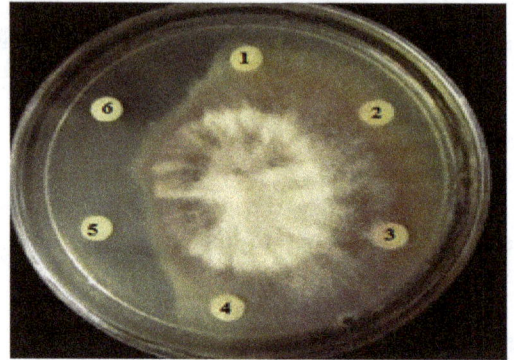

Figure 5: studying anti-fungous activity of protein extract of transgenic plant on the fungus *fusarium oxysporum* f . sp. Lycopersici, 1: 50 µg protein extract of non- transgenic plant, 2: 100 µg protein extract of non- transgenic plant , 3: 50 µl extraction buffer, 4: 100 µl extraction buffer, 5: 50 µg protein extract of transgenic plant, 6: 100 µg protein extract of transgenic plant.

Study of morphological changes in transgenic plants

Figure 6: Difference in chlorophyll values between transgenic plant and non- transgenic plant (control).

Phenotypically, transgenic plants were similar to non- transgenic plants. The only difference between these two plant types was the color of leaves. Obviously, the leaves of transgenic plants were darker (dark green) than those of non- transgenic plants (Figure 6).

To assess this alteration, the available chlorophyll in the leaves of transgenic and non- transgenic plants was measured.

In the leaves of non- transgenic and transgenic plants, the values of chlorophyll "a" were 15.41 and 20.31, respectively. On the other hand, the values of chlorophyll "b" were 5.66 and 7.71 for the leaves of non- transgenic and transgenic plants, respectively. The accumulation of PRs in the plants induces SAR (System of Acquired Resistance) genes (Ward *et al*, 1991). SAR phenomenon creates several morphological changes in the stressed plant (Ross, 1961). Some morphological and biochemical changes are created in relation to SAR phenomenon such as cellular death (Low & Merdia, 1996), increased synthesis of phytoalexins (Neuenschwader *et al*, 1996), accumulation of pathogenesis-related proteins (Jeun, 2000) and changes in chlorophyll value (Milavec *et al*, 2001). SAR phenomenon in broad bean plants increased chlorophyll value (Maggie *et al*, 2006).

Acknowledgements

The authors are grateful to Genetics and Agricultural Biotechnology Institute of Tabarestan (GABIT) for providing the instruments to perform this research.

References

1. Barone, A. and Frusciante, L. 2007. Molecular marker-assisted selection for resistance to pathogens in tomato, Marker-Assisted Selection, Current status and future perspectives in crops, livestock, forestryand fish. pp 153-164.
2. Bo, o., Han-Xia, L. and Zhi-Biao. Y. 2003. Increased resistance to fusarium wilt in transgenic tomato expressing bivalent hydrolytic Enzymes. Plant Physio & Mol Biol, 29: 179-184.
3. Bartlett, J. G., Alves, S. C., Smedley, M., Snape, J.W. and Harwood, W.A. 2008. High-throughput Agrobacterium - mediated barley transformation. Plant Meth, 4: 22.
4. Bradford, M. 1976. A rapid and sensitive method for the quantitation of micro gramquantities of protein utilizing the principle of protein dye binding. Anal Biochem, 72:248-254.
5. Chang, M. M., Culley, D., Choi, J. J. and Hadwiger, L .A. 2002. Agrobacterium- mediated co-transformation of a pea β-1, 3-glucanase and chitinase genes in potato (Solanumtuberosum L. c.v. Russet Burbank) using a single selectable marker. Plant Sci., 163: 83-89.
6. Cheng, M., Fry, J., Pang, E. S., Zhou, H., Hironaka, C. M., Duncan, D. R., Conner, T.W. and wan, Y. 1997. Genetic transformation of wheat mediated by Agrobacterium tumefaciens. Plant Physio, 115: 971-980.
7. Dellaporta, S. L., Wood, J. and Hicks, J. B. 1983. A plant DNA mini preparation: version II. Plant Mol Bio Rep, 1:19-21.
8. Dom´ınguez, A., Cervera, M., P´erez, R., Romero, J., Fagoaga, C., Cubero, J., L´opez, M.M., Ju´arez, J., Navarro, L. and Pe˜na, L. 2004. Characterisation of regenerants obtained under selective conditions after Agrobacterium- mediated transformation of citrus explants reveals production of silenced and chimeric plants at unexpected high frequencies. Mol Breeding, 14: 171–1.   183.
9. Edreva, A. 2005. Pathogensis-Related Proteins: research. Gen Appl Plant Physio, 31: 105-124.
10. Felse, P. A. and Panda, T. 1999. Regulation and cloning of microbial chitinase genes. Appl Microbiol Biotechnol, 51: 141-151.
11. Freeman, S., Minzm, O., Kolesnik, I., Barbul, O., Zveibil, A., Maymon, M., Nitzani, Y., Kirshner, B., Rav-David, D., Bilu, A., Dag, A., Shafir, S. and Elad. Y. 2004. Trichodermabiocontrol of Colletotrichumacutatum and Botrytis cinerea and survival in strawberry. E J P P, 110: 361-370.
12. Gheysen, g., Angenon, g. and Van Montagu, M. 1998. Agrobacterium mediated plant transformation: A scientifically intriguing story with significant application. In Transgenic Plant Research, Eds. Lindsey, K.,
13. Harwood Academic, A. M. Sterdam, pp:1-33.
14. Giovanni, C. D., Orco, P. D., Bruno, A., Ciccarese, F., Lotti, C. and Ricciardi, L. 2004. Identification of PCR-based markers (RAPD, AFLP) linked to a novel powdery mildew resistance gene (ol-2) in tomato. Plant Sci, 166: 41-48.
15. Hensel, G., Oleszczuk, S. S., Daghma, D. E., Zimny, J., Melzer, M. and Kumlehn,
16. J. 2012. Analysis of T-DNA integration and generative segregation in transgenic winter triticale (x TriticosecaleWittmack). BMC Plant Bio, 12:171.
17. Jach, G., Gornhardt, B., J. Mundy, J., Logemann, J., Pinsdorf, E., R leah, Schell, J. and Maas. C. 1995. Enhanced quantitative resistance against fungal disease by combinatorial expression of different barley antifungal proteins in transgenic tobacco. Plant J, 8: 97-109.
18. Jeun, Y. C. 2000. Immunolocazation of PR- pro-

tein P14 in leaves of tomato plants exhibiting systemic acquired resistance against Phytophthoriainfestanse induced by pre-treatment with 3-aminobutyric acid and preinoculation with Tobacco necrosis virus. J Plant Dis, 107: 352- 367.

19. Jones, J. P. and Woltz, S. S. 1981. Fusarium-incited disease of tomato and potato and their control. Pennsylvania State University Press, pp 157-168.

20. Li, W. L., Faris, J. D., Muthukrishnan, S.,

21. Liu, D. J., Chen, P. D. and Gill, B.S. 2001. Isolationand characterization of novel cDNA clones of acidic chitinasesand β-1,3-glucanases from wheat spikes infected by Fusariumgraminearum. Theor Appl Genet, 102: 353–362.

22. Low, P. S. and Merdia, J. R. 1996. The oxidative burst in plant defense function and signal transduction. Plant Physiol, 96: 533-542.

23. Maggie, H., Saidea, A., EI-Abbasi, I. H. and Mikhail, M.S. 2006. Inducing Resistance against Faba Bean Chocolate Spot Disease. Egypt J Phytopathol, 34: 69-79.

24. Milavec, M., Ravnikar, M. and Kovac, M. 2001. Peroxidases and photosyntthetic pigments in susceptible potato infected with potato virus YNTN. Plant Physio Biochem, 39: 891-898.

25. MohseniAzar, M., Nazeri, S., Malboobi, M. A., Ghadimzadeh, M., barzegari, M. and farokhzad. A. 2012. Optimizing Factors Affecting Agrobacterium Tumefaciens Mediated Transformation Of Dwarf Rootstock Of Apple (Malus Domestica Borkh Cv Gami Almasi). Agri biotech, 10: 7-15.

26. Mohsenpour, M., Babaeian, N. A., Tohidfar, M. and Habashi, A. A. 2008. Design and construction of four recombinant plasmid vectors containing chitinase, glucanase and BT genes, suitable for planttransformation. J Agric Sci Natural Resour, 15:41-47.

27. Murashige, T. and Skoog, F. 1962. A revised mediumfor rapid growth and bioassays with tobacco tissue cultures. Plant Physio, 15:473-497.

28. Neuenschwader, U., Lawton, K. and Ryals,

29. J. 1996. Systemic acquired resistance in plant-microbe interact. Chapman and Hall New York, 1: 81-108.

30. Nookaraju, A. and Agrawal, D. C. 2012. Enhanced tolerance of transgenic grapevines expressingchitinase and b- 1,3-glucanase genes to downy mildew. Plant Cell Tiss Organ Cult, 111:15–28.

31. Pena, L., Cervera, M., Fagoaga, C., Romero, J., Ballester, A., Soler, N., Pons, E., Rodr´ıguez, A., Peris, J., Ju´arez, J. and Navarro, L. 2010. Compendium of Transgenic Crop Plants, Volum 5: Transgenic Tropical and Subtropical Fruits and Nuts: Citrus fruit.

32. Raufi A., Tohidfar, M., Soluki, M. and Mohsenpour M. 2012. Isolation and Cloning of Two Genes from PR1 Family and Construction of Treble Plasmids Containing 3 Groups of Genes for Producing Transformed Plants Resistant to Fungal Diseases. J Agric Biotechnol, 2:27-46.

33. Selitrennikoff, C. P. 2001. Antifungal proteins. ApplEnvMicrobiol, 67: 2883-2894.

34. Ross, A. F. 1961. Systemic acquired resistance induced by localized virus in fection in plants. Virology, 14: 340-358. Simmons, C. A., Litts, J. C., Huang, N. and Rodriguez, R. L. 1992. Structure of arice β-glucanase gene regulated by ethylene, cytokinin, wounding, salicylic acid, and fungal elicitors. Plant Mol Bio, 18: 33– 45

35. Tanyolac, B. and Akkale, C. 2010. Screening of resistance genes to fusarium root rot and fusarium wilt diseases in F3 family lines of tomato (Lycopersicon esculentum)using RAPD and CAPs markers. AJB, 9: 2727-2730.

36. Van Loon, L.C. 1985. Pathogenesis-related proteins. Plant Mol Biol, 4: 111-116.

37. Van Loon, L. C., Rep, M. and Pieterse, C.

38. M. J. 2006. Significance of inducible defense-related proteins in infected plants. Phytopathol, 9: 342-431.

39. Ward, E.R., Uknes, S.J. Williams, S.C. Dincher,S.S. Widerhold, D.L. Alexander,D.C. Ahl-Goy, P. Métraux, J.P. and Ryals,J.A. 1991. Coordinate gene activity in response to agentsthat induce systemic acquired resistance. Plant Cell, 3: 1085-1094.

40. Wu, H., Sparks, C. A. and Jones, H. D. 2006. Characterisation of T-DNA loci and vector backbone sequences in transgenic wheat produced by Agrobacterium-mediated transformation. Mol Breeding, 18:195– 208.

Genetic analysis of castor (*Ricinus communis* L.) using ISSR markers

Farnaz Goodarzi[1], Reza Darvishzadeh[2,3]*& Abbas Hassani[1]

1. Department of Horticulture, Urmia University, Urmia, Iran.
2. Department of Plant Breeding and Biotechnology, Urmia University, Urmia, Iran.
3. Institute of Biotechnology, Urmia University, Urmia, Iran.
  *Corresponding Author, Email: r.darvishzadeh@urmia.ac.ir

Abstract

Castor (Ricinus communis L.) is one of the most ancient medicinal oil crops in the world. It has been vastly distributed in different parts of Iran. In the present study, the inter simple sequence repeat (ISSR) markers were used to evaluate the molecular genetic diversity among and within 12 castor accessions collected from 6 regions of Iran. Totally, 16 ISSR primers amplified 166 loci, of which 116 loci (69.89 %) were polymorph, indicating high genetic variability in castor germplasm. An accession-specific ISSR band was detected in '80-29'accession. Genetic distance among accessions ranged from 0.2 to 0.056. Analysis of molecular variance revealed a higher level of genetic variation within (80%) than between (20%) accessions. A model-based Bayesian approach subdivided 60 genotypes from 12 accessions into 6 subgroups. UPGMA dendrogram based on Nei's genetic distance classified 12 accessions into 4 groups. The result indicates that there was no association between geographical origin and ISSR patterns. The results suggest that ISSR technique is a useful tool for studying genetic diversity in castor germplasm.

Keywords: Bayesian clustering, Genetic diversity, ISSR, Ricinus communis.

Introduction

Castor is an economically and industrially important oil seed plant from Euphorbiaceae family. Castor (*Ricinus communis* L.) belongs to a monotypic genus, *Ricinus* and subtribe, *Ricininae* with chromosome number of 2n=2x=20 (62). It has a mixed mating system generating both selfed and cross fertilized offspring. Under natural conditions, cross pollination in castor can exceed 80% (54). Among the Euphorbiaceae family, castor is the only species which has lowest DNA c-value at which genome size of castor is around 350 mega base pairs (6). Castor seed contains 45-55% oil with unique properties compared to other vegetable oils. More than 80% of its oil content is an unusual hydroxyl fatty acid, ricinoleic acid (12-hydroxy-9-octadecenoic acid), which makes it acceptable for different chemical reactions (45, 30, 63). Ricinoleic acid is not found in other oil seed crops (30). Castor oil is extensively utilized in different industries such as automotive, nylons, lubricants, paints, cosmetics and medicine (43). Pharmacological researches indicated that castor oil is commonly used as a laxative and for induction of labor in pregnant women (11). Castor seed is a rich source of natural poison called ricin that makes it non-edible (4, 12). Castor is widely grown throughout tropical and subtropical regions of the world. It has been demonstrated that the greatest genetic diversity of castor is found in Ethiopia and east African regions (35, 7, 19). In western Asian, it is found in Iraq, Iran, Syria, Turkey and Armenia (3).

Availability of sufficient knowledge on genetic diversity and distributional pattern of genetic resources are necessary for plant breeding programs (28). The methods of genetic diversity study extend from analysis of morphological characters to biochemical and molecular traits (38). DNA markers have exhibited great ability

for estimating genetic diversity in plant species, because they provide direct evaluation of
distinct genetic material without environmental influences (42, 66). Different types of DNA
based molecular techniques such as polymerase
chain reaction (PCR)-based markers, hybridization-based markers and sequencing-based
markers can be used to analyze the genetic diversity of plant materials (26, 69).
Molecular markers such as SSR and AFLP (2),
SSR (5), SRAP (72), SNP (15), ISSR and RAPD
(16) have been successfully used to estimate the
genetic diversity in castor germplasm. ISSR is
a semi-arbitrary molecular marker system with
single forward primers of 16 to 18 nucleotide
length comprising repetitive units either anchored with 2 to 4 arbitrary nucleotides at the
3' or 5' end or non-anchored (18, 70, 48). As
no prior sequence knowledge is required, they
are more easily applied than SSR markers, and
are more reliable and robust than RAPD markers, possibly because the primers are longer and
PCR conditions are more stringent (17, 39, 47).
ISSR markers have been shown to be particularly useful in genetic fingerprinting and diversity analysis (1, 10, 34, 25, 67). ISSR is an ideal
method for fingerprinting and a useful alternative to single-locus or hybridization-based methods because large numbers of DNA fragments

are amplified per reaction, representing multiple
loci from across the genome (1, 10, 34).
Although Iran is considered as a center of genetic diversity for castor, unfortunately, there
is little information available on the genetic diversity of castor germplasms in Iran. The aim
of the present study was to evaluate the level
of genetic variation within and among castor
accessions originating from different regions
of country by ISSR markers. Actually, the results can generate basic information that could
be useful for castor breeding programs, such
as parental choice for developing heterotic hybrids or germplasm preservation.

Materials and methods

Plant materials
Seeds of 12 castor accessions originated from
6 different provinces were kindly provided by
Seed and Plant Improvement Institute (SPII)
Karaj, Iran (Table 1). The accessions were
planted and grown in a randomized complete
block design with three replications in Urmia
Agricultural Research Center (37° 44′ N; 45°
10′ E). Considering the mating system (54, 6)
and sampling coast, as some other studies (22,
36, 27), five plants from each accession were
randomly selected for ISSR analysis. Some
apical fresh leaves from each individual per

Table 1. List of the 12 castor (*Ricinus communis* L.) accessions collected from various regions of Iran.

Number	Gene bank code	Location	Latitude	Longitude	Altitude (m)
P1	80-23	Tafresh (Markazi province)	34° 24′	49° 43′	1735
P2	80-31	Ashtian (Markazi province)	34° 30′	50° 04′	2450
P3	80-25	Arak (Markazi province)	34° 20′	49° 49′	1753
P4	80-12-1	Sahreza (Isfahan province)	32° 11′	51° 37′	1750
P5	80-29	Toyserkan (Hamedan province)	36°30′	48° 16′	1910
P6	80-18	Taft (Yazd province)	31° 32′	54° 15′	2000
P7	80-16-1	Fasa (Fars province)	28° 58′	51° 41′	1382
P8	80-17	Ashtian (Markazi province)	32° 24′	50° 14′	1775
P9	80-7	Mehriz (Yazd province)	30° 05	54° 17′	1550
P10	80-11-1	Sahreza (Isfahan province)	32° 14′	51° 32′	1750
P11	80-4	Jiroft (Kerman province)	28° 40′	57° 44′	685
P12	80-22	Tafresh (Markazi province)	34° 27′	49° 38′	1727

accession were taken and immediately fixed in liquid nitrogen. The samples were stored at -80°C until DNA extraction.

Genomic DNA extraction

Genomic DNA was extracted from the young leaves according to Cetyltrimethylammonium bromide (CTAB) method described by Murray and Thompson (37), with some modifications. DNA quality was checked by running 1 µl DNA in 0.8% (w/v) agarose gels (Invitrogen, Carlsbad, CA, USA) in 0.5× TBE (45mM Tris base, 45mM boric acid, 1mM EDTA pH 8.0) buffer. DNA samples showing a smear on the gel were discarded. The DNA samples were diluted to 30 ng μl^{-1} in TE buffer and stored at 4°C for using as templates in polymerase chain reactions.

ISSR amplification

A total of 31 ISSR primers were examined for polymorphism on 2 individuals per accession. Sixteen primers out of 31 were selected based on clarity and reproducibility of bands for diversity analysis. PCR was carried out in a total volume of 20µl consisting of 30 ng DNA, 2µl

10× PCR buffer, 1 µM primer, 2.5 mM $MgCl_2$ (CinnaGen Co., Iran), 0.2 mM of each dNTP and 1 unit of *Taq* DNA polymerase (CinnaGen Co., Iran). Amplification was performed in a 96-well Mastercycler Gradient (Type 5331; Eppendorf AG, Hamburg, Germany). The amplification profile composed of an initial denaturation at 94°C for 4 min, followed by 36 cycles of 1 min denaturation at 94°C, annealing temperature (Table 2) for 45 s, extension at 72°C for 2 min, and a final extension at 72°C for 10 min. PCR products were electrophoresed on 1.6% agarose gels in 0.5× TBE buffer at 70 V for 3 h, stained with ethidium bromide (1.0 µg ml^{-1}) and photographed under UV light using a Gel-Doc image analysis system (Gel Logic 212 PRO, Carestream Health, Inc., Rochester, NY, USA). Molecular weight of amplified products was assessed with 100 bp DNA ladder.

Data analysis

The amplified fragments were scored independently as 1 or 0 for their presence or absence at each position, and the obtained binary

Table 2. Nucleotide sequences, annealing temperature, total number of loci, percentage of polymorphic loci and PIC per primer for ISSR primers used in the present study.

Primer name	Sequence(5'→3')	Annealing temperature(°C)	Total number of loci	Polymorphic loci	Percentage of polymorphic loci	PIC per primer
812	5'-(GA)8A-3'	48	8	4	50	0.449
816	5'-(CA)8T-3'	54	10	6	60	0.449
818	5'-(CA)8G-3'	51	7	5	71.43	0.361
822	5'-(TC)8A-3'	42.5	12	4	33.33	0.441
825	5'-(AC)8T-3'	50	9	7	77.78	0.307
834	5'-(AG)8YT-3'	50.3	11	7	63.64	0.369
836	5'-(AG)8YA-3'	51.5	14	10	71.43	0.330
840	5'-(GA)8YT-3'	42	7	3	42.8	0.298
844	5'-(CT)8RC-3'	47.5	12	8	66.67	0.336
848	5'-(CA)8RG-3'	47	10	9	90	0.516
849	5'-(GT)8CG-3'	55	10	9	90	0.386
857	5'-(AC)8YG-3'	58	13	9	69.23	0.366
859	5'-(TG)8RC-3'	44.5	14	14	100	0.394
864	5'-(ATG)6-3'	48.5	11	6	54.55	0.343
885	5'-HBH(AG)7-3'	53.5	11	9	81.82	0.388
A12	5'-(GA)6CC-3'	47	7	6	85.71	0.431
Total	-	-	166	116	69.88	-
Mean	-	-	10.375	7.25		-

Y = (C, T); R = (A, G); H = (A, C, T); B = (C, G, T). PIC: Polymorphism information content

Table 3. Characteristics of amplified ISSR loci on the studied castor (*Ricinus communis* L.) accessions.

Accession	No. Bands	No. Bands Freq. >= 5%	No. Private Bands	No. LComm Bands (<=25%)	No. LComm Bands (<=50%)	Na ± (SD)		Ne ± (SD)		He ± (SE)		H ± (SD)		I ± (SD)		Number of polymorphic loci	Percentage of polymorphic loci
P1	144	144	0	4	6	1.307	0.463	1.207	0.350	0.117	0.015	0.117	0.189	0.189	0.271	51	30.72 %
P2	148	148	0	2	6	1.301	0.460	1.228	0.379	0.125	0.015	0.125	0.200	0.181	0.284	50	30.12 %
P3	142	142	0	2	3	1.307	0.463	1.202	0.342	0.116	0.014	0.116	0.186	0.186	0.186	51	30.72 %
P4	140	140	0	0	2	1.325	0.470	1.243	0.384	0.134	0.016	0.134	0.203	0.194	0.288	54	32.53 %
P5	152	152	1	2	9	1.283	0.452	1.197	0.353	0.110	0.015	0.110	0.187	0.162	0.268	47	28.31 %
P6	143	143	0	0	5	1.247	0.433	1.177	0.338	0.099	0.014	0.099	0.181	0.144	0.260	41	24.70 %
P7	152	152	0	0	9	1.349	0.478	1.232	0.357	0.133	0.015	0.133	0.193	0.196	0.279	58	34.94 %
P8	146	146	0	0	4	1.374	0.485	1.278	0.393	0.154	0.016	0.154	0.209	0.224	0.298	62	37.35 %
P9	148	148	0	2	5	1.337	0.474	1.244	0.379	0.136	0.016	0.136	0.201	0.198	0.287	56	33.73 %
P10	142	142	0	0	2	1.325	0.470	1.243	0.385	0.134	0.016	0.134	0.203	0.194	0.289	54	32.53 %
P11	149	149	0	2	6	1.265	0.443	1.174	0.326	0.100	0.014	0.100	0.177	0.148	0.256	44	26.51 %
P12	142	142	0	0	4	1.247	0.433	1.163	0.316	0.094	0.013	0.094	0.173	0.139	0.251	41	24.70 %

No. Bands = Number of different bands. No. Bands Freq. >= 5% = Number of different bands with a frequency >= 5%. No. Private Bands = Number of bands unique to a single population. No. LComm Bands (<=25%) = Number of locally common bands (Freq. >= 5%) found in 25% or fewer populations. No. LComm Bands (<=50%) = Number of locally common bands (Freq. >= 5%) found in 50% or fewer populations. na = Observed number of alleles. ne = Effective number of alleles. He = Expected heterozygosity = 2 × p × q. H = Nei's gene diversity. I = Shannon's Information index. SD= Standard deviation. SE= Standard error.

data matrix was used for analysis. Number of markers per primer, percentage of polymorphic markers, number of marker with a frequency greater than or equal to 5%, number of private marker and number of less common marker with frequency lower than or equal to 25 and 50% (32) were calculated in the GenAlEx version 6 software (44). The PIC value for each ISSR primer was calculated as proposed by Roldan-Ruiz et al. (51): $PIC_i = 2f_i(1 - {}_if_i)$ where PIC is the polymorphism information content of marker i, fi is the frequency of marker fragments that were present, and 1-fi is the frequency of marker fragments that were absent. PIC was averaged over the fragments for each primer.

Genetic diversity within and among populations was measured as the percentage of polymorphic bands per population, Nei's gene diversity (40), Shannon's information index (56), Gst (40) and gene flow (Nm) among populations (59) and Nei's unbiased genetic distance, all of which were measured using POPGENE program, (ver. 1.32) (68). AMOVA (14) was performed to estimate the variance within and among castor accessions in the GenAlEx software.

The genetic similarity between individuals (12 accessions×5 individuals per accession= 60 individuals) was calculated using Dic similarity coefficient (41). Dendrogram was constructed by complete linkage method using NTSYS-pc software (ver. 2.11) (50). The results of cluster analysis were corroborated by principal coordinate analysis (29) visualizing the relationships among the studied genotypes. Genetic distances between accessions were calculated by Nei's genetic distance in the POPGENE software. The genetic distances were employed for construction of phylogenetic tree in the POPGENE software.

Population structure was analyzed using a model-based Bayesian approach in the software package Structure 2.3.4 (46). Five independent runs of K= 1-10 were performed, assuming an admixture model and correlated allele frequencies with 5000 Markov chain Monte Carlo (MCMC) iterations and a burn-in period of 10,000. The K value was determined by the log likelihood for each K; Ln P(D)= L(K) (52). Since the distribution of Ln P(D) did not show a clear number of true K, delta K (ΔK) based on the second order rate of change in the likelihood (ΔK) (13) was used alternatively to identify a clear peak to represent the true K value. An individual was discretely assigned to a subpopulation when more than 70% of its genome composition came from that subpopulation (9).

Results

Sixteen out of 31 ISSR primers produced clear and reproducible fragment patterns across the studied accessions. Sixteen primers amplified a total of 166 loci, of which 116 (69.88%), were polymorph. The number of polymorphic markers per primer ranged from 3 to 14 with an average of 7.25. The highest level of polymorphism was observed by UBC-859 (100%), while the primer UBC-822 produced the lowest level of polymorphism (33.33%). Amplified fragments ranged from 230 to 2800 bp in size. PIC of primers ranged from 0.298 (840) to 0.516 (848) (Table 2). Characteristics of the amplified ISSR markers in each accession were described in Table 3. The range of polymorphic loci was between 24.70% in '80-22' accession to 37.35% in '80-17'. One unique ISSR loci was observed in '80-29' accession. The mean observed number of alleles (Na) ranged from 1.247± 0.433 in '80-22' and '80-18' accessions to a maximum of 1.374± 0.485 in '80-17' accession. Values of the effective number of alleles (Ne) were less than those for (Na) with regard to every population and ranged from 1.163± 0.316 in '80-22' to 1.278 ± 0.393 in '80- 17'. The heterozygosity among the accessions ranged from 0.094 in '80-22' to 0.154 in '80-17'. The mean Nei's gene diversity (H) ranged from 0.094 ± 0.173 in '80-22' to 0.154± 0.209 in '80-17' (Table 3). The Shannon's indices (I) ranged from 0.139± 0.251 in '80-22' to 0.224± 0.298 in '80-17'. Total gene diversity (Ht) and gene diversity within population (Hs) were 0.232± 0.037 and 0.121± 0.015, respectively. The coefficient of gene differentiation (Gst) among populations was 0.479. Based on the Gst value, the mean estimated number of gene flow (Nm) between populations was found to be 0.544 (Table 4). According to AMOVA, the genetic variation was mainly within accessions (82%) rather than among accessions (Table 5). Nei's genetic distances based on ISSR analysis were

Table 4. Overall genetic variability across all studied castor (*Ricinus communis* L.) accessions.

Sample size	Na ± (SD)	Ne ± (SD)	H ± (SD)	I ± (SD)	Ht ± (SD)	Hs ± (SD)	Gst	Nm
56	1.711 (0.457)	1.402 (0.389)	0.231 (0.202)	0.347 (0.284)	0.232 (0.037)	0.121 (0.015)	0.479	0.544

na = Observed number of alleles. ne = Effective number of alleles. H = Nei's (1973) gene diversity. I = Shannon's Information index. Ht = Total genetic diversity. HS = Gene diversity within populations. Gst = Coefficient of genetic differentiation among populations. Nm = Estimate of gene flow from Gst or Gcs. e.g., Nm = 0.5(1 - Gst)/Gst.

Table 5. Analysis of molecular variance (AMOVA) for Iranian castor (*Ricinus communis* L.) accessions based on ISSR data.

Source	df	SS	MS	Est. Var.	%	Stat	Value	P(rand >= data)
Among Accessions	11	516.967	46.997	4.963	18%	PhiPT	0.183	0.001
Within Accessions	48	1064.800	22.183	22.183	82%			
Total	59	1581.767		27.146	100%			

df: degree of freedom. SS: sum of square. MS: mean of square.

computed among accessions (Table 6). The genetic distance values ranged from 0.2 ('80-29' and '80-11-1' accessions) to 0.056('80-23' and '80-25' accessions).

The dendrograms representing relationships between 60 individuals (12 accessions × 5 individuals per accession = 60 individuals) and 12 accessions are shown in Figures 1 and 2, respectively. The dendrogram of individuals (Figure 1) did not divide the individuals into distinct groups resembling the geographically-defined accessions. Generally, individuals were evenly distributed along the dendrogram, revealing high intra-accession genetic diversity. According to population's dendrogram, all the castor accessions were categorized into 4

major groups (Figure 2). Group I included '80-23', '80-31', '80-25' and '80-12-1' accessions. Group II contained '80-7', '80-11-1' and '80-22' accessions. Group III comprised '80-18', '80-16-1' and '80-17' accessions. Group IV included '80-29' and '80-4' accessions from Hamedan and Kerman provinces, respectively. Principal coordinates analysis (PCoA) showed that the first three Eigen-values explained 26.45% of the cumulative variation which were then plotted to identify the diversity of the genotypes (Figure 3). The result of principal co-ordinate analysis (PCoA) was partially in accordance with the cluster analysis. In order to understand the genetic structure of the studied panel, a model-based Bayesian approach in

Table 6. Nei's genetic identity (above diagonal) and genetic distance (below diagonal) among 12 accessions of castor (*Ricinus communis* L.) based on ISSR data.

	P1	P2	P3	P4	P5	P6	P7	P8	P9	P10	P11	P12
P1		0.898	0.946	0.923	0.824	0.844	0.859	0.851	0.860	0.853	0.826	0.844
P2	0.107		0.894	0.892	0.835	0.854	0.876	0.860	0.869	0.860	0.853	0.876
P3	0.056	0.112		0.927	0.834	0.855	0.878	0.867	0.879	0.870	0.824	0.859
P4	0.080	0.114	0.076		0.849	0.842	0.863	0.848	0.872	0.859	0.822	0.840
P5	0.194	0.180	0.181	0.164		0.843	0.845	0.835	0.831	0.819	0.893	0.837
P6	0.170	0.158	0.157	0.172	0.171		0.920	0.887	0.846	0.846	0.846	0.835
P7	0.152	0.133	0.130	0.148	0.169	0.084		0.905	0.907	0.867	0.862	0.868
P8	0.162	0.151	0.143	0.165	0.181	0.121	0.100		0.851	0.833	0.820	0.846
P9	0.151	0.141	0.129	0.137	0.185	0.168	0.097	0.162		0.913	0.860	0.887
P10	0.159	0.151	0.140	0.152	0.200	0.168	0.143	0.182	0.091		0.879	0.879
P11	0.191	0.159	0.194	0.197	0.113	0.167	0.149	0.199	0.151	0.129		0.867
P12	0.169	0.132	0.152	0.174	0.178	0.180	0.142	0.167	0.120	0.129	0.143	

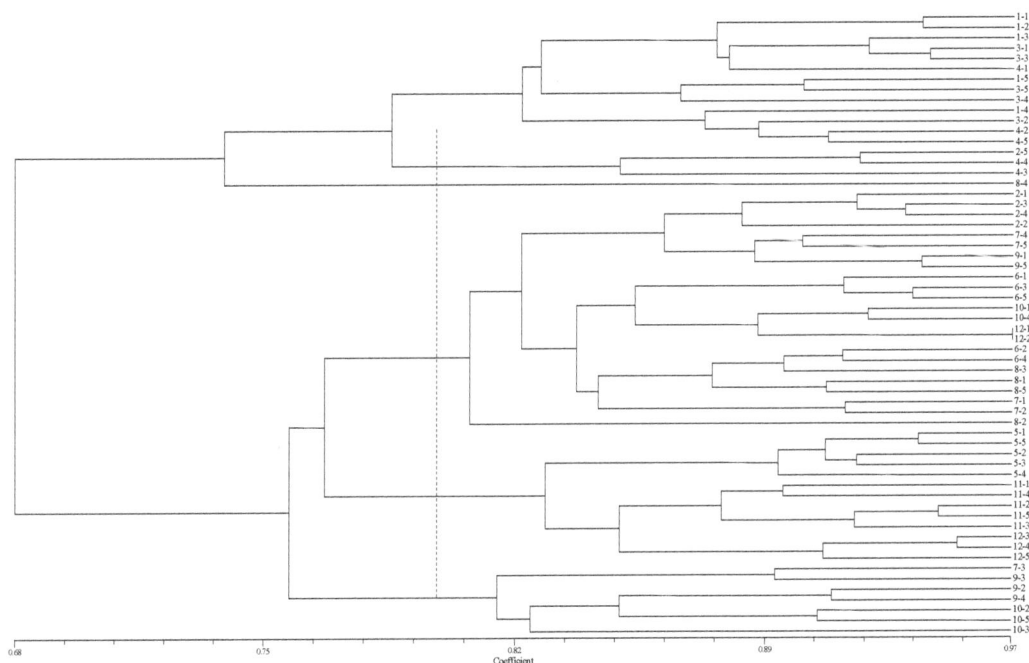

Figure 1. Dendrogram of 60 castor genotypes based on Dic similarity index.

the Structure software was used to assign each genotype to the corresponding subgroup. 21 out of 60 castor genotypes were partitioned into six subgroups and the remaining ones were categorized as mixed based on their Q values (the memberships of individuals to specific groups) (Figures 4).

Discussion

In the present study, the genetic diversity and relationships among 12 *R. communis* accessions coming from different regions of Iran were investigated using ISSR markers. The high reproducibility of ISSR markers may be because of using longer primers and higher annealing temperature than those used for RAPD. Based on its unique characters, ISSR technique can detect more genetic loci than isozyme and has higher stability than RAPD (72). ISSR markers have been used for studying genetic diversity in several plant species, such as *Coptis chinensis* (58), *Nelumbo nucifera* (8), *Asparagus acutifolius* (57), and *Solanum melongena* (24). The average level of polymorphism, revealed by 16 ISSR primers, was 69.88% indicating a high level of polymorphism. High level of genetic variability in *R. communis* germplasms

has also been revealed by SSR (5), ISSR and RAPD (16) and SRAP (71) markers. In contrast to these results, Allan *et al.* (2) by using SSR and AFLP markers and Foster *et al.* (15) by using SNP markers identified low level of genetic variation in *R. communis* gremplasms.

Based on the Shannon's information index, genetic diversity in '80-17' accession is higher than the others and that the '80-22' accession had the lower one. The high level of heterozygosity was observed in '80-17' accession (from Ashtian/Markazi province) and the low level of heterozygosity was observed in '80-22' accession (from Tafresh/Markazi province).

The accession with high level of genetic diversity can be considered as a good genetic resource for parental selection in castor breeding programs. AMOVA was used for partitioning the total genetic variation among and within *R. communis* accessions (14). Based on AMOVA analysis, genetic variation within *R. communis* accessions was higher than among accessions. Allogamous, self-incompatible and cross-pollinated species are essentially explained by more within accessions level of genetic variation (20, 53), whereas self-pollinated plants

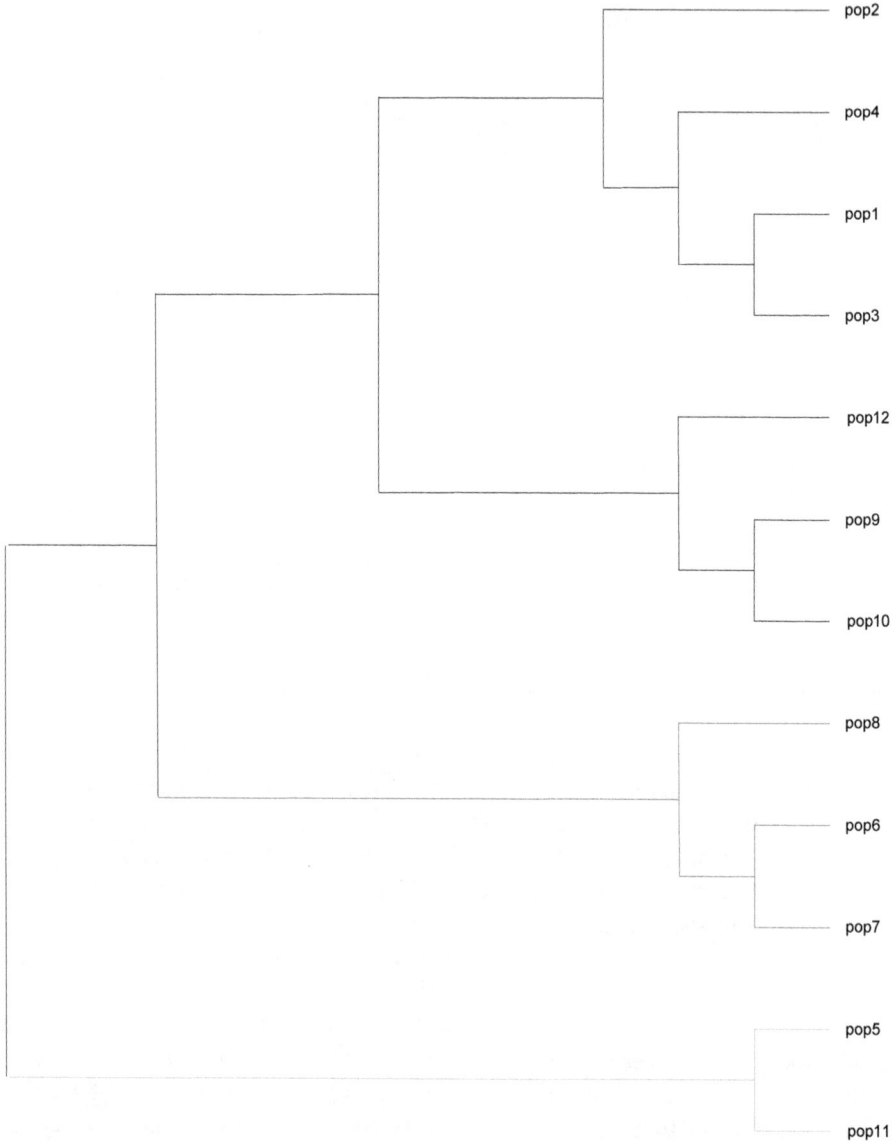

Figure 2. Dendrogram of 12 castor accessions based on Nei genetic distance calculated from ISSR data.

or plants with vegetative reproduction are often presented by more among accessions level of genetic variation (61). Similar results have been reported in other cross-pollinated species such as *Rheum tanguticum* (23) and *Murraya koenigii* (64). Genetic variation is due to numerous factors, including mating system, gene flow, genetic drift, seed and pollen scattering, human activities, long-term evolutionary history, natural selection and breeding systems (55, 21). Considering the importance of medicinal plants in our country, we believe that human activities (seed transformation) and cross pollination are possible reasons in shaping Iranian *R. communis* genetic diversity and structure. In the present study, '80-29' accession from Toyserkan/Hamedan province produced unique band. ISSR marker is one of the available techniques for identification of unique band in plant materials (69). In several studies unique ISSR band in plant species like *Dendrobium* (65), *Citrus indica* (33), *Chimonanthus praecox* (70), and *Populus cathayana* (31) has been identified. Generally, the unique bands could be

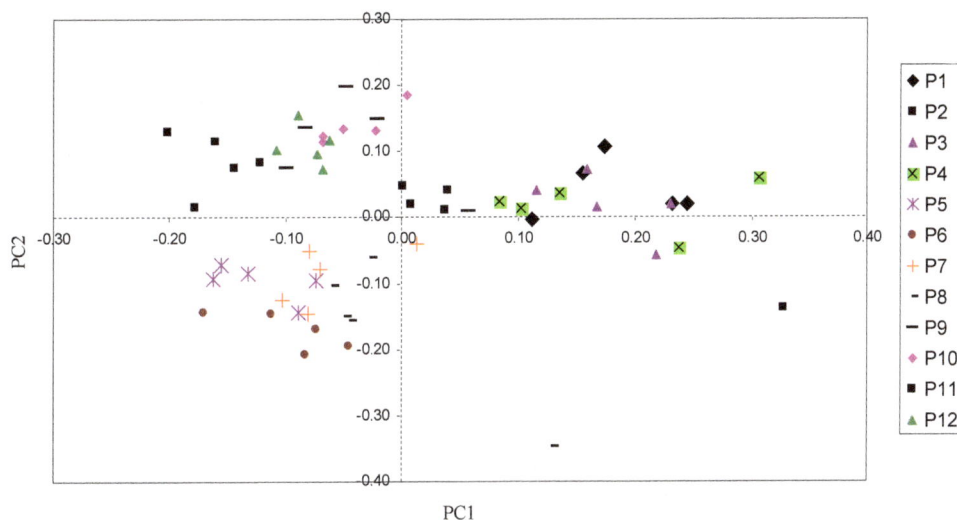

Figure 3. Two dimensional plot of the genetic relationship among 60 castor genotypes as revealed by principal co-ordinate analysis (PCoA).

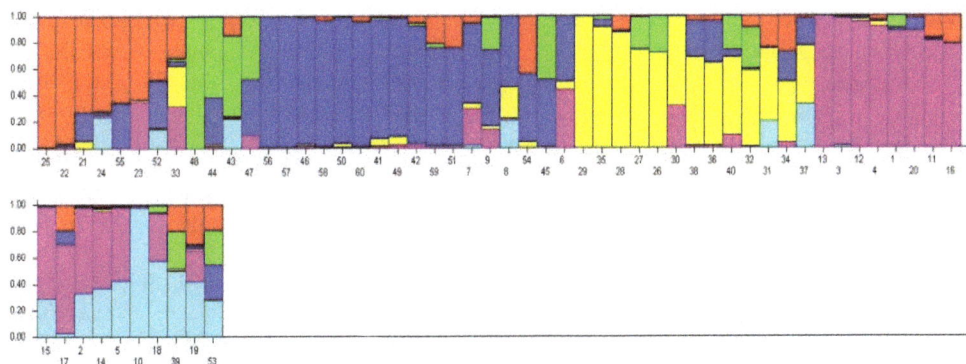

Figure 4. Genetic relatedness of 60 individuals from 12 castor (*Ricinus communis* L.) accessions with 16 inter simple sequence repeats (ISSRs) as analyzed by the Structure program. Numbers on the y-axis indicate the membership coefficient (Q) and on the x-axis indicate the individual's number. Genotypes 1-5 sampled from Tafresh accession (Markazi province); 6-10: Ashtian accession (Markazi province); 11-15: Arak accession (Markazi province); 16-20: Shahreza accession (Isfahan province); 21-25: Toyserkan accession (Hamedan province); 26-30: Taft accession (Yazd province); 31-35: Fasa accession (Fars province); 36-40: Ashtian accession (Markazi province); 41-45: Mehriz accession (Yazd province); 46-50: Shahreza accession (Isfahan province); 51-55: Jiroft accession (Kerman province); 56-60: Tafresh accession (Markazi province). Individuals with the same color belong to the same subgroup. Red: genotypes from Toyserkan accession from Hamedan province with cold and semidry climate; Green: genotypes from Shahreza accession (Isfahan province); Blue: genotypes from Tafresh accession (Markazi province with hot and dry climate); Yellow: genotypes from Taft and Fasa accessions (Yazd and Fars provinces, respectively); Pink: genotypes from Tafresh and Arak accessions (Markazi province); Aqua: genotypes from Ashtian accession (Markazi province).

transformed as distinct fingerprint into STS (sequence-tagged site) and SCAR (sequence characterized amplified regions) markers to develop the species specific marker for the best management and accurate identification of the plant materials. Reports on unique band using molecular methods are very little in the castor germplasm.

Allan *et al.* (2) reported two unique SSR alleles

but no unique AFLP allele in some castor accessions from USDA collection.

The distribution of 60 individuals from 12 *R. communis* accessions across respective groups did not reflect the geographic origins. The high degree of genetic variation revealed in dendrogram may be due to higher level of cross pollination. In view of population level, the studied castor accessions were categorized into

4 major clusters. The lowest genetic distance (0.056) was observed between '80-23' and '80-25' accessions both from Markazi province, which might be attributed to having similar ecological environment and short geographical distance.

The highest genetic distance (0.2) was observed between '80-29' accession from Toyserkan/Hamedan province and '80-11-1' accession from Shahreza/Isfahan province which might be attributed to having long geographical distance. In constructed dendrogram *R. communis* accessions coming from the same geographical location (province) were distributed in various groups. For instance, the accessions from Markazi province were distributed into 3 distinct groups. Moreover, '80-29' accession from Hamedan was classified with '80-4' accession from Kerman which is unexpected because they are geographically far distant locations. Consequently, the clustering pattern was not obvious according to geographic origin of *R. communis* accessions.

Effective analysis of the population structure and accurately classifying of individuals to appropriate subpopulations were performed by Bayesian method in the Structure software. This clustering method is based on the allocation of individual genotypes to K clusters in such a way that Hardy–Weinberg equilibrium and linkage equilibrium are valid within clusters, whereas these kinds of equilibrium are absent between clusters. Maximum value of ΔK was observed in K=6, so the studied castor genotypes probably have 6 subpopulations (Figure 4). This finding indicates existence of acceptable genetic variation and ideality of studied germplasm for association mapping in caster.

The present study revealed the utility of ISSR technique as a molecular diagnosis tool to characterize the genetic diversity, and determined the pattern of genetic structure within and among different Iranian *R. communis* accessions. The results revealed rather high level of genetic polymorphism (68.88%) and wide genetic variation within accessions, which offer valuable information for conservation and management of genetic resources and utilizing them in castor breeding programs. The genetic distance between populations is a valuable parameter to conserve and use a given germplasm in breeding activities. It was proved that crosses between unrelated parents and genetically away will show more power hybrid than crosses between genotypes closely related (49, 60). Furthermore, our results revealed population structure in present germplasm that should be taken into account to carry out unbiased association mapping in castor improvement programs.

Acknowledgements

The authors are grateful for the support provided by Institute of Biotechnology, Urmia University, Iran.

References

1. Aghaei, M., Darvishzadeh, R. and Hassani, 2012. Molecular Characterization and similarity relationships among Iranian basil (*Ocimum basilicum* L.) accessions using inter simple sequence repeat markers. Rev Ciênc Agron, 43: 312-320.
2. Allan, G., Williams, A., Rabinowicz, P.D., Chan, A., Ravel, P. and Keim, J.P. 2008. World-wide genotyping of castor bean germplasm (*Ricinus communis* L.) using ALFP and SSRs. Genet Resour Crop Evol 55: 365–378.
3. Anjani, K. 2012. Castor genetic resources: A primary gene pool for exploitation. Ind Crop Prod, 35: 1–14.
4. Audi, J., Belson, M., Patel, M., Shier, J. and Osterloh, J. 2005. Ricin poisoning— a comprehensive review. J Am Med Assoc, 294: 2342–51.
5. Bajay, M.M. 2010. Development of microsatellite markers and characterization of germplasm of castor (*Ricinus communis* L.). Dissertation, Universidade de Sao Paulo-USP, Brazil.
6. Chan, A.P., Crabtree, J., Zhao, Q., Lorenzi, H., Orvis, J., Puiu, D., Melake-Berhan, A., Jones, K.M., Redman, J., Chen, G., Cahoon, E.B., Gedil, M., Stanke, M., Haas, B.J., Wortman, J.R., Fraser-Liggett, C.M., Ravel, J. and Rabinowicz, P.D. 2010. Draft genome sequence of the oilseeds species *Ricinus communis*. Nat Biotechnol, 28(9): 951-956.
7. Carter, S. and Smith, A.R. 1987. Euphorbiaceous Flora of Tropical East Africa. A.A., Balkema Publishers, Rotterdam, Netherlands.
8. Chen, Y., Zhou, R., Lin, X., Wu, K., Qian, X. and Huang, S. 2008. ISSR analysis of genetic diversity in sacred lotus cultivars. Aquat Bot, 89: 311–316.
9. Courtois, B., Audebert, A., Dardou, A., Roques, S., Ghneim- Herrera, T., Droc G., Frouin, J., Rouan, L., Gozé, E., Kilian, A., Ahmadi, N. and Dingkuhn, M. 2013. Genome- wide association mapping of root traits in a Japonica rice panel.

PLoS ONE, 8(11): e78037. doi: 10.1371/journal. pone.0078037.

10. Dashchi, S., Abdollahi Mandoulakani, B., Darvishzadeh, R. and Bernoosi I. 2012. Molecular similarity relationships among iranian bread wheat cultivars and breeding lines using ISSR markers. Not Bot Horti Agrobo, 40(2): 254-260.

11. Duke, J.A. 1998. *Ricinus communis*. from Purdue University New Crop Resource Online Program. http: // www.hort.purdue . edu/newcrop/duke_energy/Ricinus_ communis.html.

12. Endo, Y.K., Mitsui, K., Motizuki, M. and Tsurugi, K. 1987. The mechanism of action of ricin and related toxic lectins on eukaryotic ribosomes: the site and the characteristics the modification in 28S ribosomal RNA caused by the toxins. J Biol Chem, 262: 5908–5917.

13. Evanno, G., Regnaut, S. and Goudet, J. 2005. Detecting the number of clusters of individuals using the software STRUCTURE: a simulation study. Mol Ecol, 14: 2611–2620.

14. Excoffier, L., Smouse, P.E. and Quattro, J.M. 1992. Analysis of molecular variance inferred from metric distances among DNA haplotypes: Application to human mitochondrial DNA restriction data. Genet, 131: 479-491.

15. Foster, J., Allan, T., Chan, G.J., Rabinowicz, A.P., Ravel, P.D., Jackson, J. and Keim, P.J.P. 2010. Single nucleotide polymorphism for assessing genetic diversity in castor (Ricinus communis L.). BMC Plant Biol, 10: 13.

16. Gajera, B.B., Kumar, N., Singh, A.S. and Punvar, B.S. 2010. Assessment of genetic diversity in castor (*Ricinus communis* L.) using RAPD and ISSR markers. Ind Crop Prod, 32: 491-498.

17. Godwin, L.A., Aitken, E.A. and Smith L.A. 1997. Application of inter simple sequence repeat (ISSR) markers to plant genetics. Electrophor, 18: 1524-1528.

18. Gupta, M., Chyi, Y.S., Romero-Severson, J. and Owen, J.L. 1994. Amplification of DNA markers from evolutionarily diverse genomes using single primers of simple-sequence repeats. Theor Appl Genet, 89: 998–1006.

19. Govaerts, R., Frodin, D.G. and Radcliffe-Smith, A. 2000. World checklist and bibliography of Euphorbiaceae (with Pandaceae). Redwood Books Limited, Trowbridge, Wiltshire.

20. Hamrick, J.L. and Godt, M.J.W. 1989. Allozyme diversity in plants species. In Plant Population Genetics, Breeding and Genetic Resources, pp. 43–63. Eds A.H.D. Brown, M.T. Clegg, A.L. Kahler and B.S. Weir. Sunderland, MA, USA: Sinauer Associates.

21. Hamrick, J.L. and Godt, M.J.W. 1996. Effects of life history traits on genetic diversity in plant species. Philos Trans R Soc Lond B, 351: 1291-1298.

22. He, F., Kang, D., Ren, Y., Qu, L-J., Zhen, Y., and Gu, H. 2007. Genetic diversity of the natural populations of Arabidopsis thaliana in China. Hered, 99: 423-431.

23. 23. Hu, Y., Wang, L., Xie, X., Yang, J., Li Y. and Zhang, H. 2010. Genetic diversity of wild populatio ns of Rheum tanguticum endemic to China as revealed by ISSR analysis. Biochem Sys Ecol, 38: 264–274.

24. Isshiki, S., Iwata, N. and Mizanur Rahim Khan, M.D. 2008. ISSR variations in eggplant (Solanum melongena L.) and related Solanum species. Sci Hort, 117: 186–190.

25. Jabbarzadeh, Z., Khosh-Khui, M., Salehi, H. and Saberivand, A. 2010. Inter simple sequence repeat (ISSR) markers as reproducible and specific tools for genetic diversity analysis of rose species. Afr J Biotechnol, 9: 6091- 6095.

26. Joshi, K., Chavan, P., Warude, D. and Patwardhan, B. 2004. Molecular markers in herbal drug technology. Curr Sci, 87(2):159-165.

27. Kholghi, M., Darvishzadeh, R., Bernousi I., Pirzad, A. and Laurentin H. 2012. Assessment of genomic diversity among and within Iranian confectionery sunflower (Helianthus annuus L.) populations by using simple sequence repeat markers. Acta Agri Scand Sect B Soil Plant Sci, 62: 488-498.

28. Khush, G.S. 2002. Molecular genetics-plant breeder's perspective. In: Jain, S.M., Brar, D.S., Ahloowalia, B.S. (Eds.), Molecular Techniques in Crop Improvement. Kluwer Academic.

29. Kovach, W. 1999. MVSP-A Multivariate Statistical Package for Windows, ver. 3.1. Kovach Computing Services, Pentraeth, Wales, UK.

30. Lobb, K. 1992. Fatty acid classification and nomenclature. In: Chow, C.K. (Ed.), Fatty Acids in Foods and their Health Implications. Marcel Dekker, New York, pp. 1–15.

31. Lu, Z., Wang, Y., Peng, Y., Korpelainen,

32. H. and Li, C. 2006. Genetic diversity of populus cathayana Rehd populations in southwestern china revealed by ISSR markers. Plant Sci, 170: 407–412.

33. Lynch, M. and Milligan B. 1994. Analysis of population-genetic structure using RAPD markers. Mol Ecol, 3: 91-99.

34. Marak, C.K. and Laskar, M.A. 2010. Analysis of phenetic relationship between Citrus indica Tanaka and a few commercially important citrus species by ISSR markers. Sci Hort, 124: 345–348.

35. Modareskia, M., Darvishzadeh, R., Hassani, A. and Kholghi M. 2012. Molecular diversity within and between Ajowan (Carum copticum L.) populations based on inter simple sequence repeat (ISSR) markers. J Plant MB, 1: 51-62.

36. Moshkin, V.A. 1986. Castor. Amerind Publishing Co. PVT Ltd, New Delhi.

37. Muirhead, J. R., Gray, D. K., Kelly, D. W., Ellis, S. M., Heath, D. D., and Macisaac, H.

38. J. 2008. Identifying the source of species invasions: sampling intensity vs. Genet Divers Mol Ecol, 17: 1020-1035.

39. Murray, M.G. and Thompson, W.F. 1980. Rapid isolation of high molecular-weight plant DNA. Nucleic Acids Res, 8: 4321-4325.

40. Muthusamy, S., S. Kanagarajan and S. Ponnusamy, 2008. Efficiency of RAPD and ISSR marker system in accessing variation of rice bean (*Vigna umbellate*) landrace. Electron J Biotechnol, Vol. 11.

41. Nagaoka, T. and Ogihara, Y. 1997. Applicabilty of inter-simple sequence repeat polymorphisms in wheat for use as DNA markers in comparison to RFLP and RAPD markers. Theor Appl Genet, 94: 597–602.

42. Nei, M. 1973. Analysis of gene diversity in subdivided populations. Proc Natl Acad Sci USA, 70: 3321-3323.

43. Nei, N. M. and Li, W. 1979. Mathematical model for studying genetic variation in terms of restriction endonucleases. Proc Natl Acad Sci USA, 76: 5269-5273.

44. Nybom, H. 1994. DNA fingerprinting – a useful tool in fruit breeding. Euphytica, 77: 59-64.

45. Ogunniyi, D.S. 2006. Castor oil: a vital industrial raw material. Bioresour Technol, 97: 1086–1091.

46. Peakall, R. and Smouse, P.E. 2006. GENALEX 6: genetic analysis in Excel. Population genetic sof tware for teaching and research. Mol Ecol Notes, 6: 288-295.

47. Piazza, G.J. and Farrell, H.M. 1991. Generation of ricinoleic acid from castor oil using the lipase from ground oat (*Avena sativa* L.) seeds as a catalyst. Biotechnol Lett, 13:179–184.

48. Pritchard, J.K., Stephanes, M., Rosenberg, N.A. and Donnelly, P. 2000. Association mapping in structured populations. Am J Hum Genet, 67:170–181.

49. Qian, W., Ge S. and Hong, D.Y. 2001. Genetic variation within and among populations of a wild rice Oryza granulata from China detected by RAPD and ISSR markers. Theor Appl Genet, 102: 440–449.

50. Ratnaparkhe, M.B., Tekeoglu, M. and Muehlbauer, F.J. 1998. Inter-simple- sequence-repeat (ISSR) polymorphisms are useful for finding markers associated with disease resistance gene clusters. Theor Appl Genet, 97: 515–519.

51. Reif, J.C., Gumpert, F., Fischer, S. and Melchiger, A.E. 2007. Impact of genetic divergence on additive and dominance variance in hybrid populations. Genet, 176: 1931-1934.

52. Rohlf, F.J. 1998. NTSYS-pc: Numerical Taxonomy and Multivariate Analysis System. version 2.02. Exter Software, Setauket, New York, NY, USA.

53. Roldan-Ruiz, I., Dendauw, J., VanBockstaele, E., Depicker, A., and De Loose, M. 2000. AFLP markers reveal high polymorphic rates in ryegrasses (*Lolium spp.*). Mol Breed, 6: 125-134.

54. Rosenberg, N.A., Pritchard, J.K., Weber, J.L., Cann, H.M., Kidd, K.K., Zhivotovsky, L.A. and Feldman, M.W. 2002. The genetic structure of human populations. Sci, 298: 2381-2385.

55. Rossetto, M., Weaver, P.K. and Dixon, K.W. 1995. Use of RAPD analysis in devising conservation strategies for the rare and endangered *Grevillea scapigera* (Proteaceae). Mol Ecol, 4: 321–329.

56. Savy Filho, A. 2005. Castor bean breeding. In: Borém A. (Ed.) Improvement of Cultivated Species. Viçosa: Federal University of Viçosa.

57. Schall, B.A., Hayworth, D.A., Olsen, K.M., Rauscher, J.T. and Smith, W.A. 1998. Phylogeographic studies in plants: problems and prospects. Mol Ecol, 7: 465– 474.

58. Shannon, C.E. and Weaver, W. 1949. The Mathematical theory of communication, Urbana (Illinois), University of Illinois Press.

59. Sica, M., Gamba, G., Montieri, S., Gaudio, L. and Aceto, S. 2005. ISSR markers show differentiation among Italian populations of *Asparagus acutifolius* L. BMC Genet, 6: 17.

60. Shi, W., Yang, C.-F., Chen, J.-M. and Guo, Y.-H. 2007. Genetic variation among wild and cultivated populations of the Chinese medicinal plant *Coptis chinensis* (Ranunculaceae). Plant Biol, 10: 485-491.

61. Slatkin, M., and Barton, N. H. 1889. A comparison of three methods for estimating average levels of gene flow. Evol, 43: 1349-1 368.

62. Solomon, K.F., Labuschagne, M.T. and Viljoen, C.D. 2007. Estimates of heterosis and association of genetic distance with heterosis in durum wheat under different moisture regimes. J Agric Sci, 145: 239-248.

63. Tachida, H. and Yoshimaru, H. 1996. Genetic diversity in partially selfing populations with the stepping-stone structure. Heredity, 77(5): 469–475.

64. Tomar Rukam, S., Parakhia, M.V., Kavani, R.H., Dobariya, K.L., Thakkar, J.R., Rathod, V.M., Dhingani, R.M. and Golakiya B.A. 2014. Characterization of castor (*Ricinus communis* L.) genotypes using different markers. Res J Biotech, 9(2): 6-13.

65. Velasco, L., Rojas-Barros, P. and Fernández-Martínez, J.M. 2005. Fatty acid and tocopherol accumulation in the seeds of a high oleic acid castor mutant. Ind Crop Prod, 22: 201–206.

66. Verma, S. and Rana, T.S. 2011. Genetic diversity within and amon g the wild populations of Murraya koenigii (L.) Spreng., as revealed by ISSR analysis. Biochem Syst Ecol, 39: 139–144.

67. Wang, H.Z., Feng, S.G., Lu J.J., Shi, N.N. and

Liu, J.J. 2009. Phylogenetic study and molecular identification of 31 DendrObium species using inter-simple sequence repeat (ISSR) markers. Sci Hort, 122: 440–447.

68. Weising, K., Nybom, H., Wolff, K. and Meyer, W. 1995. DNA fingerprinting in plants and fungi. CRC Press, Inc Boca Raton, FL. 322 pp.

69. Yao, M.Z., Chen, L. and Liang, Y.R. 2008. Genetic diversity among tea cultivars from China, Japan and Kenya revealed by ISSR markers and its implication for parental selection in tea breeding programmes. Plant Breed, 127: 166 -172.

70. Yeh, F.C., Yang, R.C., Boyle, T.J., Ye, Z.H. and Mao, J.X. 1999. Popgene ver. 1.32, the user-friendly shareware for population genetic analysis. Edmonton, Canada: Molecular Biology and Biotechnology Centre, University of Alberta.

71. Yip, P.Y., Chau, C.F., Mak, C.Y. and Kwan, H.S. 2007. DNA methods for identification of Chinese medicinal materials. Chin Med, 2(9): 1-19.

72. Zhao K.G., Zhou M.Q., Chen L.Q. 2007. Genetic diversity and discrimination of *Chimonanthus praecox* (L.) Link germplasm using ISSR and RAPD markers. Hortscience, 42(5): 1144–1148.

73. Zheng L., Qi J., Fang P., Su J., Xu J., Tao 2010. Genetic diversity and phylogenetic relationship of castor germplasm as revealed by SRAP analysis. Plant Sci J, 28(1): 1-6.

74. Zietkiewicz, E., Rafalski, A. and Labuda, D. 1994. Genome fingerprinting by simple sequence repeat (SSR)-anchored polymerase chain reaction amplification. Genomics, 20: 176–183.

3

The role of microRNAs and phytohormones in plant immune system

M. Alizadeh & H. Askari*

Department of Biotechnology, Faculty of New Technologies Engineering, Shahid Beheshti University, G. C., Evin, Tehran, Iran.
*Corresponding Author, Email: H_askari@sbu.ac.ir

Abstract

The plant-pathogen interaction is a multifactor process that may lead to resistance or susceptible responses of plant to pathogens. During the arms race between plant and pathogens, various biochemical, molecular and physiological events are triggered in plant cells such as ROS signaling, hormone activation and gene expression reprogramming. In plants, microRNAs (miRNAs) are key post-transcriptional regulators of gene expression and are involved in several cellular processes including response to environmental stress. In recent years, plant pathologists have presented a logical approach of plant immune system as zigzag based model that includes two phases of immunity, PTI and ETI in which miRNA molecules are determinant regulators. Here, we present an overview of miRNA biology, a brief explanation of plant immune systems in zigzag model, the role of phytohormones and miRNAs in plant immunity with a main focus on *Arabidopsis-Pseudomonas* interactions and finally we discuss our results on miRNA expression in lemon-*Xanthomonas* interactions.

Keywords: Effector-Triggered immunity, miRNAs, PAMP-Triggered immunity, Plant immune systems.

Introduction

Crop plants are often exposed to various environmental stress factors which severely affect crop production (Board and Kahlon, 2011). Plant responses to different stresses are highly complex and involve changes at the transcriptome, cellular, and physiological levels. Through an evolutionary process, plants have evolved specific mechanisms that allow them to detect precise environmental changes and respond to the stress condition, minimizing damage while conserving valuable resources for growth and reproduction. (Atkinson and Urwin, 2012).

Under conditions generated by pathogen attack, host plants must be able to orchestrate adaptive responses according to these circumstances in order to survive (Dodds and Rathjen, 2010). Plant immunity is controlled by a complex signaling network depending on cell-autonomous events. Indeed, plants rely on the innate immunity of each cell and on systemic signals emanating from infection sites (Ausubel, 2005; Dangl and Jones, 2001). Some parts of plant immunity systems may be established on natural or modified mineral (Hassabi *et al.*, 2014a), organic (Hassabi *et al.*, 2014b) and biochemical (Hassabi *et al.*, 2014c) compositions of plant tissues. The sensing of biotic stress conditions induces signaling cascades that activate ion channels, kinase cascades, production of reactive oxygen species (ROS) and accumulation of hormones such as salicylic acid (SA), ethylene (ET), jasmonic acid (JA) and abscisic acid ABA (Bari and Jones, 2009; Jones and Dangl, 2006). These signals ultimately induce expression of specific subsets of defense genes that lead to the assembly of the overall defense reaction (Jones and Dangl, 2006). In an attempt to reduce the damage of stress and adapt to their environment, plants have evolved multiple gene regulatory mechanisms

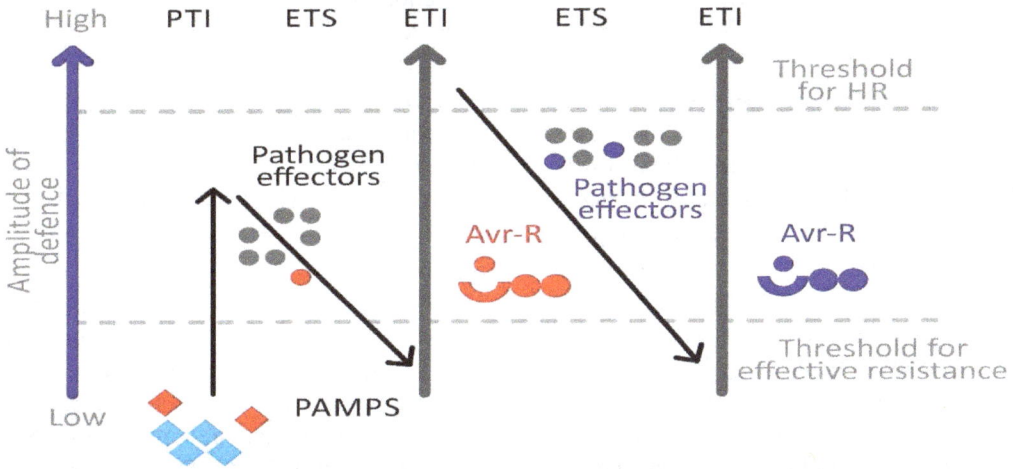

Figure 1. The recommended zigzag model in plant immunity (Jones and Dangl, 2006). In phase 1, plants detect PAMPs (red diamonds) via PRRs to trigger PTI. In phase 2, successful pathogens deliver effectors that interfere with PTI, or otherwise enable pathogen nutrition and dispersal, resulting in effector-triggered susceptibility (ETS). In phase 3, one effector (indicated in red) is recognized by an NB-LRR protein, activating ETI, an amplified version of PTI that often passes a threshold for induction of HR. In phase 4, pathogen isolates are selected that have lost the red effector, and perhaps gained new effectors through horizontal gene flow (in blue)—these can help pathogens to suppress ETI.

involving transcriptional, post-transcriptional and post-translational regulation (Hirayama and Shinozaki, 2010).

Small non-coding RNAs (ncRNAs), which consist of 20–24 nucleotides (nt), have been increasingly investigated as important regulators of protein-coding gene expression; these small RNAs function by causing either transcriptional (TGS) or post-transcriptional gene silencing (PTGS) (Baulcombe, 2004). Our understanding of the complexity of plant's responses to stress has been enhanced by the discovery of ncRNA species which play crucial regulatory roles (Ruiz-Ferrer and Voinnet, 2009). MicroRNAs (miRNAs) are a class of ncRNAs that exist in most eukaryotic genomes. Over the past decade, miRNA molecules have emerged as critical post-transcriptional regulators of animal and plant genomes (Bartel, 2004; Carrington and Ambros, 2003). miRNAs are involved in development, signal transduction, protein degradation, response to environmental stress and pathogen invasion, and regulate their own biogenesis (Unver *et al.*, 2010; Dugas and Bartel, 2004). Plants miRNAs were initially described in *Arabidopsis thaliana* (Ehrenreich and Purugganan, 2008) and since then, an increasing number of miRNAs has been identified in plants (Jones-Rhoades and Bartel, 2004). The levels of conserved and species-specific miRNAs change in response to different pathogens in plants, providing new avenues for the investigation of plant signalling in biotic stresses (Ruiz-Ferrer and Voinnet, 2009). Recently, authors successfully detected and analyzed three conserved miRNAs (mir159, mir167 and mir398) in *Citrus×Limon* (lemon) infected by *Xanthomonas* using stem-loop qRT-PCR (Alizade *et al.*, 2014).

This review explains in detail the miRNA biogenesis and function in plants. Subsequently, plant immune system and the role of phytohormones and plant miRNAs in this system will be discussed with a focus on bacteria- responsive miRNAs.

Plant immune systems

Over the last 25 years, researches have led to an increasingly clear conceptual understanding of the molecular components of the plant immune system (Dangl *et al.*, 2013). Currently, the evolutionary development of the plant immune system is represented as a zigzag model (Figure 1) (Jones and Dangl, 2006). In accordance with this model, plant pathologists discriminate two phases of plant immunity: PTI (PAMP-Triggered Immunity) and ETI (Effector-Triggered Immunity).

PTI is induced where the first level of microbe recognition is performed by membrane proteins termed pattern recognition receptors (PRRs), which perceive molecular signatures charac- teristic of a whole class of microbes, termed pathogen-associated (or microbe- associated) molecular patterns (MAM- Ps/PAMPs) (Medzhitov and Janeway., 1997). ETI as a second phase of plant immunity is mediated by intracellular nucleotide-binding leucine-rich repeat receptors (NLR) that recognize the presence or the activity of specific microbial effectors (García and Hirt, 2014). Although PTI and ETI employ distinct immune receptors, they seem to use a similar signaling network (Tsuda et al., 2009) and activate a largely overlapping set of genes (Zipfel et al., 2006; Navarro et al., 2004), with the paradigm that activated immune responses in ETI occur quicker and are more prolonged and more robust than those in PTI (Jones and Dangl, 2006; Tao et al., 2003). ETI amplifies PTI responses and is normally associated with the appearance of localized cell death lesions known as hypersensitive response (HR) (Figure 1) (Heidrich et al., 2012). In plants, HR is defined as a rapid cell death that causes necrosis to restrict the growth of a pathogen (Morel and Dangl, 1997). Following PAMPs perception, a series of downstream defense responses are triggered including ion fluxes, MAPK (mitogen- activated protein kinase) cascade activation, ROS (ROS) production, defense gene expression, callose (β-1->3 glucose polymer) deposition, stomatal closure, hormone activation and gene silencing (Nicaise et al., 2009).

Recent progresses have been made in understanding the complex hormone network that governs plant immunity. Downstream of PTI or ETI activation, diverse plant hormones act as central players in triggering of the plant immune signaling network (Pieterse et al., 2009; Bari and Jones, 2009).

The role of phytohormones in plant immunity

The plant hormones ethylene, jasmonic acid and salicylic acid play a central role in the regulation of plant immune responses (Robert-Seilaniantz et al., 2011; Vlot et al., 2009). In addition, other plant hormones, such as auxins, ABA, cytokinins, gibberellins and brassinosteroids that have been thoroughly described to regulate plant development and growth, have recently emerged as key regulators of plant immunity (Kazan and Manners, 2009; Ton et al., 2009).

SA plays a crucial role in plant defense and is generally involved in the activation of defense responses against biotrophic and hemibiotrophic pathogens as well as the establishment of systemic acquired resistance (SAR) (Grant and Lamb, 2006). HR development is usually accompanied by an increase in SA and an accumulation of defense related proteins such as the pathogenesis related (PR) proteins (Vlot et al., 2008). By contrast with SA, JA and ET are involved in resistance to necrotrophic pathogens and herbivorous insects (Beckers and Spoel, 2006). Although SA and JA/ET defense pathways are mutually antagonistic, evidences of synergistic interactions have also been reported (Mur et al., 2006; Kunkel and Brooks, 2002).

The phytohormone ABA plays regulatory functions in many aspects of plant growth and development including seed germination, embryo maturation, leaf senescence, stomatal aperture and adaptation to environmental stresses (Wasilewska et al., 2008). In general, ABA is shown to be involved in the negative regulation of plant defense against various biotrophic and necrotrophic pathogens (Thaler and Bostock, 2004; Audenaert et al., 2002). ABA was shown to attenuate SA-mediated resistance at later infection stages and can also suppress callose deposition in response to PAMPs (De Torres-Zabala et al., 2007).

Many biotrophic pathogens could synthesize auxin or auxin-like molecules to promote disease symptoms in many plants (Navarro et al., 2006). Treatments with the auxin analogs 2,4-dichlorophenoxyacetic acid (2,4-D) or 1- naphthalacetic acid (NAA) enhance disease symptoms in *Arabidopsis* infected by *Pseudomonas syringae* pv. *tomato* (*Pst*) *DC3000* (Chen et al., 2007). Gene expression analysis of sweet orange leaves treated with auxin analogs suggested that auxin affects GA synthesis in citrus as it occurs in numerous plant species (Cernadas and Benedetti, 2009). Auxin transport inhibitor, naphthylphthalamic acid can attenuate canker development of sweet orange infected by *Xanthomonas citri* pv. *citri*, but NAA can provoke more serious disease symptoms (Cernadas and Benedetti, 2009). As auxin and gibberellin hormones are core signals in cell division and growth, they are suggested to play key roles in contributing to citrus canker symptoms (Cernadas and Benedetti, 2009).

A comprehensive overview of miRNA

History of discovery

miRNAs were first found and characterized in a worm, *Caenorhabditis elegans*; *Lin-4*, a mutant worm which lost many adult structures and developmental plasticity (Lee and Ambors, 2001; Lau *et al.*, 2001). It was observed that no protein sequences were encoded by this gene but were transcribed into RNA in wild-type worms (Lee *et al.*, 1993). Another worm mutant, *let-7*, followed a similar pattern in gene expression (Moss, 2000). In both cases, the primary transcripts were sliced into smaller RNA fragments and finally into a sRNA with about 21nt in length, which is now known to be a miRNA. Lin-4 or let-7 miRNAs act as negative regulators of gene expression by annealing with their target mRNAs (Moss, 2000), resulting in time-dependent regulation of developmental phase change.

The importance of miRNA in plants was first demonstrated by Palatnik *et al* (2003). They showed that the gene locus responsible for the mutation in *Arabidopsis* mutants (*JAW* mutants), did not encode any protein. The transcript generated from this locus, had the potential to produce a miRNA. They showed that the miRNA was produced in the wild-type plants but not in *jaw* mutants (Palatnik *et al.*, 2003). The miRNA partially complemented to mRNA sequence encoding the so-called TCP proteins, which are a class of transcription factors (TCP) (Palatnik *et al.*, 2003).

Genomic location of miRNA-encoding genes (MIR genes)

In the recent years, the vast majority of conserved and novel microRNAs have been discovered by small RNA deep sequencing. These technologies are making it quickly possible to identify novel microRNAs as well as they are published and submitted to a database. Among the databases exist for miRNA information, miRBase (www .mirbase .com) is the most valid source for the biological studies. miRBase is the source for miRNA information includes databases of sequences and predicted targets, as well as an official name registry for new miRNA genes. In the *Arabidopsis* plant whose genome has been fully sequenced, over 100 miRNA encoding loci have been identified (Ehrenreich and Purugganan, 2008; Bonnet *et al.*, 2004). miRNA-encoding (*MIR*) genes are frequently expressed individually, but many exist in clusters of 2–7 genes with small intervening sequences. Experimental results suggest that they are expressed co- transcriptionally, which indicates that they are under the control of common regulatory sequences (Lee *et al.*, 2002; Lau *et al.*, 2001). Other miRNA genes are usually located in intergenic regions, some in the introns of known genes, and even within the expressed sequence tags (ESTs) (Lim *et al.*, 2003). In addition, *MIR* genes are excised from the introns and exons of non-coding genes (Rodriguez *et al.*, 2004), or even from the 3'-UTR of protein-coding genes (Cai *et al.*, 2004). In mammalian genomes, it is also possible to find miRNAs in repetitive regions, and some studies suggest that transposable elements may be involved in the creation of new miRNAs (Smalheiser and Torvik, 2005).

miRNAs Biogenesis

Most characterized eukaryotic *MIR* genes are RNA polymerase II (Pol II) transcription units that generate a primary miRNA transcript called a pri- miRNA, therefore pri-miRNAs can be subjected to elaborate transcriptional control (Lee *et al.*, 2004). miRNA biogenesis in animals is a two-step process (Figure 2) (Lee *et al.*, 2002). In the first step, pri-miRNAs, which are several hundred nucleotides long, are processed by a nuclear multiprotein complex (Microprocessor) containing an enzyme called Drosha (nuclear RNase III type) into a 70~90nt hairpin long precursor miRNA (pre-miRNA) which is then exported to the cytoplasm (Lee *et al.*, 2003). This cleavage event is important because it predetermines mature miRNA sequence and generates optimal substrate for the subsequent events (Lund *et al.*, 2004; Lee *et al.*, 2003). The nuclear export is elicited by a complex of Exportin 5 (Exp5) and Ran- GTP which selectively bind pre- miRNAs and protect them from exonucleolytic digestion (Lund *et al.*, 2004). In the cytoplasm, the second step takes place where the pre-miRNA is cleaved by cytoplasmic RNase III Dicer into ~22nt miRNA duplex (miRNA: miRNA∗ duplex), with each strand originating from opposite arms of the stem–loop (Hutvágner and Zamore, 2002). The duplex strand with the weakest 5' end base pairing is then selected as the mature miRNA and the remaining strand, called miRNA*, is degraded (Tomari *et al.*, 2004). In general, the miRNA strand is then integrated in a ribonucleoprotein complex known as the (mi)RNA- induced silencing complex (miRISC or RISC) or miRNA-containing ribonucleoprotein particles (miRNPs) (Lau *et al.*, 2001).

miRNA biogenesis in plants differs from animal biogenesis mainly in the steps of nuclear processing and export (Figure 2) (Millar and Waterhouse, 2005). All maturation steps of plant miRNAs are processed by Dicer-like proteins (Jones- Rhoades *et al.*, 2006). In plants, miRNAs seem to be fully matured into a single stranded miRNA before being exported to the cytoplasm by a homologue of Exp5 termed HASTY (HST) and integrated into the silencing complex (Park *et al.*, 2005; Bartel, 2004). The enzymes for miRNA biogenesis are under feedback regulation by miRNAs (Jones-Rhoades *et al.*, 2006) and this feedback regulatory

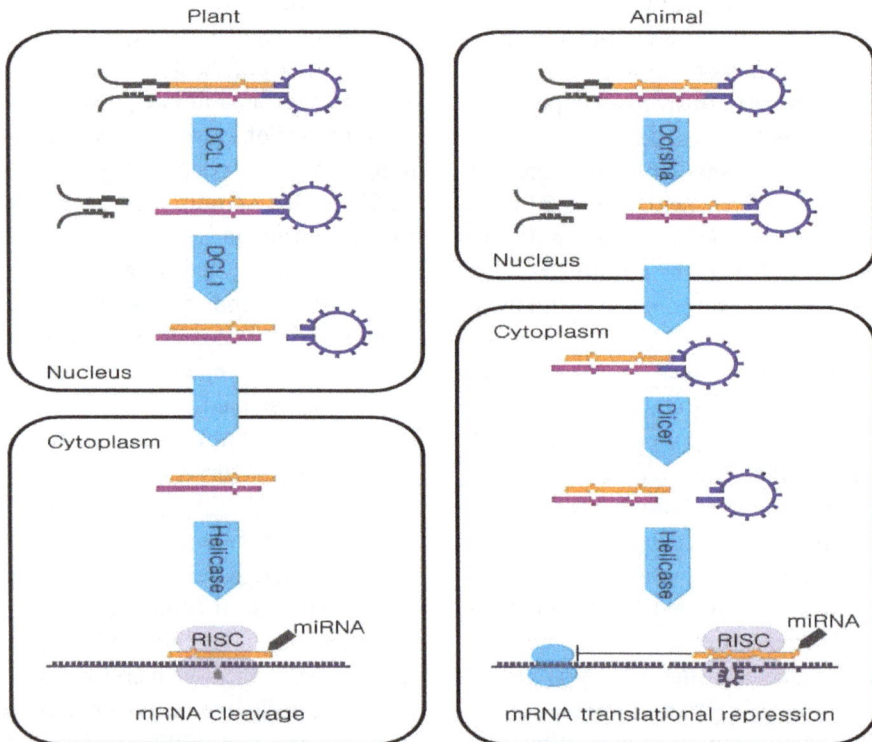

Figure 2. A comparative view of miRNA biogenesis and action in plant and animal. The Drosha gene that is responsible for processing of pri-miRNA to pre-miRNA in animals is absent from plant genomes; this function is performed by the plant Dicer-like 1 (DCL1). In animals, miRNA/miRNA* duplex is formed in the cytoplasm by Dicer endonucleolytic activity. In contrast, all maturation steps of plant miRNAs occur in the nucleus. Depending on the level of miRNA-mRNA complementarity, miRNA in animals acts as translational repressor whereas plant miRNA is considered for its mRNA decay activity.

mechanism is deeply conserved among diverse plant species (Xie *et al.*, 2010).

miRNAs Function

In an Overall view, miRNAs regulate gene expression by inhibiting mRNA translation and/or facilitating mRNA degradation (Voinnet, 2009). Post-transcriptional control of gene silencing by miRNAs is a ribonucleoprotein-driven process, which involves specific RNA binding proteins, miRNAs and their mRNA targets (Cava *et al.*, 2014). To this end, mature miRNA assembles into RISC, activating the complex to target mRNA specified by the miRNA (Pratt and MacRae, 2009). Members of the Argonaute (AGO) protein family are central to RISC function (Pratt and MacRae, 2009). A key component in the miRNA pathway is AGO1, which predominately binds mature miRNAs to cleave the target mRNA or represses translation depending on the level of miRNA-mRNA complementarity (Okamura *et al.*, 2004). AGOs contain four characteristic domains: the N-terminal domain; the PAZ domain, which binds the 2nt overhang of the 3' end of the mature miRNA; the MID domain, which provides a binding pocket for the 5' phosphate of mature miRNAs; the PIWI domain, which adopts an RNase H fold and has endonucleolytic activity in some, but not all, AGOs (Ma *et al.*, 2005; Parker *et al.*, 2005).

miRNAs and their targets seem to constitute remarkably complex regu- latory networks since a single miRNA can bind to and regulate many different mRNA targets and, conversely, several different miRNAs can bind to and cooperatively control a single mRNA target (Lewis *et al.*, 2003). In animals, miRNAs are considered to act mainly as translational repressors by their partially complementary binding to specific 3'- UTR regulatory elements on target mRNAs (Lai, 2002), although target sites in the coding region and 5'-UTR can also be functional (Lytle *et al.*, 2007; Kloosterman *et al.*, 2004). On the other hand, plant miRNAs frequently cleave and thus induce immediate degradation of the target mRNAs and are often almost perfectly complementary to sites in the coding region (Ehrenreich and Purugganan, 2008), as well as in the 3'- UTR (Sunkar and Zhu, 2004), and even in the 5'-UTR (Millar

and Waterhouse, 2005). The functions of plant miRNAs are highly diverse and have essential roles in regulating plant growth, organogenesis, pattern formation, organ polarity, and hormone homeostasis (Voinnet, 2009).

miRNAs involvement in plant immunity

In dealing with pathogens, host plants can establish defense responses against pathogens which involve rapid changes in gene expression, hormone and metabolite levels (Sunkar *et al.*, 2012). Plant small RNAs have been demonstrated as critical regulators in gene expression reprogramming during both PTI and ETI establishment (Padmanabhan *et al.*, 2009; Voinnet, 2008). In *Arabidopsis*, the first reported miRNA contributing to antibacterial resistance was miR393 which plays a role in PTI response by regulating the auxin signaling pathway (Navarro *et al.*, 2006). It has been shown that bacterial PAMP flg22 rapidly induces the miR393 expression which targets receptors of auxin (AFBs receptors) (Figure 3) (Navarro *et al.*, 2006; Jones-Rhoades and Bartel, 2004). Perception of auxin by AFBs leads the degradation of the AUX/IAA protein, and subsequently activates auxin response genes by derepressing the auxin-response factor (ARF) transcription factors (Figure 3) (Chapman and Estelle, 2009). Fahlgren *et al.*, (2007) reported that miR393 can be significantly induced at 3h post- inoculation (hpi) by nonpathogenic *Pst* DC3000 *hrcC*, a strain responsible for the induction of initiate immunity. In addition, miR160 and miR167 upregulated by *Pst* DC3000 *hrcC* at 3-hpi rather than mir393 (Rhoades *et al.* 2002). mir160 and mir167 target the members of Auxin-responsive factor (ARF) family that are involved in auxin signaling pathway (Figure 3) (Li Y *et al.*, 2010). Thus, three bacteria-responsive miRNAs (mir160, mir167 and mir393) suppress the auxin signaling and contribute to the PTI in plants. Auxin is a plant hormone which has growth- promoting role and is antagonistic to SA-mediated resistance (Wang *et al.*, 2007). Upon perceiving the pathogen PAMPs, these miRNAs are induced to rapidly repress the auxin signaling and shift the energy from plant growth to defense responses.

In *Arabidopsis-pseudomonas* interaction model, some miRNAs are induced which target negative defense response regulators and a group of miRNAs targeting positive regulators (e.g. resistance genes) are repressed upon bacterial infection (Ruiz-Ferrer and Voinnet, 2009). miR398 is down- regulated in response to avirulent strains of *Pst* DC3000 (*avrRpm1*) or *Pst* DC3000 (*avrRpt2*) at 12-hpi and continued until 24-hpi (Jagadeeswaran *et al.*, 2009). The targets of miR398 are Cu/Zn superoxide dismutases 1 and 2 (CSD_1 and CSD_2) (Figure 3) (Bonnet *et al.*, 2004). These enzymes decrease superoxide (as a form of ROS) levels by converting it to H_2O_2 and O_2 (Figure 3) (Draper, 1997). It has been found that miR398 negatively regulated PAMP induced callose deposition (Li *et al.*, 2010).

The repertoire of known bacterial- responsive miRNAs has increased and includes several families, such as miR159 involved in ABA signaling and miR319 (Zhang *et al.*, 2011; Fahlgren *et al.*, 2007). miR159 is down-regulated by *Pst* DC3000 (*EV*) and *Pst* DC3000 (*avrRpt2*) at 6-hpi, but up-regulated by *Pst* DC3000 (*avrRpt2*) at 14-hpi (Zhang *et al.*, 2011). miR159 targets transcription factors MYB33, MYB65 and MYB101, the homologous genes of the barley GAMYB that activates Gibberellin (GA)-signaling pathways (Figure 3) (Reyes and Chua, 2007; Millar and Gubler, 2005). MYB33 and MYB101 act as positive regulators of ABA signaling pathways in *Arabidopsis* (Figure 3) (Reyes and Chua, 2007).

Northern blot analysis showed that mir319 is induced by *Pst* DC3000 *hrcC* and *Pst* DC3000 (*avrRpt2*) at 14-hpi (Zhang *et al.*, 2011). miR319 targets TCP (*TEOSINTE BRANCHED/CYCLOIDEA/PCF*) transcription factor family genes which directly regulate *LIPOXYGENASE2* (*LOX2*) (Figure 3) (Schommer *et al.*, 2008). *LOX2* encodes a chloroplast-localized enzyme that is responsible for the first step in the JA biosynthesis pathway. JA signaling is usually antagonistic to SA signaling (Overmyer *et al.*, 2003), while SA signaling is important for plant defense against biotrophic pathogens, including *Pst*.

A case study on miRNA time-dependent expression in lemon-Xanthomonas interaction

Xanthomonas citri subsp. *citri* strain A (Xc) with a broad host range is causal agent of citrus canker disease and is considered as one of the most devastating biotic stresses affecting the

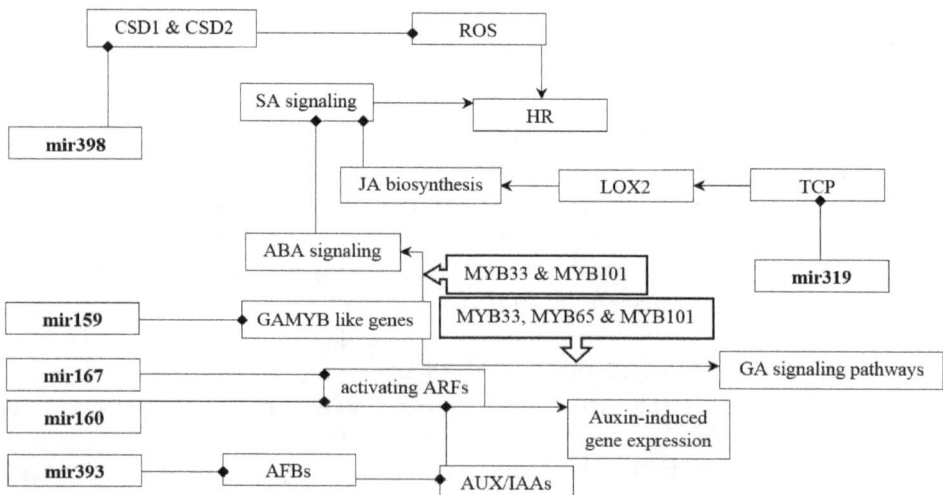

Figure 3. The regulatory role of responsive miRNAs in Arabidopsis-pseudomonas interaction. Arrows indicate positive regulations and diamond arrows indicate inhibitions. It has been shown that microRNAs inhibit protein production of their target genes and consequently lead plant biosystem toward regulation of hormone signaling and HR.

citrus industry (Brunings and Gabriel, 2003). Citrus canker is characterized by pustule-like lesions that raise on both surfaces of the leaf and which later become corky and surrounded by a watersoaked margin with a yellow halo (Schubert *et al.*, 2001). Canker lesions can also develop on stems and fruits (Schubert *et al.*, 2001) and are thought to be the result of intense cell division (hyperplasia) and expansion (hyper- trophy) that occurs in the host tissues after pathogen infection (Brunings and Gabriel, 2003). *Xanthomonas fuscans* subsp. *aurantifolii* strain C (XaC) has a narrower range of citrus hosts which are restricted to some citrus-producing areas in South America (Schubert *et al.*, 2001). In addition, XaC induces HR in various citrus species including *Citrus × Limon* (lemon) (Brunings and Gabriel, 2003).

Expression analysis of conserved miRNAs including mir159 involved in gibberellin and ABA signaling, mir167 involved in auxin signaling and mir398 involved in detoxification of ROS demonstrates a time-dependent expression regulation during seven hours (0.5, 3, 6, 12, 24, 48 and 72) after lemon leaves infection by Xc and XaC (Alizadeh *et al.*, 2014). It seems that the expression patterns of the miRNAs follow a rather zigzag model in lemon- *Xhanthomonas* interaction. According to the results, all three miRNAs are significantly induced at 6-hpi (Table 1). mir159 and mir167 gene expression follow a similar pattern upon both strains infection. After induction at 6-hpi, the high levels of mir159 and mir167 expression are reduced upon Xc infection whereas abundance of transcripts maintained at high levels in response to XaC.

The expression patterns in response to Xc suggest that mir159 and mir167 may contribute to inhibition of disease development through their down- regulatory roles in gibberellin and auxin signaling, respectively. Upon XaC infection, the expression patterns of mir159 and mir167 suggest probable roles in HR induction for both miRNAs. On the other hand, a stable level is observed after 6-hpi induction for mir398 gene expression in response to both strains. Opposite regulation patterns of mir398 gene expression in this study compared to previous studies mention different strategy in mir398 regulation in various plant-pathogen interaction systems. The study eventually concludes that mir159 and mir167 can be investigated in future as major nods in lemon gene regulation network in order to develop the resistance to citrus canker and also proposes 6-hpi as a critical time for future studies to develop a model of gene expression regulatory network in lemon-*Xanthomonas* interaction.

Future perspectives

In recent years, Identification and characterization of plant miRNAs and their targets in biotic stresses have demonstrated the importance of small RNAs machinery in plant immunity. miRNAs have central roles in gene expression reprogramming and balancing the host immune responses and fitness costs during host-microbial interaction. Despite the many experimental methods and computational approaches developed in order to solve the mystery of miRNAs involved networks, there is a need for a global and comprehensive understanding of the functions of miRNAs to

Table 1. The fold change (FC) values of selected miRNAs after three post inoculation times in lemon-*Xanthomonas* strains interaction. The FC value was reported as Log $_2$ (Ratio). The Ratio was calculated using efficiency-based mathematical model (Pfaffl, 2001).

Strain	Xc			XaC		
Time (h)	6	12	48	6	12	48
mir159	6.985327	-3.44701	-0.27441	1.819573	2.463257	3.065113
mir167	5.628179	-0.18154	-0.26342	1.460915	1.94406	1.965262
mir398	3.183962	-1.71882	1.546557	2.724107	2.803568	0.266999

provide adequate insights for conferring plant resistance to pathogens.

References

1. Alizadeh, M., Askari, H., Najafabadi, M.S., Rajaei, E. 2014. Time-dependent regulation of conserved putative micro- RNAs in response of *Citrus × Limon* to *Xanthomonas citri* subsp. *citri* and *Xanthomonas fuscans* subsp. *aurantifolii*. *Mol. Biol. Rep.*, in press.
2. Atkinson, N.J. and Urwin, P.E. 2012. The interaction of plant biotic and abiotic stresses: from genes to the field. *J. Exp. Bot*, 63:3523-3543.
3. Audenaert, K., De Meyer, G.B. and Höfte, M.M. 2002. Abscisic acid determines basal susceptibility of tomato to *Botrytis cinerea* and suppresses salicylic acid- dependent signaling mechanisms. *Plant Physiol*, 128:491-501.
4. Ausubel, F.M. 2005. Are innate immune signaling pathways in plants and animals conserved? *Nat. Immunol*, 6:973-979.
5. Bari, R. and Jones, J.D.G. 2009. Role of plant hormones in plant defence responses. *Plant Mol. Biol*, 69:473-488.
6. Bartel, D.P. 2004. MicroRNAs: genomics, biogenesis, mechanism, and function. *Cell*, 116:281-297.
7. Baulcombe, D. 2004. RNA silencing in plants. *Nature*, 431:356-363.
8. Beckers, G.J.M. and Spoel, S.H. 2006. Fine- Tuning Plant Defence Signalling: Salicylate versus Jasmonate. *Plant Biol*, 8:1-10.
9. Board, J.E. and Kahlon, C.S. 2011. Soybean yield formation: what controls it and how it can be improved. *Soybean physiology and biochemistry. Intech Publ. InTech Open Access, Rijeka, Croatia*, 1-36.
10. Bonnet, E., Wuyts, J., Rouzé, P. and Van de Peer, Y. 2004. Detection of 91 potential conserved plant microRNAs in *Arabidopsis thaliana* and *Oryza sativa* identifies important target genes. *Proc. Natl. Acad. Sci. U. S. A*, 101:11511- 11516.
11. Brodersen, P., Malinovsky, F.G., Hématy, K., Newman, M.A. and Mundy, J. 2005. The role of salicylic acid in the induction of cell death in Arabidopsis *acd11*. *Plant Physiol*, 138:1037-1045.
12. Brunings, A.M. and Gabriel, D.W. 2003. *Xanthomonas citri*: breaking the surface. *Mol. Plant Pathol*, 4:141-157.
13. Cai, X., Hagedorn, C.H. and Cullen, B.R. 2004. Human microRNAs are processed from capped, polyadenylated transcripts that can also function as mRNAs. *RNA*, 10:1957-1966.
14. Carrington, J.C. and Ambros, V. 2003. Role of microRNAs in plant and animal development. *Science*, 301:336-338.
15. Cava, C., Bertoli, G., Ripamonti, M., Mauri, G., Zoppis, I., Della Rosa, P.A., Gilardi, M.C. and Castiglioni, I. 2014. Integration of mRNA Expression Profile, Copy Number Alterations, and microRNA Expression Levels in Breast Cancer to Improve Grade Definition. *PloS One*, 9:e97681.
16. Cernadas, R.A. and Benedetti, C.E. 2009. Role of auxin and gibberellin in citrus canker development and in the transcriptional control of cell-wall remo- deling genes modulated by *Xanthomonas axonopodis* pv. *citri*. *Plant Sci*, 177:190- 195.
17. Chandra-Shekara, A.C., Gupte, M., Navarre, D., Raina, S., Raina, R., Klessig, D. and Kachroo, P. 2006. Light-dependent hypersensitive response and resistance signaling against Turnip Crinkle Virus in *Arabidopsis*. *Plant J*, 45:320-334.
18. Chapman, E.J. and Estelle, M. 2009. Mechanism of auxin-regulated gene expression in plants. *Annu. Rev. Genet*, 43:265-285.
19. Chen, Z., Agnew, J.L., Cohen, J.D., He, P., Shan, L., Sheen, J. and Kunkel, B.N. 2007. *Pseudomonas syringae* type III effector AvrRpt2 alters *Arabidopsis thaliana* auxin physiology. *Proc. Natl. Acad. Sci. U. S. A*, 104:20131-20136.
20. Chisholm, S.T., Coaker, G., Day, B. and Staskawicz, B.J. 2006. Host-microbe interactions: shaping the evolution of the plant immune response. *Cell*, 124:803- 814.
21. Dangl, J.L., Horvath, D.M. and Staskawicz, B.J. 2013. Pivoting the plant immune system from dissection to deployment. *Science*, 341:746-751.
22. Dangl, J.L. and Jones, J.D.G. 2001. Plant pathogens and integrated defence responses to infection. *Nature*, 411:826- 833.
23. De Torres-Zabala, M., Truman, W., Bennett, M.H., Lafforgue, G., Mansfield, J.W., Rodriguez Egea, P., Bögre, L. and Grant, M. 2007. *Pseudomonas syringae* pv. *tomato* hijacks the *Arabidopsis* abscisic acid signalling pathway to cause disease. *EMBO J*, 26:1434-1443.
24. Delledonne, M., Xia, Y., Dixon, R.A. and Lamb, C. 1998. Nitric oxide functions as a signal in plant disease resistance. *Nature*, 394:585-588.
25. Depuydt, S. and Hardtke, C.S. 2011. Hormone signalling crosstalk in plant growth regulation. *Curr. Biol*, 21:365- 373.
26. Dodds, P.N. and Rathjen, J.P. 2010. Plant immunity: towards an integrated view of plant-pathogen interactions. *Nat. Rev. Genet*, 11:539-548.
27. Draper, J. 1997. Salicylate, superoxide synthesis and cell suicide in plant defence. *Trends Plant Sci*, 2:162-165. Dugas, D.V. and Bartel, B. 2004. MicroRNA regulation of gene expression in plants. *Curr. Opin. Plant Biol*, 7:512-520.
28. Ehrenreich, I.M. and Purugganan, M. 2008. MicroRNAs in plants. *Plant Signaling Behav*, 3:829-830.
29. Fahlgren, N., Howell, M.D., Kasschau, K.D., Chapman, E.J., Sullivan, C.M., Cumbie, J.S., Givan, S.A., Law, T.F., Grant, S.R. and Dangl, J.L.

2007. Highthroughput sequencing of *Arabidopsis* microRNAs: evidence for frequent birth and death of *MIRNA genes*. *PloS One*, 2:e219.

30. García, A.V. and Hirt, H. 2014. *Salmonella enterica* induces and subverts the plant immune system. *Front. Microbiol*, 5:141.

31. Grant, M. and Lamb, C. 2006. Systemic immunity. *Curr. Opin. Plant Biol*, 9:414- 420.

32. Hasabi, V., Askari, H., Alavi, S.M., Goodarzi, T., Najafabadi, M.S., Zamanizadeh, H. 2014a. Inhibitory impact of plant nutritional compounds on Xanthomonas citri subsp. citri, the causal agent of bacterial canker of citrus. *J. Plant Pathol.*, 96:369-375.

33. Hasabi, V., Askari, H., Alavi, S.M., Najafabadi, M.S. 2014b. In Vitro and In Vivo Antibacterial Activity of Some Organic and Inorganic Salts against Asiatic Citrus Canker Agent *Xanthomonas* Citri Subsp. Citri. Turkish. J Sci. Food Agr, 2:296-300.

34. Hasabi, V., Askari, H., Alavi, S.M., Zamanizadeh, H. 2014c. Effect of amino acid application on induced resistance against citrus canker disease in lime plants. *J. Plant Prot. Res.*, 54:144-149.

35. Heidrich, K., Blanvillain-Baufumé, S. and Parker, J.E. 2012. Molecular and spatial constraints on NB-LRR receptor signaling. *Curr. Opin. Plant Biol*, 15:385-391.

36. Hirayama, T. and Shinozaki, K. 2010. Research on plant abiotic stress responses in the post-genome era: past, present and future. *Plant J*, 61:1041-1052.

37. Hutvágner, G.r. and Zamore, P.D. 2002. A microRNA in a multiple-turnover RNAi enzyme complex. *Science*, 297:2056- 2060.

38. Jagadeeswaran, G., Saini, A. and Sunkar, R. 2009. Biotic and abiotic stress downregulate miR398 expression in *Arabidopsis*. *Planta*, 229:1009-1014.

39. Jiang, C.J., Shimono, M., Sugano, S., Kojima, M., Yazawa, K., Yoshida, R., Inoue, H., Hayashi, N., Sakakibara, H. and Takatsuji, H. 2010. Abscisic acid interacts antagonistically with salicylic acid signaling pathway in rice- *Magnaporthe grisea* interaction. *Mol. Plant-Microbe Interact*, 23:791-798.

40. Jones, J.D.G. and Dangl, J.L. 2006. The plant immune system. *Nature*, 444:323- 329.

41. Jones-Rhoades, M.W. and Bartel, D.P. 2004. Computational identification of plant microRNAs and their targets, including a stress-induced miRNA. Mol Cell, 14:787-799.

42. Jones-Rhoades, M.W., Bartel, D.P. and Bartel, B. 2006. MicroRNAs and their regulatory roles in plants. *Annu. Rev. Plant Biol*, 57:19-53.

43. Kazan, K. and Manners, J.M. 2009. Linking development to defense: auxin in plantpathogen interactions. *Trends Plant Sci*, 14:373-382.

44. Kloosterman, W.P., Wienholds, E., Ketting, R.F. and Plasterk, R.H.A. 2004. Substrate requirements for let-7 function in the developing zebrafish embryo. *Nucleic* The role of microRNAs and phytohormones in plant immune system 39 Acids Res, 32:6284-6291.

45. Kunkel, B.N. and Brooks, D.M. 2002. Cross talk between signaling pathways in pathogen defense. *Curr. Opin. Plant Biol*, 5:325-331.

46. Lai, E.C. 2002. Micro RNAs are complementary to 3' UTR sequence motifs that mediate negative posttranscriptional regulation. *Nat. Genet*, 30:363.

47. Lau, N.C., Lim, L.P., Weinstein, E.G. and Bartel, D.P. 2001. An abundant class of tiny RNAs with probable regulatory *roles in Caenorhabditis elegans*. *Science*, 294:858-862.

48. Lee, R.C. and Ambros, V. 2001. An extensive class of small RNAs in *Caenorhabditis elegans*. *Science*, 294:862-864.

49. Lee, R.C., Feinbaum, R.L. and Ambros, V. 1993. The C. elegans heterochronic *gene lin-4* encodes small RNAs with antisense complementarity to lin-14. Cell, 75:843- 854.

50. Lee, Y., Ahn, C., Han, J., Choi, H., Kim, J., Yim, J., Lee, J., Provost, P., Rådmark, O. and Kim, S. 2003. The nuclear RNase III Drosha initiates microRNA processing. *Nature*, 425:415-419.

51. Lee, Y., Jeon, K., Lee, J.T., Kim, S. and Kim, V.N. 2002. MicroRNA maturation: stepwise processing and subcellular localization. *EMBO J*, 21:4663-4670.

52. Lee, Y., Kim, M., Han, J., Yeom, K.H., Lee, S., Baek, S.H. and Kim, V.N. 2004. MicroRNA genes are transcribed by RNA polymerase II. *EMBO J*, 23:4051- 4060.

53. Lewis, B.P., Shih, I.h., Jones-Rhoades, M.W., Bartel, D.P. and Burge, C.B. 2003. Prediction of mammalian microRNA targets. *Cell*, 115:787-798.

54. Li, Y., Zhang, Q., Zhang, J., Wu, L., Qi, Y. and Zhou, J.M. 2010. Identification of microRNAs involved in pathogenassociated molecular pattern-triggered plant innate immunity. *Plant Physiol*, 152:2222-2231.

55. Lim, L.P., Lau, N.C., Weinstein, E.G., Abdelhakim, A., Yekta, S., Rhoades, M.W., Burge, C.B. and Bartel, D.P. 2003. The microRNAs of *Caenorhabditis elegans*. *Genes Dev*, 17:991-1008.

56. Lund, E., Güttinger, S., Calado, A., Dahlberg, J.E. and Kutay, U. 2004. Nuclear export of microRNA precursors. *Science*, 303:95-98.

57. Lytle, J.R., Yario, T.A. and Steitz, J.A. 2007. Target mRNAs are repressed as efficiently by microRNA-binding sites in the 5' UTR as in the 3' UTR. *Proc. Natl. Acad. Sci. U. S. A*, 104:9667-9672.

58. Ma, J.B., Yuan, Y.R., Meister, G., Pei, Y., Tuschl, T. and Patel, D.J. 2005. Structural basis for 5'-end-specific recognition of guide RNA by the *A. fulgidus Piwi protein*. *Nature*, 434:666- 670.

59. Medzhitov, R. and Janeway, C.A. 1997. Innate im-

munity: the virtues of a nonclonal system of recognition. *Cell,* 91:295-298.

60. Millar, A.A. and Gubler, F. 2005. The *Arabidopsis GAMYB-like Genes, MYB33* and MYB65, are microRNA-regulated genes that redundantly facilitate anther development. *Plant Cell,* 17:705-721.

61. Millar, A.A. and Waterhouse, P.M. 2005. Plant and animal microRNAs: similarities and differences. *Funct. Integr. Genomics,* 5:129-135.

62. Moss, E.G. 2000. Non-coding RNAs: lightning strikes twice. *Curr. Biol,* 10:436-439.

63. Morel, J.B. and Dangl, J.L. 1997. The hypersensitive response and the induction of cell death in plants. *Cell Death Differ,* 4:671-683.

64. Mur, L.A.J., Kenton, P., Atzorn, R., Miersch, O. and Wasternack, C. 2006. The outcomes of concentration-specific interactions between salicylate and 40 Molecular Plant Breeding jasmonate signaling include synergy, antagonism, and oxidative stress leading to cell death. *Plant Physiol,* 140:249-262.

65. Navarro, L., Dunoyer, P., Jay, F., Arnold, B., Dharmasiri, N., Estelle, M., Voinnet, O. and Jones, J.D.G. 2006. A plant miRNA contributes to antibacterial resistance by repressing auxin signaling. *Science,* 312:436-439.

66. Navarro, L., Zipfel, C., Rowland, O., Keller, I., Robatzek, S., Boller, T. and Jones, J.D.G. 2004. The transcriptional innate immune response to flg22. Interplay and overlap with Avr gene-dependent defense responses and bacterial pathogenesis. *Plant physiol,* 135:1113-1128.

67. Nicaise, V., Roux, M. and Zipfel, C. 2009. Recent advances in PAMP-triggered immunity against bacteria: pattern recognition receptors watch over and raise the alarm. *Plant Physiol,* 150:1638-1647.

68. Okamura, K., Ishizuka, A., Siomi, H. and Siomi, M.C. 2004. Distinct roles for Argonaute proteins in small RNAdirected RNA cleavage pathways. *Genes Dev,* 18:1655-1666.

69. Overmyer, K., Brosché, M. and Kangasjärvi, J. 2003. Reactive oxygen species and hormonal control of cell death. *Trends Plant Sci,* 8:335-342.

70. Padmanabhan, C., Zhang, X. and Jin, H. 2009. Host small RNAs are big contributors to plant innate immunity. Curr. Opin. *Plant Biol,* 12:465-472.

71. Palatnik, J.F., Allen, E., Wu, X., Schommer, C., Schwab, R., Carrington, J.C. and Weigel, D. 2003. Control of leaf morphogenesis by microRNAs. *Nature,* 425:257-263.

72. Park, M.Y., Wu, G., Gonzalez-Sulser, A., Vaucheret, H. and Poethig, R.S. 2005. Nuclear processing and export of microRNAs in *Arabidopsis. Proc. Natl. Acad. Sci. U. S. A,* 102:3691-3696.

73. Parker, J.S., Roe, S.M. and Barford, D. 2005. Structural insights into mRNA recognition from a PIWI domain-siRNA guide complex. *Nature,* 434:663-666.

74. Pfaffl, M.W. 2001. A new mathematical model for relative quantification in realtime RT-PCR. *Nucleic Acids Res,* 29: e45. Pieterse, C.M.J., Leon-Reyes, A., Van der Ent, S. and Van Wees, S.C.M. 2009. Networking by small-molecule hormones in plant immunity. *Nat. Chem. Biol,* 5:308-316.

75. Pratt, A.J. and MacRae, I.J. 2009. The RNAinduced silencing complex: a versatile gene-silencing machine. J. *Biol. Chem,* 284:17897-17901.

76. Reyes, J.L. and Chua, N.H. 2007. ABA induction of miR159 controls transcript levels of two MYB factors during Arabidopsis seed germination. *Plant J,* 49:592-606.

77. Rhoades, M.W., Reinhart, B.J., Lim, L.P., Burge, C.B., Bartel, B. and Bartel, D.P. 2002. Prediction of plant microRNA targets. *Cell,* 110:513-520.

78. Robert-Seilaniantz, A., Grant, M. and Jones, J.D.G. 2011. Hormone crosstalk in plant disease and defense: more than just jasmonate-salicylate antagonism. *Annu. Rev. Phytopathol,* 49:317-343.

79. Rodriguez, A., Griffiths-Jones, S., Ashurst, J.L. and Bradley, A. 2004. Identification of mammalian microRNA host genes and transcription units. *Genome Res,* 14:1902-1910.

80. Ruiz-Ferrer, V. and Voinnet, O. 2009. Roles of plant small RNAs in biotic stress responses. *Annu. Rev. Plant Biol,* 60:485- 510.

81. Schommer, C., Palatnik, J.F., Aggarwal, P., Chételat, A., Cubas, P., Farmer, E.E., Nath, U. and Weigel, D. 2008. Control of jasmonate biosynthesis and senescence by miR319 targets. *PLoS Biol,* 6:e230.

82. Schubert, T.S., Rizvi, S.A., Sun, X., Gottwald, T.R., Graham, J.H. and Dixon, The role of microRNAs and phytohormones in plant immune system 41 W.N. 2001. Meeting the challenge of eradicating citrus canker in Florida- Again. *Plant Dis,* 85:340-356.

83. Smalheiser, N.R. and Torvik, V.I. 2005. Mammalian microRNAs derived from genomic repeats. *Trends Genet,* 21:322- 326.

84. Sunkar, R., Li, Y.F. and Jagadeeswaran, G. 2012. Functions of microRNAs in plant stress responses. *Trends Plant Sci,* 17:196-203.

85. Sunkar, R. and Zhu, J.K. 2004. Novel and stress-regulated microRNAs and other small RNAs from *Arabidopsis. Plant Cell,* 16:2001-2019.

86. Tao, Y., Xie, Z., Chen, W., Glazebrook, J., Chang, H.S., Han, B., Zhu, T., Zou, G. and Katagiri, F. 2003. Quantitative nature of Arabidopsis responses during compatible and incompatible interactions with the bacterial pathogen *Pseudomonas syringae. Plant Cell,* 15:317-330.

87. Thaler, J.S. and Bostock, R.M. 2004. Interactions between abscisic-acidmediated responses and

plant resistance to pathogens and insects. *Ecology,* 85:48- 58.

88. Tomari, Y., Matranga, C., Haley, B., Martinez, N. and Zamore, P.D. 2004. A protein sensor for siRNA asymmetry. *Science,* 306:1377-1380.

89. Ton, J., Flors, V. and Mauch-Mani, B. 2009. The multifaceted role of ABA in disease resistance. *Trends Plant Sci,* 14:310-317.

90. Tsuda, K., Sato, M., Stoddard, T., Glazebrook, J. and Katagiri, F. 2009. Network properties of robust immunity in plants. *PLoS Genet,* 5:e1000772.

91. Unver, T., Bakar, M., Shearman, R.C. and Budak, H. 2010. Genome-wide profiling and analysis of *Festuca arundinacea* miRNAs and transcriptomes in response to foliar glyphosate application. *Mol. Genet. Genomics,* 283:397-413.

92. Vlot, A.C., Dempsey, D.A. and Klessig, D.F. 2009. Salicylic acid, a multifaceted hormone to combat disease. *Annu. Rev. Phytopathol,* 47:177-206.

93. Vlot, A.C., Klessig, D.F. and Park, S.W. 2008. Systemic acquired resistance: the elusive signal (s). *Curr. Opin. Plant Biol,* 11:436-442.

94. Voinnet, O. 2008. Post-transcriptional RNA silencing in plant-microbe interactions: a touch of robustness and versatility. *Curr. Opin. Plant Biol,* 11:464-470.

95. Voinnet, O. 2009. Origin, biogenesis, and activity of plant microRNAs. *Cell,* 136:669-687.

96. Wang, D., Pajerowska-Mukhtar, K., Culler, A.H. and Dong, X. 2007. Salicylic acid inhibits pathogen growth in plants through repression of the auxin signaling pathway. *Curr. Biol,* 17:1784-1790.

97. Wasilewska, A., Vlad, F., Sirichandra, C., Redko, Y., Jammes, F., Valon, C., Frei dit Frey, N. and Leung, J. 2008. An update on abscisic acid signaling in plants and more... *Mol. Plant,* 1:198-217.

98. Xie, Z., Khanna, K. and Ruan, S. 2010. Expression of microRNAs and its regulation in plants. *Semin Cell Dev Biol,* 21:790-797.

99. Zeier, J., Delledonne, M., Mishina, T., Severi, E., Sonoda, M., Lamb, C. 2004. Genetic elucidation of nitric oxide signaling in incompatible plant-pathogen interactions. *Plant Physiol,* 136:2875- 2886.

100. Zhang, W., Gao, S., Zhou, X., Chellappan, P., Chen, Z., Zhou, X., Zhang, X., Fromuth, N., Coutino, G. and Coffey, M. 2011. Bacteria-responsive microRNAs regulate plant innate immunity by modulating plant hormone networks. *Plant Mol. Biol,* 75:93-105.

101. Zipfel, C., Kunze, G., Chinchilla, D., Caniard, A., Jones, J.D.G., Boller, T. and Felix, G. 2006. Perception of the Bacterial PAMP EF-Tu by the Receptor EFR Restricts *Agrobacterium*-Mediated Transformation. Cell, 125:749-760.

4

Marker assisted selection for the improvement of Sarjoo-52 for drought tolerance by introgression of MQTL1.1 from the source Nagina–22

S. Awasthi* & J. P. Lal

Department of Genetics and Plant Breeding, Institute of Agricultural Sciences, Banaras Hindu University, Varanasi, India.
*Corresponding Author, Email: sand.saumya@gmail.com

Abstract

Literatures have reported that a lot of drought related genes were cloned and individual gene showed positive effects under controlled stress experiments, but were not much effective in the field. Although, the progresses by conventional breeding approaches were achievable as some drought varieties have been released to the farmers in the recent years but this is not adequate to cope up with the future demand of high yield for rice, as drought seems to spread to more regions and seasons. Therefore, marker assisted selection came into lime light for accelerating and giving pace to plant breeding.From the cross (Sarjoo- 52× Nagina-22) × Sarjoo- 52, plants were selected on the basis of presence of gene $MQTL_{1.1}$ responsible for the drought tolerance. These lines have been subjected to further breeding and trial tests. Agronomic performances and physiological behavior of these lines are also under track. The results showed that the variety Sarjoo 52 could be efficiently converted to a drought tolerant variety in a backcross generation followed by selfing and selection, involving a time of two to three years. Polymorphic markers for foreground and background selection were identified for the high yielding variety to develop a wider range of drought tolerant variety to meet the needs of farmers in the drought-prone regions. This approach demonstrates the effective use of marker assisted selection for a major QTL in a molecular breeding program.

Keywords: Background Selection, Drought, Foreground Selection, Marker Assisted Selection, Oryza Sativa, QTLs.

Introduction

More than three billion people in the world depends upon rice (*Oryza sativa* L.) and most of them comes from Asia. Surprisingly, ninety percent of world's total rice production is cultivated and consumed in Asia. Rice is a crop which is being cultivated under diverse ecologies, from irrigated to rainfed upland to rainfed lowland to deep water. Irrigated rice accounts for 55% of world area and about 75% of total rice production. Rainfed lowland represents about 25% of total rice area, accounting for 17% of world rice production. Upland rice covers 13% of the world rice area and accounts for 4% of global rice production. Rice being a water loving crop, is more prone to drought as compared to other cereal crops which shows better adaptation to lesser water availability.

Studies on the plant response to water stress are becoming increasingly important, as most of climatic change scenarios suggest an increase in aridity in many areas of the globe (Petit *et al.*, 1989). On a global basis, drought in conjunction with high temperature and radiation,is known to be the most important environmental constraints to plant survival and to crop productivity. As irrigation water is not adequate as per crop requirement, the possible solutions to improve field productivity are i) environment control development i.e. improve plant living environment to fit the needs of crop, this includes technologies which reduce soil and water loss, decrease soil water evaporation, increase and maximize the use of soil water storage, collect non cultivate field run offs and use them as irrigation supplement. ii) Approach of biological water saving i.e. modify plant to adapt the dry environment; this includes genetic modification of plant, physiological regulation and application of crop complementary effort. As a matter of fact, management practices can contribute to increase yield in moisture stress environments but major progress will be realized through genetic improvement and therefore through plant breeding and molecular breeding, it would be better to develop drought tolerant varieties than to irrigate drylands.

Despite the importance of drought as a constraint, little effort has been devoted to developing drought-tolerant rice cultivars. In drought years, high yielding varieties inflict high yield losses, leading to a sudden decline in the country's rice production. Farmers of drought-prone areas require varieties that provide them with high yield in years of good rainfall and sustainable good yield in years with drought. The earlier approach of improving grain yield under drought through selection on secondary traits such as root architecture, leaf water potential, panicle water potential, osmotic adjustment, and relative water content (Fukai et al., 1999; Price and Courtois, 1999; Jongdee et al., 2002; Pantuwan et al., 2002) did not yield the expected results to improve yield under drought. Breeders and physiologists practiced selection for secondary traits as several earlier studies reported low selection efficiency for direct selection for grain yield under drought stress (Rosielle and Hamblin, 1981; Blum, 1988; Edmeades et al., 1989). Similarly, at the molecular level, initial efforts in rice were devoted to mapping of QTLs for secondary drought-related traits such as root morphology and osmotic adjustment (Yadav et al. 1997; Kamoshita et al. 2002; Babu et al. 2003). Because QTLs for secondary traits are not linked to direct yield increase under drought, marker-assisted selection for such QTLs has not been successfully used to improve yield under drought stress in rice.

Upto now, a lot of drought related genes were cloned and individual gene showed positive effects under controlled stress experiments, but were not effective in the field. However, the progresses by conventional breeding approaches were achievable as some drought varieties have been released to the farmers in the recent

years. Although, this is not adequate to cope up with the future demand of high yield for rice, as drought seems to spread to more regions and seasons across the country. Therefore, marker assisted selection came into lime light for accelerating and giving pace to plant breeding.

Mapping studies are performed to detect linkage of a molecular marker to a gene affecting a trait of interest. It then becomes possible to select for the desirable allele of those genes based on marker genotype rather than, or in addition to, field phenotype (Jongdee et al., 2002). This technique, known as marker assisted selection (MAS), is theoretically more reliable than selection based solely on phenotype, as a marker tightly liked to the desirable gene would represent selection with a heritability of near unity for that specific gene (Bernardo, 2002). Marker-assisted selection may be useful to improve traits that are either controlled by a few genes or where phenotypic evaluation is difficult/costly to perform. The relative difficulty associated with drought- resistance phenotyping suggests that there is scope for the use of MAS in breeding for drought resistance (Bernardo, 2002).

The effectiveness of MAS depends on the availability of closely linked markers and/or flanking markers for the target locus, the size of the population, the number of backcrosses and the position and number of markers for background selection (Frisch et al. 1999a; Frisch and Melchinger 2005). MAS has previously been used in rice breeding to incorporate the bacterial blight resistance gene Xa21 (Chen et al. 2000) and waxy gene (Zhou et al. 2003) into elite varieties.

The identification of QTLs with a major effect on grain yield raises a new hope of improving grain yield under drought through marker assisted breeding. The availability of the major effect QTL for drought tolerance, a theoretical framework for marker assisted selection and the existence of intolerant varieties that are widely accepted by farmers provides an opportunity to develop cultivars that would be suitable for larger areas of drought prone rice (Mackill, 2006).

Considering the above aspects, the present study was proposed to exploit the gene action and variability through marker assisted selection with the objective of isolating promising high yielding drought tolerant recombinants through conventional and marker assi- sted selection (MAS) approaches.

Materials and method

The present investigation was conducted during three seasons i.e. 2010, 2011 and 2012 at Agricultural Research Farm, Institute of Agricultural Sciences, Banaras Hindu University, Varanasi and off season, 2010- 2011 at Central Rice Research Institute (C.R.R.I.) Cuttack, Odisha.Sarjoo –52, derived from T(N)1×Kashi was notified in 1982 for general cultivation. It takes approximately 130- 133 days after sowing DAS to harvest. Sarjoo–52 is an irrigated, semi dwarf (98cm) and erect type. Grains are long, bold, abdominal white present (AWP) and white. It is moderately resistant to bacterial leaf blight BLB. It is reported to yield 50-60 q/ha. It is mainly grown in Uttar Pradesh. Nagina 22, a selection from Rajbhog was used as donar parent. Nagina–22 was notified in 1978. It takes around 85-102 days. Grains are short, bold and white. The variety is susceptible to blast, BLB and resistant to drought. It gives yield of about 20-25 Q/ha and grown well in Uttar Pradesh as upland crop.

The seeds of drought tolerant variety (Nagina-22) and drought susceptible variety (Sarjoo–52) were sown in raised nursery beds in the last week of June, 2010 at the Research Farm, Banaras Hindu University. Twenty one days old seedlings were transplanted in the well puddled field at a spacing of 30 × 15 cm between row to row and plant to plant, respectively with row length of 3 m in a crossing block on three different dates at the interval of seven days in three replications. Standard agronomic practices were followed to raise good crop. Five rows of Nagina-22 (donar) were transplanted in separate blocks on different dates at the interval of 7 days to synchronize the flowering for making the crosses.

Half of the F_1 seeds of the cross with their parents were transplanted at the Research Farm at Central Rice Research Institute (C.R.R.I.), Cuttack, at the spacing of 30 × 15 cm between row to row and plant to plant, respectively with a row length of 3.0m in three replications.

Screening of the F_1s was done at different growth stages i.e., seedling stage and vegetative stage. Depending upon the screening test, backcross with drought susceptible parents was done.

Seeds from thebackcross (BC_1) and F_1 plants were harvested separately. Marker validation and hybrid confirmation were done before commencing marker assisted selection. F_1 plant progeny, along with the parents were grown to raise F_2 generation. Compact family randomized block design with three replications was followed. Drought susceptible and drought tolerant plants were screened on the basis of leaf rolling and were harvested separately. Fresh crosses were also made to get F_1 seeds.

BC_1F_1 along with the parents were also grown in compact family randomized block design in three replications for phenotypic study.20 plants in parents F_1s and 50 plants in F_2 per replication were selected for MAS. Heterozygous plants selected on the basis of marker assisted selection were allowed to self to produce BC_1F_2 generation.BC_1F_2 plants were again subjected to second round of marker assisted selection.

Young leaves were collected from 20-25 days old seedlings and immediately stored in -20°C till further processing. The DNA was extracted following CTAB extraction method (Doyle and Doyle, 1987). Polymerase chain reaction was performed to selectively amplify *in vitro* a specific segment of the total genomic DNA to a billion fold (Mullis *et al.,* 1986). The most essential requirement of PCR is the availability of a pair of short (typically 20-25bp nucleotides) primers having sequence complementary to either end of the target DNA segment (called template DNA) to be synthesized in large amount. The components of the PCR reaction were first added in a sterilized 1.5ml microcentifuge tube and then mixed thoroughly by vortexing. To each PCR tubes (0.2ml), 14 μl of reaction mixture was distributed, and finally template DNA of individual rice genotpyes was added. The tubes containing reaction mixture were placed in the wells of the thermal cycler block (Eppendorf Thermo–cycler, USA) and amplification reaction was carried out with the thermalcycler programme. 40 cycles of denaturation, annealing and extension was programmed for the study. The amplified DNA fragments generated through SSR primers were resolved through electrophoresis in 2.5% agarose gel prepared in TAE [242g Tris– base; 57.1ml glacial acetic acid and 100ml 0.5 M EDTA (pH 8.0) bring final volume to 1000ml] buffer. Ethidium bromide solution at a final concentration of 0.03 ng/μl was added to the agarose solution.

For electrophoresis, 15μl of the PCR product was mixed with 2μl of 6x loading dye (0.25% bromophenol blue in 30% glycerol) and loaded in the slot of the agarose gel. In order to determine the molecular size of the amplified products, each gel was also loaded with 6 μl of 50 bp DNA size marker (Fermentas, USA). Gel electrophoresis was performed at a constant voltage of 65V for about 3.5 hours. Finally, the gels were visualized under a UV light source in a gel documentation system (Gel Doc^{TM} XR +, BIO RAD, USA) and the images of amplification products were captured and stored in a computer for further analysis and future use.

19 SSR markers linked to the QTLs for drought tolerance on various linkage groups were used for foreground selection to select the individuals presumably having the donor allele. Particular target (drought tolerance QTL) was flanked by these markers. The tighter the markers are linked to the QTL, the greater the chance that the QTL mapped between a pair of flanking marker has indeed been transferred. Therefore, phenotypic testing of final products of the MAS exercise needs to be performed in order to confirm the transfer of drought tolerance QTL. At the same time selected markers unlinked to drought tolerance have been used to select those individuals with minimal linkage drag (background selection).

Results:

Parental polymorphism survey

Initially, parental polymorphism survey was performed among parental genotypes, Sarjoo–52 (drought susceptible) and Nagina-22 (drought tolerant). On the basis of parental polymorphism survey, 19 SSR markers were used to validate for the drought tolerance in the back-cross population. Flanking markers RM 212- RM 3825, produced reproducible and polymorphic bands. These markers clearly distinguished drought susceptible and tolerant parents. In rice, hundreds of microsatellite markers were developed which are publicly available and are being used for MAS, gene tagging, mapping and phylogenetic studies.

Hybrid confirmation in F_1's

Oryza sativa is basically a self-pollinated crop, with limited degree of outcrossing (< 0.5%). The factors limiting the receptivity of rice flowers to outcrossing include a short style and stigma (1.5 to 4 mm in combined length), short anthers, limited pollen viability and brief period between opening of florets and release of pollen (between 30 seconds and 9 minutes) (Morishima, 1984; Oka, 1988). It therefore became essential to confirm the true type hybrid condition in the F_1s. SSR markers RM 212- RM 3825 (linked with MQTL$_{1.1}$) which are co - dominant in nature, gave promising results. Heterozygous plants were selected for further generation advancement and backcrossing.

Production of BC_1F_1 generation

Five plants in F_1 generation of the cross Sarjoo-52 × Nagina-22 were back crossed with the recurrent parent Sarjoo-52 to produce around 250 seeds at Agricultural farm, Institute of Agricultural Sciences, Banaras Hindu University, Varanasi.

Marker assisted foreground selection in BC_1F_1 and production of BC_1F_2 generation

BC_1F_1 plants were screened for the presence of *MQTL$_{1.1}$* with the linked and validated SSR markers RM 212–RM 3825, respectively from the cross (Sarjoo-52 × Nagina-22) × Sarjoo-52. SSR markers used in the study are co-dominant in nature therefore, in BC_1F_1 population, two types of banding patterns were amplified i.e., homozygous susceptible type and heterozygous types.

The segregation of BC_1F_1 plants into drought tolerant and susceptible can be seen clearly in the representative gel picture of screening of 107 BC_1F_1 plants for *MQTL$_{1.1}$* with linked molecular marker RM 212–RM 3825 (Figure1).

RM 212

RM 3825

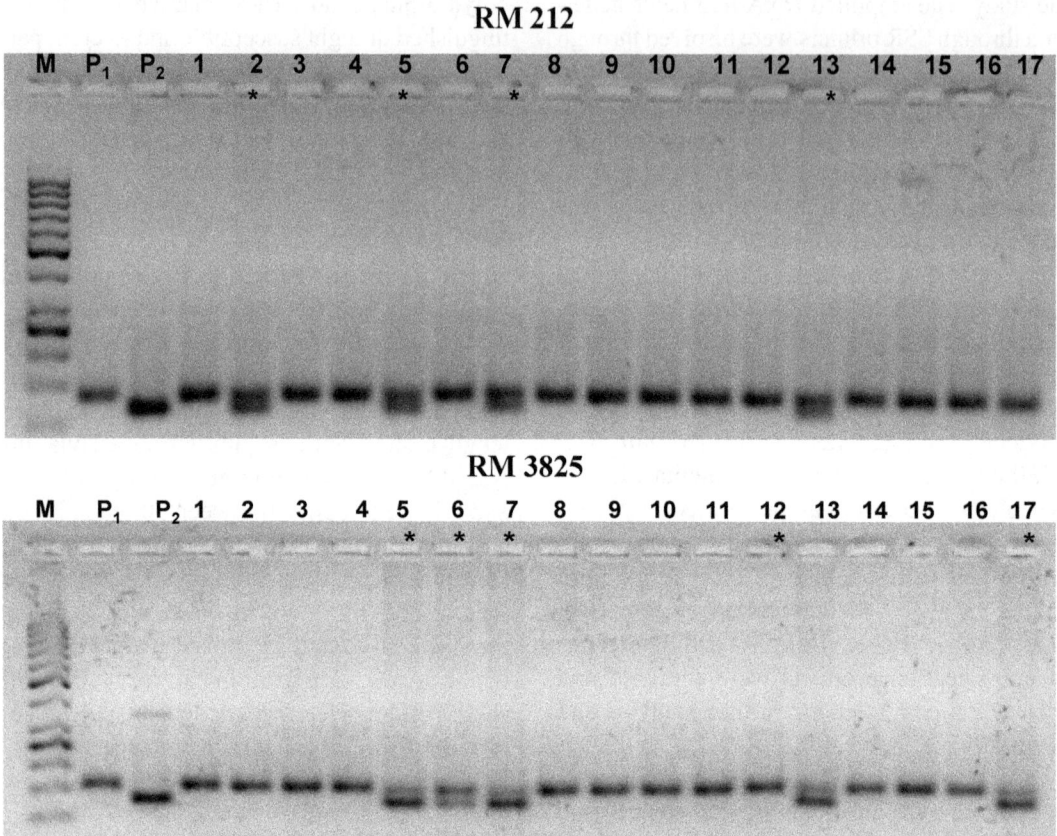

Figure 1. Representative gel picture of foreground selection for MQTL1.1 in BC1F1 generation of (Sarjoo-52×Nagina–22) × Sarjoo-52).

Production of BC₁F₂ generation

Since, flanking markers were used in the study, emphasis was given on the selection of only those plants which exhibited heterozygous banding pattern for both the markers i.e. RM 212–RM 3825. Thus plant number SA-C-5, SA– C–7, SA–C–24, SA–C–25, SA–C–31, SA–C–39 and SA–C–65 from the cross (Sarjoo-52 × Nagina-22) × Sarjoo-52 were selected. Based on agronomic performance and drought related traits, three top performers were selected and allowed to self to produce BC_1F_2 seeds.

Marker assisted foreground and background selection in BC₁F₂ generation and production of BC₁F₃ seeds

Foreground selection: Selected BC_1F_2 plants from the cross (Sarjoo-52 × Nagina-22) × Sarjoo-52 was screened for $MQTl_{1.1}$ with the linked and validated markers (Figure 2).

Background selection:

The gene positive plants in BC_1F_2 generation were taken for background selection with polymorphic primers (Figure 3). After, background selection in BC_1F_2, 70 loci became homozygous out of 73 polymorphic loci between Sarjoo–52 and Nagina-22+MQTL1.1 in plant no. SA–C–10–23 of the cross (Sarjoo–52 × Nagina-22) × Sarjoo–52. Maximum genome recovery in BC_1F_2 with $MQTL_{1.1}$ was about 83.5%.

Discussion

The present study indicated that MAS strategy is an effective means of utilizing QTLs with large effects in rice breeding programs. Sarjoo 52 is an elite Indian rice variety that was first introduced in India way back in 1982 and still widely grown in many areas of the country due to it stable yield, aromatic and good quality. Hence, Sarjoo 52 was

RM 212

M P₁ P₂ 1 2 3 4 5 6 7 8 9 10 11 12 13 14 15 16 17

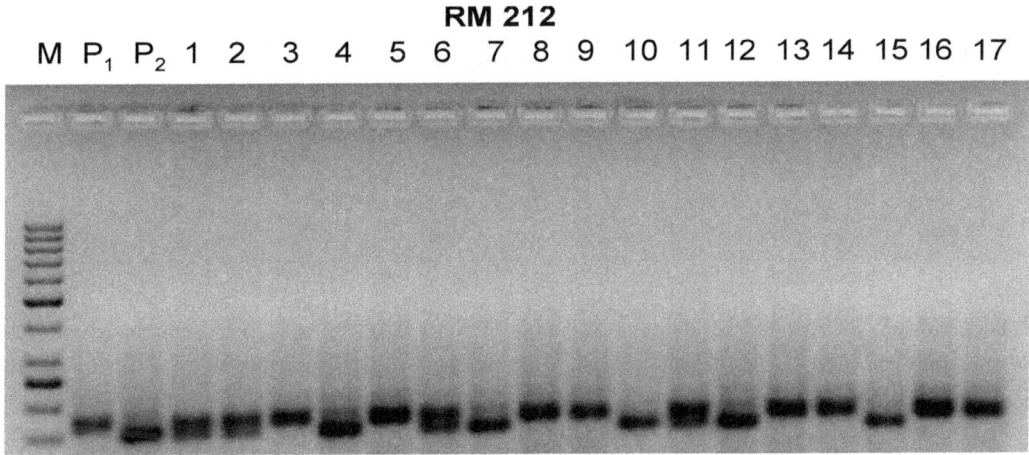

RM 3825

M P₁ P₂ 1 2 3 4 5 6 7 8 9 10 11 12 13 14 15 16 17

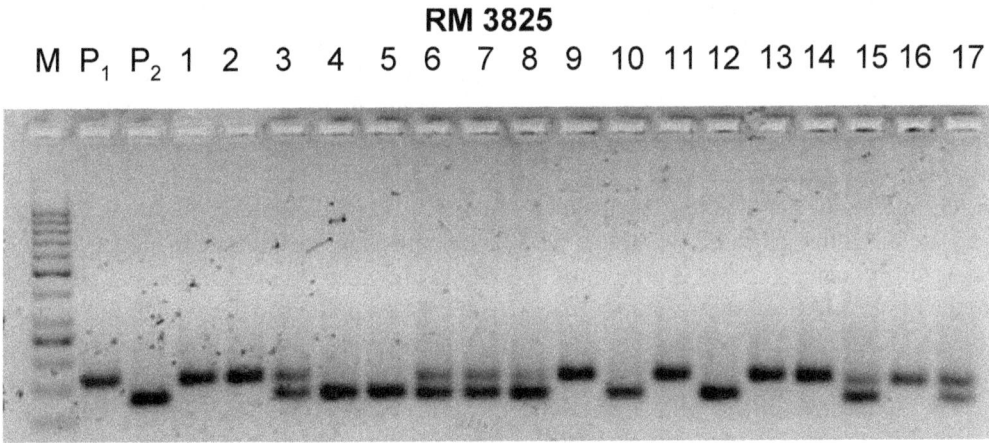

Figure 2. Representative gel picture of foreground selection for MQTL1.1 in BC1F2 generation of (Sarjoo-52 × Nagina–22) × Sarjoo 52).

M P₁ P₂ 1   2   3   4   5   6   7   8   9   10 11 12 13 14 15 16 17 18 19

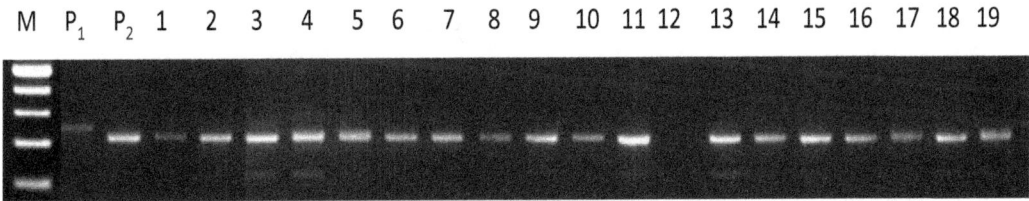

Figure 3. Representative gel picture of background selection in BC_1F_2 of (Sarjoo–52× Nagina-22) × Sarjoo–52 using polymorphic markers.

selected as a recipient parent. Marker assisted foreground selection was proposed by Tanksley (1983) and investigated in the context of introgression of tolerant genes by Melchinger (1990). If in BC_1 generation more than one individual satisfying the strongest condition is found, selection between them can be performed on the basis of analysis of other marker loci (located either on the carrier or on non carrier chromosome) to determine the most desirable individual for producing BC2 (Tanksley et al., 1989).

Table 1. Details of SSR primers associated with drought QTL *MQTL$_{1.1}$*.

QTL	Flanking markers	LG	Forward sequence	Reverse sequence	R^2	Tm	Reference
Meta QTL	RM 212	1	CCACTTTCAGCTACTACCAG	CACCCATTTGTCTCTCATTATG	12.1	55°C	Salunke *et al.*, 2011
	RM 3825		AAAGCCCCCAAAAGCAGTAC	GTGAAACTCTGGGGTGTTCG		60°C	Salunke *et al.*, 2011

The success of marker assisted backcross breeding (MAB) depends upon several factors, including the distance between the closest markers and the target gene, the number of target genes to be transferred, the genetic base of the trait, the number of individuals that can be analyzed and the genetic background in which the target gene has to be transferred, the type of molecular marker (s) used and available technical facilities (Weeden *et al.*, 1992; Francia *et al.*, 2005). Identification of molecular markers that should co- segregate or be closely linked with the desired trait (if possible, physically located beside or within genes of interest) is a critical step for the success of MAB. The most favourable case for MAB is when the molecular marker is located directly within the gene of interest (direct markers). MAB conducted using direct markers is called gene assisted selection (Dekkers, 2003). Alternatively, the marker is genetically linked to the trait of interest. Before a breeder can utilize linkage–based associations between a trait and markers, the associations have to be assessed with a certain degree of accuracy so that marker genotypes can be used as indicators or predictors of trait genotypes and phenotypes.

The lower the genetic distance between the marker and the gene, the more reliable is the application of the marker in MAS. However only in few cases will the selected marker allele be separated from the desired trait due to a recombination event (appearance of false positives). The presence of a tight linkage between desirable trait(s) and a molecular marker(s) may be useful in MAS to increase gain from selection. Based on the studies by Lee (1995) and Ribaut *et al.* (2002), it could be generalized that whenever a target gene is introduced for the first time from either wild or unadapted germplasm, flanking markers as close as 2cM is considered an ideal option, while in the transfer of the same target gene in subsequent phases from elite into elite lines, positioning the flanking markers near by might be effective in reducing the required size of the backcross population.

MAS have generated a good deal of expectations, which in some cases has leaded to over optimism and in others to disappointment because many of the expectations have not yet been realized. Although documentation is limited the current impact of MAS on products delivered to farmers seems to be small. New developments and improvements in marker technology, the integration of functional genomics with QTL mapping and the availability of more high density maps are the other factors that will greatly affect the efficiency and effectiveness of QTL mapping and MAS in the future. The development of high density maps that incorporate new marker types, such as single nucleotide polymorphism (SNPs) and expressed sequence tags (EST) will provide researchers with a great arsenal of tools for QTL mapping and MAS.

Conclusion

Cross (Sarjoo–52 × Nagina-22) × Sarjoo–52 exhibited polymorphism for the presence of a single gene for drought tolerance i.e. MQTL1.1. The selected lines have been subjected to further breeding and trial tests. Agronomic performances and physiological behavior of these lines are also under track. The results showed that the variety Sarjoo-52 could be efficiently converted to a drought tolerant variety in a backcross generation followed by selfing and selection, involving a time of two to three years. Polymorphic markers for foreground and background selection. This study could have a good impact in rice breeding and it is applicable for the introduction of important agronomic traits into the genomes of popular rice cultivars.

References

1. Babu, RC., Nguyen, B.D., Chamarerk, V., Shanmugasundaram, P., Chezhian, P., Jeyaprakash, P., Ganesh, SK., Palchamy, A., Sadasivam, S., Sarkarung, S., Wade, LJ. and Nguyen, HT. 2003. Genetic analysis of drought resistance in Rice by molecular markers, Association between Secondary Traits and Field Performance, *Crop sci*, 43, 1457- 1469.

2. Bernardo, R. 2002. Breeding for quantitative traits in plants, Stemma Press, Wood- burn, Minnesota.

3. Blum, A. 1988. Plant Breeding for Stress Environments, CRC Press, Boca Raton, FL.

4. Boyer, J.S. 1982. Plant productivity and environment.Science.218: 443.

5. Chen, S., Lin, X.H., Xu, C.G. and Zhang, Q. 2000. Improvement of bacterial blight resistance of "Minghui 63', an elite restorer line of hybrid rice, by molecular marker assisted selection. *Crop Sci.*40 (1): 239- 244.

6. Dekkers, JCM. 2003. commercial appli- cation of marker and gene assisted selection in livestock: stategies and lessons, Paper presented at the 54th annual meeting of the European associa- tion for animal production, Rome, Italy, 31st August–3rd September.

7. Edmeades, GO., Bolanos, J., Lafitte, HR., Rajaram, S., Pfeiffer, W. and Fischer, RA. 1989. Traditional approaches to breeding for drought resistance in cereals, In, Baker, F,W,G, (ed,), Drought Resistance in Cereals, CAB International, Wallingford, Oxon, UK, 27-52.

8. Francia, E., Tacconi, G., Crosatti, C., Barabaschi, D., Bulgarelli, D., Dall'Aglio, E. and Vale, G. 2005. Marker assisted selection in crop plants. *Plant Cell Tissue Organ Cult,*82, 317- 342.

9. Frisch, M., Bohn, M. and Melchinger, A.E. 1999a. Minimal sample size and optimal positioning of flanking markers in markr assisted backcrossing for transfer of a target gene. *Crop sci.*39: 967- 975.

10. Frisch, M. and Melchinger, A.E. 2005. Selection theory for Marker assisted backcrossing. *Genet.*170: 909- 917.

11. Fukai, S., Pantuwan, G., Jongdee, B. and Cooper, M. 1999. Screening for drought resistance in rainfed lowland rice, *Field Crop Res,*64, 61-74.

12. Jongdee, B., Fukai, S. and Cooper, M. 2002. Leaf water potential and osmotic adjustment as physiological traits to improve drought tolerance in rice, *Field Crop Res*, 76, 153-163.

13. Kamoshita, A., Zhang, J., Siopongco, J., Sarkarung, S., Nguyen, HT. and Wade, LJ. 2002.

14. Lee, M. 1995. DNA markers and plant bree- ding programs, *AdvAgron*, 55, 265- 344.

15. Mackill, D.J. 2006. Breeding for resistance to abiotic stresses in rice: the value of QTLs. Plant breeding: The Arnel R Hallauer International Symposium.

16. Melchinger, AE. 1990. Use of molecular markers in breeding for oligogenic disease resistance, *Pl Breeding,* 104(1), 1- 19.

17. Morishima, H. 1984. Wild plants and domestication, In, Tsunoda, S, and Takahashi, N, (ed) Biology of rice, Elsevier, Amsterdam, 3-30.

18. Oka, HI. 1988. Origin of cultivated rice, Elsevier, Amsterdam.

19. Pantuwan, G., Fukai, S., Cooper, M., Rajataseeekul, S. and O' Toole, JC. 2002. Yield response of rice (Oryza sativa L,) genotypes to drought under rainfed lowland, Plant factors contri- buting to drought resistance, *Field Crop Res*, 73, 181- 200.

20. Petit, J.R., Jouzel, J. and Raynaud, D. 1999. Climate and atmospheric history of the past 420 000 years from the Vostok ice core, Antarctica. Nature. 399: 429–436.

21. Price, A. and Courtois, B. 1999. Mapping QTLs associated with drought resistance in rice, Progress, problems and prospects, *Plant Growth Regul,* 29, 123-133.

22. Ribaut, J.M., Banziger, M., Betran, J., Jiang, C., Edmeades, GO., Dreher, K. and Hoisington, D. 2002. Use of molecular markers in plant breeding: drought tolerance improvement in tropical maize , in Kang MS(ed) Quantitative genetics, genomics and plant breeding, Wallingford, CABI, 85- 99.

23. Rosielle, A.A., and Hamblin, J. 1981. Theoretical aspects of selection for yield in stress and non-stress environments, *Crop Sci*, 21, 943-946.

24. Salunkhe, AKS., Poornima, R., Prince, KSJ., Kanagaraj, P., Sheeba, JA., Amudha, K., Suji, KK., Senthil, A. and Babu, RC. 2011. Fine mapping QTL for drought resistance traits in rice (Oryza sativa L,) using bulk segregant analysis, *Mol-Biotechnol,*49, 90- 95.

25. Singh, MP. 2009. Rice productivity in India under variable climates, www, niaes, affrc,go,jp/marco/marco2009/english/„,/ W2-02_Singh_P.pdf.

26. Tanksley, SD. 1983. Molecular markers in plant breeding.*Plant MolBiol,* 1: 3–8.

27. Tanksley, SD., Young, ND., Paterson, AH. and Bonierbale, MW. 1989. RFLP mapping in plant breeding, new tools for an old science, *Biotechnol,* 7, 257–264.

28. Vikram, P., Swamy, BPM., Dixit, S., Ahmed, HA., Cruz, MTS., Singh, AK., Yec, G. and Kumar, A. 2012. Bulk segre- gant analysis, "An effective

Effects of phenotyping envi- ronment on identification of quantitative trait loci for rice root morphology under anaerobic conditions, *Crop Sci,*42, 255- 265.

approach for mapping consistent-effect drought grain yield QTLs in rice", *Field Crops Res,* 134, 185–192.

29. Weeden, NR., Muehlbauer, FJ. and Ladi- zinky, G. 1992. Extensive conservation of linkage relationships between pea and lentil genetic maps, *J. Hered,* 83, 123- 129.

30. Yadav, R., Courtois, B., Huang, N. and McLaren, G. 1997. Mapping genes con- trolling root morphology and root distribution in a double-haploid popul- ation of rice. *Theor Appl Genet.* 94, 619-632.

31. Zhou, P., Tan, Y., He, Y., Xu, C. and Zhng, Q. 2003. Simultaneous improvement for four quality traits of Zhenshan 97, an elite parent of hybrid rice, by molecular marker-assisted selection. Theor Appl Genet.106(2): 326- 331.

5

Identification of QTLs for grain yield and some agro-morphological traits in sunflower (*Helianthus annuus* L.) using SSR and SNP markers

N. Eyvaznejad[1] & R. Darvishzadeh[1,2,*]

1. Department of Plant Breeding and Biotechnology, Urmia University, Urmia, Iran.
2. Institute of Biotechnology, Urmia University, Urmia, Iran.
   *Corresponding author: r.darvishzadeh@urmia.ac.ir

Abstract

Many agriculturally important traits are complex, affected by many genes and the environment. Quantitative trait loci (QTL) mapping is a key tool for studying the genetic structure of complex traits in plants. In the present study QTLs associated with yield and agronomical traits such as leaf number, leaf length, leaf width, plant height, stem and head diameter were identified by using 70 recombinant inbred lines (RILs) from the cross (♀) PAC2 × RHA266(♂). RILs and their parents were evaluated in a rectangular 8×9 lattice design with two replications. High genetic variability and transgressive segregation were observed in all studied traits. Genetic gain representing the difference between 10% of selected RILs and their parents was significant for most of the studied traits. Positive and significant genotypic and phenotypic correlations were observed among the studied traits. QTL analysis was performed using a recently developed SSR and SNP sunflower linkage map. The map consists of 210 SSRs and 11 SNP markers placed in 17 linkage groups (LGs). The total map length is 1,653.1 cM with a mean density of 1 marker per 7.44 cM. Composite interval mapping (CIM) procedure detected 21 QTLs involved in genetic control of studied traits. The phenotypic variance explained by the identified QTLs varied from 1.13 to 73.70%. QTLs such as HMBPP associated with the expression of more than one trait could increase the efficiency of marker-assisted selection (MAS) and genetic progress in sunflower.

Keywords: Genetic variation, linkage map, molecular markers, QTL mapping, sunflower, yield-related traits.

Abbreviations: AFLP: amplified fragment length polymorphism, BIO: total dry mater, CIM: composite interval mapping, DNA: deoxyribonucleic acid, DHs: doubled haploids, F2: second filial generation, Gb: giga base pairs = 1,000,000,000 bp, GLM: general linear model, GYP: grain yield per plant, HD: head diameter, INDEL: Insertion-deletion polymorphisms as genetic markers in natural populations, LAF: leaf area at flowering, LAD: leaf area duration, LL: leaf length, LGs: linkage groups, LOD: logarithm (base 10) of odds, LW: leaf width, NL: number of leaf, PL: petiole length, PH: plant height, OP: osmotic potential; QTL: quantitative trait loci, RAPD: random amplified polymorphic DNA, RFLP: restriction fragment length polymorphism, RILs: recombinant inbred lines, SD: stem diameter, SF: days from sowing to flowering, SSR: simple sequence repeat, SNP: single nucleotide polymorphism, TRAP: target region amplification polymorphism.

Introduction

Sunflower is an annual and diploid (2n = 34) species from North America with an estimated genome size of 3.5 Gb (Baack *et al.* 2005). The crop is grown worldwide and performs well in most temperate climates of the world (Hu *et al.* 2010). It is the second largest hybrid crop and the fifth largest edible oilseed crop worldwide (Hu *et al.* 2010). Between 2000/01 and 2010/11, sunflower has been cultivated on 19-23 million hectares in 60 countries, producing 23-33 million tons of seed and 7.4-12.2 million tons of oil. The areas planted in Iran in 2012 were 68000 hectare with production of 90000 ton grain yield (http:// faostat.fao. org/ site/567/DesktopDefault.aspx?PageID=5

67#ancor). Sunflower seed contains high oil content ranging from 35-48% with some types yielding up to 50% (Skoric and Marinkovic 1986), 20-27% protein (Nazir *et al.* 1994) and high percentage of poly-unsaturated fatty acids (60%) including oleic acid (16.0%) and linoleic acid (72.5%), which control cholesterol in blood (Satyabrata *et al.* 1988). The majority of edible oil requirements of Iran are met through imports. Any breeding efforts that increase oily plants production can be effective in reducing the country's dependence on incoming oil.

Traits are classified into two categories including quantitative and qualitative. Quantitative traits are naturally complex and controlled by many genes (Kearsey and Pooni 1996). Understanding the genetic structure of quantitative traits is a long-term challenge for quantitative geneticists and plant breeders, who wish to design efficient breeding programs. Conventionally, the genetic properties of traits can be deciphered by partitioning the total variation into variation components caused by specific genetic effects (Kearsey and Pooni 1996). With recent advances in molecular genotyping and high-throughput technology, unrevealing the genetic architecture of complex traits has become possible via quantitative trait loci (QTL) analysis (Collard *et al.* 2005; De Vienne 2003). Molecular markers linked to QTLs/major genes of traits are being routinely identified in many crops by utilizing genetic linkage map developed from F_2, back cross families, recombinant inbred lines (RILs) or doubled haploids (DHs) population (Collard *et al.* 2005). DHs and RILs are genetically homozygous in genetic loci and can be mostly multiplied without genetic changes. These populations provide more accurate phenotypic data by testing multiple plants per line to minimize environmental effects. The seeds of DHs and RILs can be transferred from a laboratory to another place for collaborative works. However, only additive effects of genes are estimated in DHs and RILs populations (Kearsey and Pooni 1996; Darvishzadeh *et al.* 2007). A DH population is produced by doubling the gametes of F_1 or F_2 population. RILs are developed by single-seed selections from individual plants of an F_2 population. DH population is quicker to generate than RILs but

the production of DHs is only possible for species with well established protocols for pollen grains culture and chromosome doubling.

In sunflower, several molecular genetic linkage maps have been constructed on F_2 or RIL mapping populations using restriction fragment length polymerphism (RFLP), random amplified polymorphic DNA (RAPD), amplified fragment length polymorphism (AFLP), simple sequence repeat (SSR), insertion / deletion (INDELs) polymorphism, single nucleotide polymorphism (SNP) and target region amplification polymorphism (TRAP) markers (Berry *et al.* 1995; Gentzbittel *et al.* 1995, 1998; Jan *et al.* 1998; Flores-Berrios *et al.* 2000; Burke *et al.* 2002; Mokrani *et al.* 2002; Bert *et al.* 2003, 2004; Langar *et al.* 2003; Yu *et al.* 2003; Rachid Al- Chaarani *et al.* 2004; Lai *et al.* 2005; Hu *et al.* 2007; Poormohammad Kiani *et al.* 2007a; Yue *et al.* 2008).

Recently microsatellites or simple sequence repeat (SSR) and single nucleotide polymorphism (SNP) markers have gained substantial attention in sunflower studies (Tang *et al.* 2002). SSR are tandem repeats of short (1 to 6 bp) DNA sequences. The desirable features of SSR markers include their easy use, high information content, codominant inheritance pattern, even distribution along chromosomes, reproducibility, and locus specificity (Kalia *et al.* 2011). SNPs widely distributed in the genome are the most abundant type of DNA variation currently used as a genetic marker (Brooks 1999). Compared to markers based on size discrimination or hybridization, SNPs directly interrogate sequence variation and can reduce genotyping errors (Oliver *et al.* 2011).

Up to date, several QTL mapping studies were published in sunflower (Poormohammad Kiani and Sarrafi 2010). An RFLP/isoezymes F_2:F_3 mapping population (162 F3 plants) from a cross between two inbred lines (GH and PAC2) developed by INRA-France with 82 markers was used for mapping days from sowing to flowering, and three QTLs were detected on linkage group LG A, LG G and LG L (Mestries *et al.* 1998). The three QTLs explained 30% of the total phenotypic variation (R^2) and the type of gene action observed was consistent with partial dominance on LG L and overdominance on LG A and LG G. For the QTLs on LG A and LG G, the parent, GH, contributed to positive alleles and for the QTL on LG L, PAC2 alleles increased the trait. Leon *et al.* (1995) used a genetic map of 201 RFLP markers and identified six QTLs associated with 57% of genetic variation for oil content among the F2 progeny. Rachid Al-Chaarani *et al.* (2004) identified five QTLs for each of plant height and shoot diameter, and four QTLs for head diameter on different linkage groups in the RIL population of sunflower. LOD scores for these QTLs ranged from 4.35 to 13.66 and total phenotypic variances explained by the QTLs were 71% for plant height, 50% for shoot diameter and 25% for head diameter. Identification of molecular markers associated with genes controlling traits facilitates pyramiding multiple QTL alleles in the same genotype background and also progeny selection in segregation population via marker-assisted selection (MAS) (Sarrafi and Gentzbittel 2004; Mohler and Singrun, 2004). The objective of the present research was to identify the chromosomal regions involved in genetic variation of different agro- morphological traits using a linkage map developed by SSR and SNP markers on RILs population from the cross (♀) PAC2×RHA266(♂). This public RILs population has been widely used for genetic analysis of complex traits in sunflower (Rachid Al-Chaarani *et al.* 2004, 2005; Abou Al Fadil *et al.* 2007; Darvishzadeh *et al.* 2007; Poor- mohammad Kiani *et al.* 2007a, 2007b, 2008, 2009).

Materials and methods

Plant materials and experimental design

Seventy RILs developed via single seed descent (SSD) method from the cross (♀) PAC2 × RHA266 (♂) was used for QTL analysis (Flores Berrios *et al.* 2000). RHA266 is an American inbred line developed from a across between the wild species *H. annuus* and peredovik. PAC2 is an INRA-France inbred line developed from a cross between *H. petiolaris* and HA61 (Gentzbittel *et al.* 1995). RHA266 is a branched and high yielding line in comparison with PAC2 (Rachid Al-Chaarani *et al.* 2004). PAC2 is a line with high total dry matter per

plant and head weight per plant compared to RHA266. PAC2 showed high net photosynthesis and relative water content in water stressed condition compared to RHA266.

The experiment was conducted in 2011-2012 at the research farm of Shahid Beheshti Agricultural high school with the latitude and longitude of 37°/32' north and 45°/5' east and the height of 1313 m above sea level. Climate of the region was cold and semidry and the average rainfall and temperature according to 16 years statistics were 184 mm and 12°C, respectively. Seeds of RILs and their two parents, kindly provided by INRA (France), were cultivated in rectangular 8×9 lattice design with two replications. Each plot comprised 3 lines 3 m long spaced 65 cm. Plants spaced 25 cm in each raw. Irrigations were carried out when an amount of evaporated water (from Class 'A pan' evaporation) reached to 60 mm (Akbari *et al.* 2008). Weeds were mechanically controlled. 100 kg per hectare urea was distributed between rows at 8 leaf growth stage and irrigation was performed immediately. Any particular pest or disease was not seen during vegetative and reproductive growth stages. In order to prevent sparrows' damage during seed filling stage, the heads were covered by white envelopes. Seven traits including plant height (PH; cm), stem diameter (SD; cm), head diameter (HD; cm), number of leaf (NL), leaf length (LL; cm), leaf width (LW; cm) and petiole length (PL; cm) were measured on 5 plants per plot selected randomly at flowering stage. Grain yield per plant (GYP; g) was measured at maturity stage by harvesting five plants per plot.

Statistical and QTL analyses

Normality test were done by Proc Univariate in the SAS 9.2 software (SAS Institute Inc., Cary, N.C.). Variance analysis of phenotypic data was performed with Proc GLM in the SAS software. Estimation of phenotypic and genotypic correlations and their standard errors were realized by using multivariate restricted maximum likelihood with SAS Proc MIXED (Holland 2006). QTL analysis was performed using a recently developed sunflower linkage map (Haddadi *et al.* 2012; Amouzadeh *et al.* 2013) via composite interval mapping (CIM) in

Windows QTL Cartographer Version 2.5 (Wang *et al.* 2005). The map consists of 210 SSRs and 11 SNP markers placed in 17 linkage groups (LGs). The total map length is 1,653.1 cM with a mean density of 1 marker per 7.44 cM. Significant LOD for each trait was determined by permutation test (n=1,000 permutations) (Churchill and Doerge 1994). The genome was scanned at 2 cM intervals with a window size of 15 cM and up to 15 background markers were used as cofactors in the CIM analysis that was identified by the module Srmapqtl (model 6) of QTL Cartographer. The percentage of phenotypic variance (R^2) explained by each QTL was estimated at the peak of the LOD curve in Windows QTL Cartographer. MapChart 2.2 (Voorrips 2002) was used to draw a graphical presentation of linkage groups and the map of QTLs.

Results

Phenotypic variation

The results of ANOVA for rectangular lattice design showed that inter-block variance was smaller than residual variance in all studied traits, then these two variances were pooled (Table 1). Analysis of variance revealed significant differences among genotypes for studied traits (Table 1). The genetic parameters and phenotypic variation observed among RILs and their parents are presented in Table 2. The performance of RHA266 was better than that of PAC2 in GYP and PH. The difference between the mean of RILs ($\overline{X}$ RILS) and their parents ($\overline{X}$ p) was only significant for GYP, PH and PL (Table 2). Genetic gain, calculated as difference between the mean values of 10% selected RILs ($\overline{X}_{10\%bestRILs}$) and the mean of parents ($\overline{X}$ p) (Rachid Al-Chaarani *et al.* 2004; Poormohammad Kiani *et al.* 2007a, b), was significant for all the studied traits (Table 2). Frequency distribution of RILs and their parents for most of the studied traits showed continuous patterns, suggesting that studied agromorphological traits were controlled by a polygenic system (Figure

1, Table 2). The amounts of studied traits in some RILs were less than their parents, whereas in some others, the values were higher than their parents (Figure 1).

Correlation analysis

Genotypic and phenotypic correlations among studied traits are summarized in Table 3. The highest genotypic correlation coefficients were observed between LL and LW (0.91±0.02), SD and LW (0.78± 0.05), LL and SD (0.76±0.05), LW and PL (0.75±0.05) and GYP and HD (0.72±0.05).

Significantly positive genotypic correlations were observed between LL, PL, LW, HD and SD with GYP (Table 3).

QTLs

The map and the characteristics of QTLs associated with the studied traits are presented in Table 4. QTL names were constructed using the trait abbreviations suffixed with numbers presenting the linkage group and the order of QTL on the linkage group. For an easier overview of overlapping QTLs between traits, an image of QTL regions was presented in Figure 2. A total number of 21 QTLs was identified for the studied traits.

The QTLs corresponding to various traits were located throughout the genome except on linkage groups 1, 3, 4, 7, 9, 13 and 16 (Figure 2). LOD thresholds determined after 1,000- permutation LR tests ranged from 2.85 to 4.73 depending on the trait with the mean of 3.55. Individual QTLs explained 1.13 to 73.70% of phenotypic variation of the studied traits. The sign of additive gene effects showed that favorable alleles for the studied traits come from both parental lines. Overlapping QTLs were found on linkage groups 6, 11, 12 and 17.

For PH, 2 QTLs were identified on linkage groups 6 and 17. The phenotypic variance explained by PH.6.1 QTL (LOD=4.30) was 73.70% and the positive alleles came from the maternal line 'PAC2'. The phenotypic variance explained by PH.17.1 QTL (LOD=4.21) was 1.80% and the positive alleles came from the paternal line 'RHA266'. Two QTLs were identified for HD, on linkage groups 8 and 11,

Table1. Mean squares of yield and some agro-morphological traits in sunflower recombinant inbred lines (RILs) and their two parents.

Source of variation	LN df	LN MS	LL df	LL MS	PL df	PL MS	^{1}PH df	^{1}PH MS	LW df	LW MS	SD df	SD MS	HD df	HD MS	^{1}GYP df	^{1}GYP MS
Rectangular lattice design																
Replication	1	34.00**	1	27.74**	1	3.02*	1	1.24**	1	12.04**	1	8.80*	1	17.47**	1	2.26ns
Genotype	70	17.79**	68	18.43**	68	12.08**	71	1.06**	68	26.40**	71	26.48**	69	32.87**	68	3.31**
Block/ Replication	16	1.49ns	16	0.29ns	16	0.10ns	16	0.05ns	16	0.25ns	16	1.11ns	16	2.629ns	16	0.54ns
Error	49	2.21	46	0.53	47	0.06	50	0.06	47	0.36	47	1.90	47	2.623	45	1.22
Normality test on residuals																
BDT		0.99ns		0.99ns		0.99ns		0.95**		0.99ns		0.99ns		0.98ns		0.95**
ADT		-		-		-		0.07ns		-		-		-		0.05ns
%CV		11.42		4.16		2.89		2.45		4.21		7.42		10.58		23.42
Pooling																
Replication	1	34.0**	1	27.74**	1	3.02**	1	1.24**	1	12.04**	1	8.80*	1	17.47**	1	2.26**
Genotype	70	17.79**	68	18.43**	68	12.08**	71	1.06**	68	26.40**	71	26.48**	69	32.87**	68	3.31**
Error	65	2.08	62	0.47	63	0.07	66	0.06	63	0.33	63	1.71	63	2.62	61	1.04

^{1}Square root transformation, BDT: before data transformation, ADT: after data transformation, CV: coefficient of variation, df: degree of freedom, MS: mean of square, LN: leaf number, LL: leaf length, PL: petiole length, PH: plant height, LW: leaf width, SD: stem diameter, HD: head diameter, GYP: grain yield per plant, ns: non significant, * and ** significant at 0.05 and 0.01 probability level, respectively.

Table 2. Genetic parameters and gain for traits in sunflower recombinant inbred lines (RILs) and their two parents.

Item	Traits							
	GYP	HD	SD	LW	PH	PL	LL	LN
PAC2 (P1)	10.56	12.33	19.17	14.76	95.56	7.48	18.22	14.53
RHA266 (P2)	16.40	15.50	17.99	12.84	106.84	7.48	16.59	13.73
P1- P2	-5.84^*	-3.17^{ns}	1.18^{ns}	1.93^*	-11.28^*	0^{ns}	1.63^*	0.80^{ns}
$\overline{X}_P$	13.48	13.92	18.58	13.8	101.20	7.48	17.4	14.13
Max	98.56	24.87	28.57	26.52	152.27	16.1	24.91	22.47
Min	5.13	6.72	10.42	7.74	87.52	3.84	11.44	7.51
$\overline{X}_{RIL}$	21.10	15.40	18.57	14.40	115.82	8.83	17.60	13.08
$\overline{X}_{RIL} - \overline{X}_P$	7.61^*	1.48^{ns}	-0.01^{ns}	0.60^{ns}	14.62^*	1.35^*	0.20^{ns}	-1.05^{ns}
$\overline{X}_{10\% \text{ best RIL}}$	58.91	22.98	25.21	21.48	143.28	13.74	23.10	19.06
$\overline{X}_{10\% \text{ best RIL}} - \overline{X}_P$	45.43^*	9.06^*	6.63^*	7.68^*	42.08^*	6.26^*	5.7^*	4.93^*
STDEV	1.47	3.84	4.43	3.76	0.74	2.45	3.10	2.96
LSD	2.21	3.24	2.76	1.20	0.49	0.49	1.46	2.97
Normality test on original data	0.15^{**}	0.06^{ns}	0.04^{ns}	0.10^{**}	0.06^{ns}	0.09^{**}	0.12^{**}	0.12^{**}

$\overline{X}_P$: mean of parents; $\overline{X}$ RILs: mean of recombinant inbred lines; 10%SRILs: mean of the 10% of selected recombinant inbred lines; GG10%: genetic gain when the mean of 10% of selected recombinant inbred lines are compared with the mean of parents; STDEV: standard deviation. LN: leaf number; LL: leaf length; PL: petiole length; PH: plant height; LW: leaf width; SD: stem diameter; HD: head diameter; GYP: grain yield per plant.

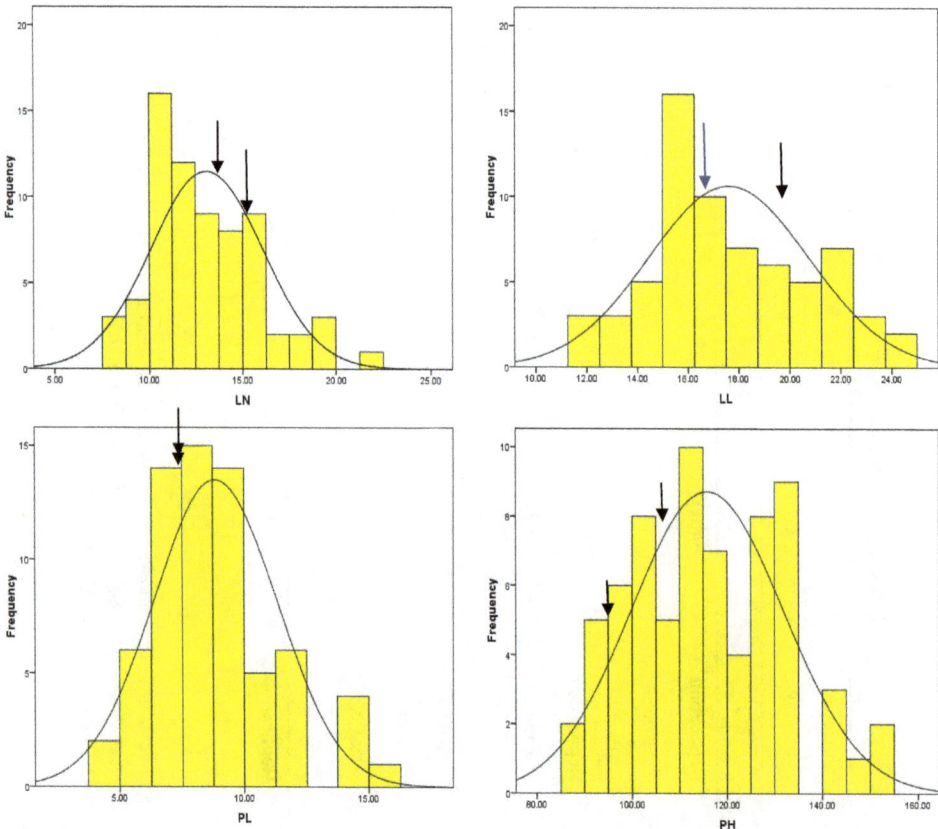

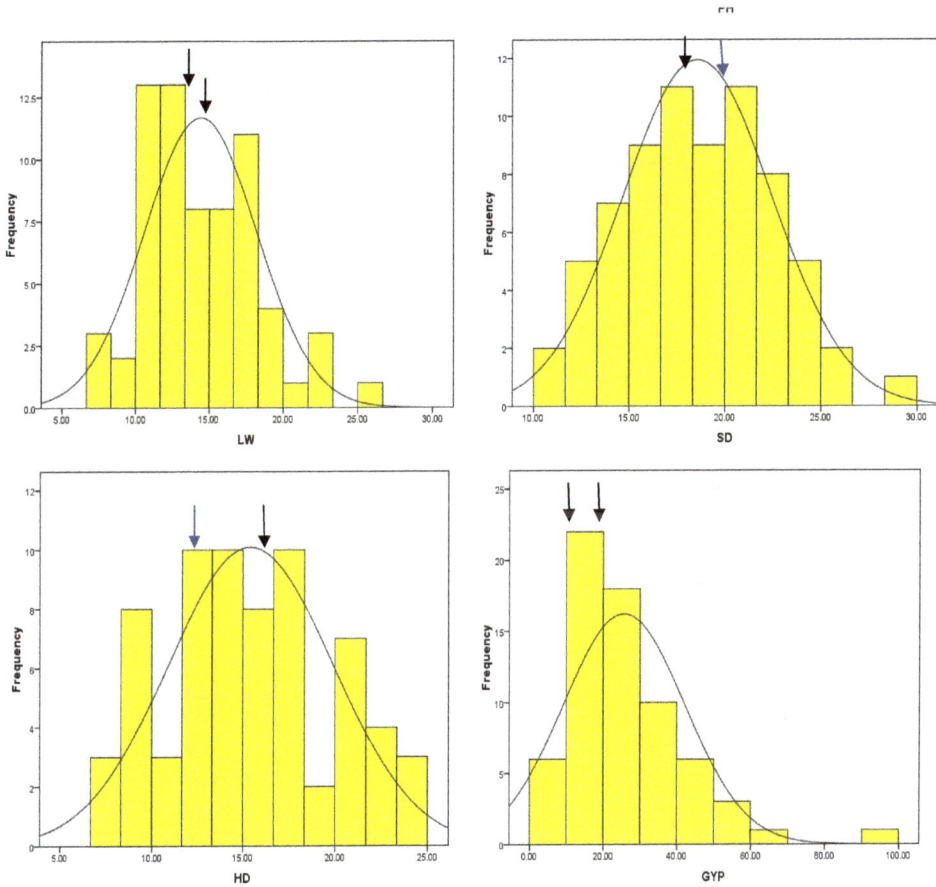

Figure 1. The frequency distribution of sunflower recombinant inbred lines (RILs) and their parents for agro-morphological traits. Red and blue arrows show the phenotypic value of PAC2 (maternal line) and RHA266 (paternal line). LN: leaf number; LL: leaf length; PL: petiole length; PH: plant height; LW: leaf width; SD: stem diameter; HD: head diameter; GYP: grain yield per plant.

Table 3. Correlation among traits in sunflower recombinant inbred lines (RILs).

Variable	LN	LL	PL	PH	LW	HD	SD
Genotypic correlation							
LL	0.15±0.11						
PL	0.28±0.11	0.65±0.07					
PH	0.21±0.11	0.48±0.09	0.32±0.11				
LW	0.25±0.11	0.91±0.02	0.75±0.05	0.41±0.10			
HD	0.23±0.11	0.54±0.08	0.60±0.08	0.27±0.11	0.63±0.07		
SD	0.40±0.09	0.76±0.05	0.71±0.06	0.54±0.08	0.78±0.05	0.66±0.06	
GYP	0.22±0.10	0.32±0.10	0.41±0.09	0.13±0.10	0.42±0.09	0.72±0.05	0.44±0.09
Phenotypic correlation							
LL	0.17±0.13						
PL	0.32±0.12	0.67±0.07					
PH	0.22±0.13	0.49±0.10	0.34±0.11				
LW	0.28±0.12	0.94±0.02	0.77±0.05	0.41±0.11			
HD	0.25±0.14	0.60±0.09	0.64±0.08	0.31±0.12	0.70±0.07		
SD	0.50±0.11	0.82±0.05	0.76±0.06	0.61±0.09	0.85±0.04	0.75±0.07	
GYP	0.26±0.18	0.53±0.14	0.58±0.13	0.19±0.16	0.66±0.12	0.89±0.07	0.59±0.12

LN: leaf number; LL: leaf length; PL: petiole length; PH: plant height; LW: leaf width; SD: stem diameter; HD: head diameter; GYP: grain yield per plant. $r \geq |\pm 0.232|$ is significant at 0.05 probability level (http://www.gifted.uconn.edu/siegle/research/correlation/corrchrt.htm).

Table 4. Map position and effect of QTLs detected for agro-morphological traits in sunflower recombinant inbred lines (RILs) population.

Traits	QTL	LG	Position (cM)	LOD	Additive effects	R^2	Traits	QTL	LG	Position (cM)	LOD	Additive Effects	R^2 (%)
LN	NL.8.1	8	102.71	3.78	-0.81	11.53	SD	SD.2.1	2	10.01	3.27	-1.14	13.13
	NL.12.1	12	64.01	3.08	-0.22	1.85		SD.2.3	2	41.01	3.50	-1.16	10.15
LL	LL.11.1	11	10.01	3.48	-3.30	25.31		SD.5.1	5	16.01	3.16	-1.02	10.05
PL	PL.11.1	11	0.01	3.33	-1.05	11.37		SD.5.2	5	49.01	3.14	1.65	18.91
	PL.14.1	14	36.01	3.39	1.85	4.99		SD.6.1	6	28.01	3.77	1.25	53.88
LW	LW.12.1	12	67.01	3.67	0.95	2.11		SD.6.2	6	47.01	3.16	1.32	51.45
PH	PH.6.1	6	23.01	4.30	8.29	73.70		SD.10.1	10	12.01	3.50	-1.61	1.62
	PH.17.1	17	29.01	4.21	7.81	1.80		SD.11.1	11	12.01	3.64	-3.85	17.65
HD	HD.8.1	8	53.21	2.87	-4.90	60.86		SD.15.1	15	106.01	4.16	1.59	7.65
	HD.11.1	11	0.01	2.85	-1.88	13.19		SD.17.1	17	29.01	3.60	1.07	11.54
GYP	GYP.17.1	17	11.01	4.73	2.35	1.13							

cM: centi Morgan; LG: linkage group; LOD: log10 likelihood ratio (likelihood that the effect occurs by linkage/likelihood that the effect occurs by chance); QTL: quantitative trait loci; R^2: percentage of phenotypic variance explained by the individual QTLs. The positive additive effect shows that PAC2 allele increase the trait and negative value shows that RHA266 allele increases the trait. LN: leaf number; LL: leaf length; PL: petiole length; PH: plant height; LW: leaf width; SD: stem diameter; HD: head diameter; GYP: grain yield per plant. Ns: non significant; *, ** significant at 0.05, 0.01 probability level. QTLs with LOD $\geq$ 3 and $R^2 \geq$ 10% are considered major QTLs. The relatively high R^2 values found for some QTL with lower LOD scores may be influenced by large distances between flanking markers (Balyejusa Kizito et al., 2007).

accounting for 60.86 and 13.19% of phenotypic variation, respectively. The positive alleles for these QTLs came from PAC2. Ten QTLs were identified for SD. The phenotypic variance explained by QTLs ranged from 1.62 to 53.88%. The positive alleles for the identified QTLs came from both parental lines. One QTL (LOD=4.73) was identified for GYP on linkage group 17 that explained 1.13% of the phenotypic variance. The positive allele for detected QTL came from RHA266. Two QTLs were identified for NL on linkage groups 8 and 12, accounting for 1.85 and 11.53% of phenotypic variance, respectively. The positive allele for QTL located on linkage group 8 (R2 =11.53%) derived from RHA266, and for other the positive allele came from PAC2. Two QTLs were identified for PL on linkage groups 11 and 14. The phenotypic variance explained by PL QTLs was 4.99 and 11.37%. The positive alleles for the detected QTLs came from both parental

lines. One QTL (LOD=3.67) was detected for LW that explained 2.11% of phenotypic variance of trait and the positive allele came from RH266. A QTL (LOD=3.48) was detected for LL on linkage group 11 accounting for 25.31% of phenotypic variance. Positive allele for QTL came from PAC2. QTLs with LOD $\geq$ 3 and R2 $\geq$ 10% are considered major QTLs. The relatively high R2 values found for some QTL with lower LOD scores may be influenced by large distances between flanking markers (Balyejusa Kizito *et al.* 2007).

Discussion

Phenotypic variation

Significant differences among genotypes for the studied traits provide necessary genetic variation for QTL analysis. Any significant differences between the means of RILs and their parents for most of the studied traits indicate that RILs used in this study were representative of possible recombination of the cross 'PAC2×RHA266' (Rachid Al-Chaarani *et al.* 2004; Poormohammad Kiani *et al.* 2007a). The frequency distribution of RILs and their parents for most of the studied traits showed a continuous pattern suggesting polygenic control. Continuous variation was also observed for early seedling vigour traits in sunflower (Davar *et al.* 2011). Transgressive segregation that would be the result of the accumulation of positive alleles from both parental lines was observed for most of the studied traits. The positive and negative signs of additive effect at different loci indicate the contribution of both parental lines and confirm the transgressive segregation observed at the phenotypic level. Transgressive segregation for morphological and agronomical traits as well as for water status traits under well- watered and water-stressed conditions has been also reported by Rachid Al- Chaarani *et al.* (2004) and Poormohammad Kiani *et al.* (2007a, b) in sunflower.

Positive and significant genotypic correlations were observed among the studied traits (Table 3). Genotypic correlation analysis indicated that leaf length, petiole length, head diameter, leaf width and stem diameter positively influenced the grain yield of sunflower. A strong correlation between yield and agro-morphological

traits has been reported in Rachid Al-Chaarani *et al.* (2004) research on sunflower. Significant correlations among traits may be resulted from either pleiotropic effect of single gene or tight linkage of several genes that individually influence specific traits (Veldboom *et al.* 1994).

Co-localization of identified QTLs for some traits confirms the observed significant correlations among traits. Yield and yield associated traits are complex quantitative traits controlled by multiple genes and are highly influenced by environmental conditions (Shi *et al.* 2009). Yield associated traits that are less environmentally sensitive and have higher heritabilities than grain yield (Cuthbert *et al.* 2008) could be crucial for sustained sunflower improvement.

The range of identified QTLs for the studied traits was from 1 to 10. The number of QTLs detected in a given study depends on different factors, including type and size of mapping population used, trait investigated, the number of environments used for phenotyping, and genome coverage. The larger the environmental effect on the character (low heritability), the less likely a QTL will be detected. The low number of identified QTLs for some studied traits such as GYP could be improved by replicate phenotyping in different environments in order to controlling environmental error.

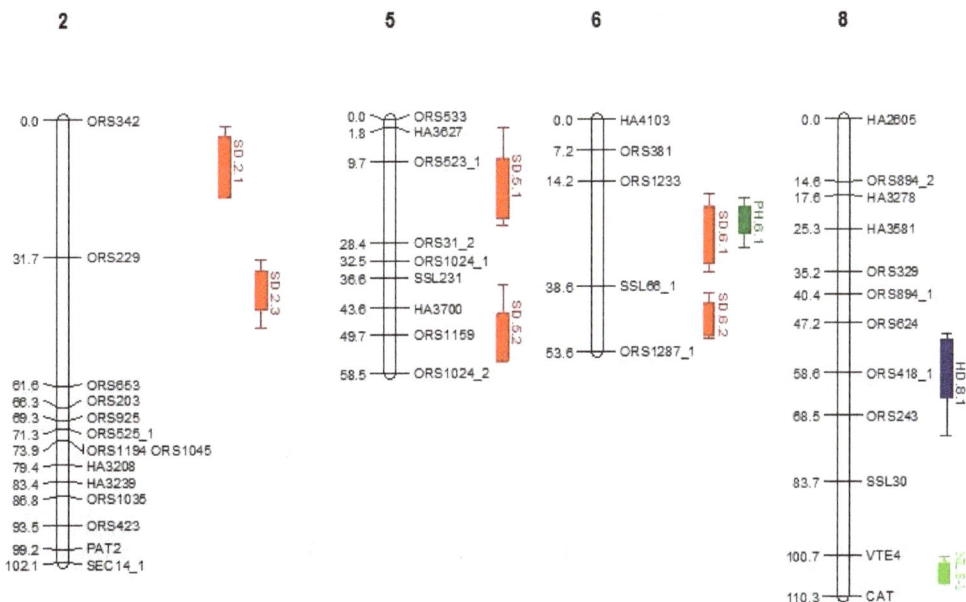

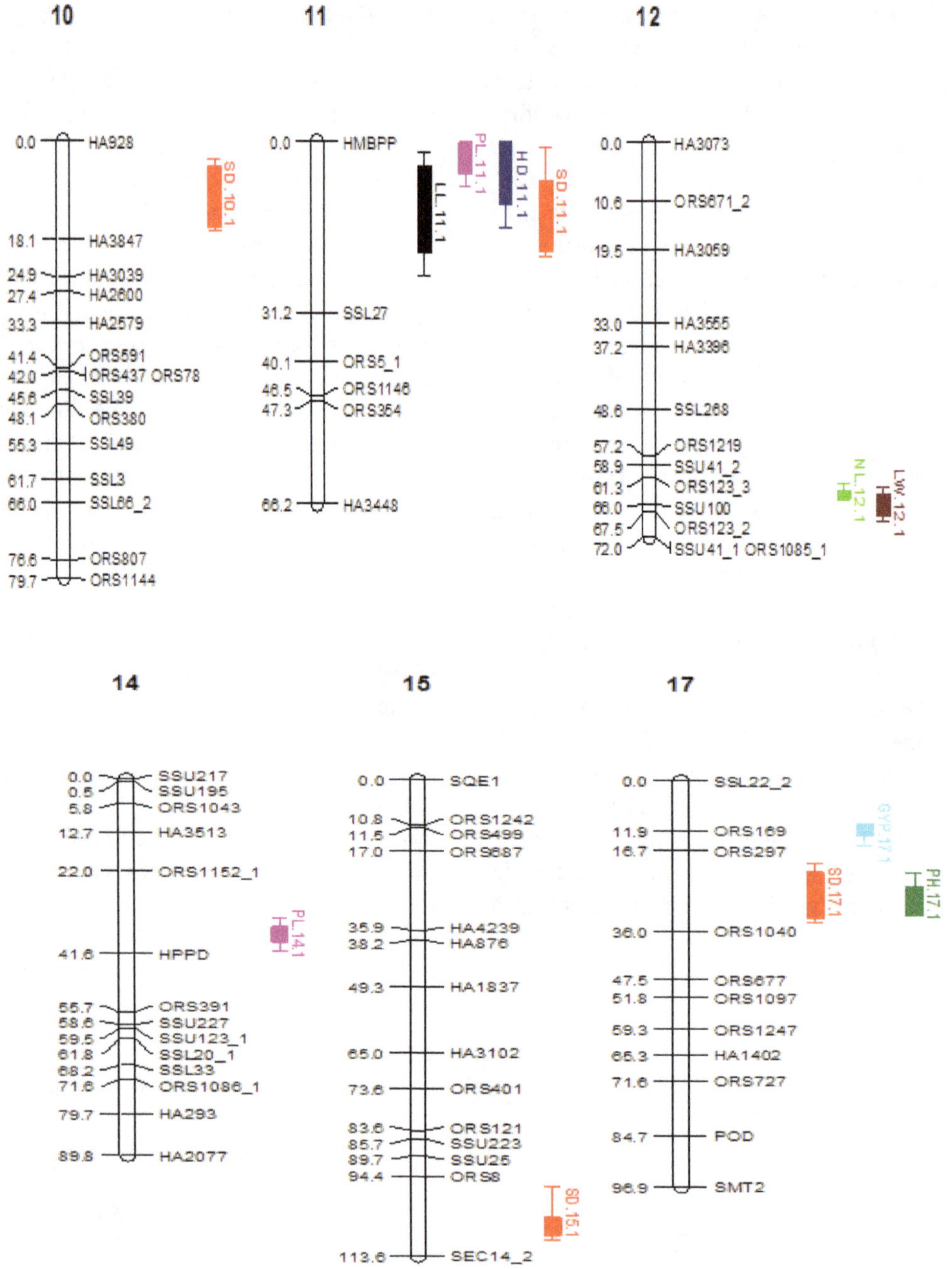

Figure 2. Sunflower linkage groups and QTLs for agro-morphological characteristics in recombinant inbred lines (RILs) population. Linkage groups are named as 1 to 17 according to reference linkage map of sunflower (Tang *et al.* 2002). LN: leaf number; LL: leaf length; PL: petiole length; PH: plant height; LW: leaf width; SD: stem diameter; HD: head diameter; GYP: grain yield per plant.

Individual QTLs explained 1.13 (GYP.17.1) to 73.70% (PH.6.1) of phenotypic variation of the studied traits. As anticipated, because of the quantitative and complex nature of GYP, the magnitude of phenotypic variance explained by identified QTL was low, and low phenotypic variation is consistent with results also from other crop species.

Co-localization of QTLs for GYP and other agro-morphological traits

QTL analysis identified several putative genomic regions involved in the expression of the studied traits. Some identified QTLs associated with the expression of more than one trait. For example, PH.6.1 (LOD= 4.30) and SD.6.1 (LOD= 3.77) co-localized on the linkage group 6 at 23.01 and 28.01 cM distances involved in plant height and stem diameter variations; LL.11.1 (LOD=3.48), PL.11.1 (LOD=3.33), HD.11.1 (LOD=2.85) and SD.11.1 (LOD=3.64) co-localized on linkage group 11 at 0.01-12.01 cM interval associated with leaf length, petiole length, head diameter and stem diameter phenotypes; NL.12.1 (LOD=3.08) and LW.12.1 (LOD=3.67) co-located on the linkage group 12 at 64.01 and 67.01 cM associate with number of leaf and leaf width phenotypes. QTLs: PH.17.1 (LOD=4.21) and SD.17.1 (LOD=3.60) co-localized on the linkage group 17 at 11.01-29.01 cM associate with plant height and stem diameter phenotypes (Table 4, Figure 2). An important QTL was detected on linkage group 11 in interval 0.01-12.01 near marker HMBPP for head diameter, stem diameter, petiole length and leaf length traits. The positive allele for this QTL comes from PAC2. The findings were supported by the results of correlation analysis. Identification of co-localized QTLs could be the reason for high correlation coefficients among the traits. The co- locality of QTLs for different traits implies the presence of pleiotropic or the close linkage of QTLs. Co-localized QTLs significantly increases the efficiency of selection in breeding programs (Tuberosa et al. 2002a; Hittalmani et al. 2003).

Some QTLs identified herein for agro- morphological traits showed co-locality with QTLs identified for water status traits in Poormohammad Kiani et al. (2007a; 2009) studies (Table 5). For example, the petiole length QTL on linkage group 14 (PL.14.1, LOD=3.39) was found in the same interval for the osmotic potential QTL under water- stressed condition in Poormohammad Kiani et al. (2007a, 2009) studies (Table 5). QTL identified herein for stem diameter on linkage group 10 co- localized with QTL identified for total dry mater (BIO) and leaf area duration (LAD) in Poormohammad Kiani et al. (2007a) studies (Table 5). One of the identified QTLs for stem diameter on linkage group 2 was co-localized with QTL identified for grain yield per plant (GYP) (Poormohammad Kiani et al. 2009) (Table 5).

In conclusion, we have detected several specific and nonspecific QTLs for agro- morphological traits. The detection of QTLs influencing various traits such as HMBPP on LG 11 could increase the efficiency of marker-assisted selection and genetic progress.

Table 5. Co-localized QTLs for agro-morphological traits as compared with published QTLs in sunflower.

Linkage Group	In present study	In published works	Overlapped QTLs
LG2	SD	GYP (Poormohammd Kiani et al. 2009)	SD.2.1, GYPN.2.1
LG5	SD	LN, GYP (Poormohammd Kiani et al. 2009)	SD.5.1, LND.5.1, GYPI.5.1
LG8	LN	HD (Poormohammd Kiani et al. 2009)	LN.8.1, HD-W-8-1
LG10	SD	BIO, LAD (Poormohammd Kiani et al. 2007a, 2009)	SD.10.1, BIOW.10.1, LADW.10.2
LG12	PH	LAF (Poormohammd Kiani et al. 2007a, 2009)	PH.12.1, LAF-W-12-1
LG14	PL	OP (Poormohammd Kiani et al. 2007a)	PL.14.1, OP.WW.14.1
LG17	PH, SD	OP (Poormohammd Kiani et al. 2007a)	NAW.17.1, OPF.WS.17.1

Acknowledgements

The authors would like to thank the Institute of Biotechnology, Urmia University, Iran, for financial support. The reviewers are kindly acknowledged for their helpful suggestions.

References

1. Abou Al Fadil, T.A., Kiani, S.P., Dechamp- Guillaume, G., Genzbittel, L. and Sarrafi, 2007. QTL mapping of partial resistance to Phoma basal stem and root necrosis in sunflower (*Helianthus annuus* L.). Plant Science, 172: 815-823.
2. Akbari, G.H.A., Jabbari, H., Daneshian, J., Alahdadi, I. and Shahbazian, N. 2008. The effect of limited irrigation on seed physical characteristics in sunflower hybrids. Journal of Crop Production and Processing, 12: 513-523.
3. Amouzadeh, M., Darvishzadeh, R., Haddadi, P., Abdollahi Mandoulakani, B. and Rezaee Danesh, Y. 2013. Genetic analysis of partial resistance to basal stem rot (*Sclerotinia sclerotiorum*) in sunflower. Genetika, 45: 737-748.
4. Baack, E.J., Whitney, K.D. and Rieseberg, L.H. 2005. Hybridization and genome size evolution: timing and magnitude of nuclear DNA content increases in *Helianthus* homoploid hybrid species. New Phytologist, 167: 623-630.
5. Balyejusa Kizito, E., Rönnberg-Wästljung, A.-C., Egwang, T., Gullberg, U., Fregene, M. and Westerbergh, A. 2007. Quantitative trait loci controlling cyanogenic glucoside and dry matter content in cassava (*Manihot esculenta* Crantz) roots. Hereditas, 144: 129-136.
6. Berry, S.T., Leon, A.J., Hanfrey, C.C., Challis, P., Burkolz, A., Barnes, S.R., Rufener, G.K., Lee, M. and Caligari, P.D.S. 1995. Molecular-marker analysis of *Helianthus annuus* L. 2. Construction of an RFLP map for cultivated sunflower. Theoretical and Applied Genetics, 9l: 195-199.
7. Bert, P.F., Dechamp-Guillaume, G., Serre, F., Jouan, I., de Labrouhe, D.T., Nicolas, P. and Vear, F. 2004. Comparative genetic analysis of quantitative traits in sunflower (*Helianthus annuus* L.) 3. Characterisation of QTL involved in resistance to *Sclerotinia sclerotiorum* and *Phoma macdonaldii*. Theoretical and Applied Genetics, 109: 865–874.
8. Brookes, A.J. 1999. The essence of SNPs. Gene, 234: 177-86.
9. Burke, J.M., Tang, S., Knapp, S.J. and Rieseberg, L.H. 2002. Genetic analysis of sunflower domestication. Genetics, 6l: 1257-1267.
10. Churchill, G.A. and Doerge, R.W. 1994. Empirical threshold values for quantitative trait mapping. Genetics, 138: 963-971.
11. Collard, B.C.Y., Jahufer, M.Z.Z., Brouwer, J.B. and Pang, E.C.K. 2005. An introduction to markers, quantitative trait loci (QTL) mapping and marker-assisted selection for crop improvement: The basic concepts. Euphytica, 142: 169- 196.
12. Cuthbert, J.L., Somers, D.J., Brule-Babel, A.L., Brown, P.D. and Crow, G.H. 2008. Molecular mapping of quantitative trait loci for yield and yield components in spring wheat (*Triticum aestivum* L.). Theoretical and Applied Genetics, 117: 595-608.
13. Darvishzadeh, R., Poormohammad Kiani, S., Dechamp-Guillaume, G., Gentzbittel, L. and Sarrafi, A. 2007. Quantitative trait loci associated with isolate specific and isolate nonspecific partial resistance to *Phoma macdonaldii* in sunflower. Plant Pathology, 56: 855-861.
14. Davar, R., Majd, A., Darvishzadeh, R. and Sarrafi, A. 2011. Mapping quantitative trait loci for seedling vigour and development in sunflower (*Helianthus annuus L.*) using recombinant inbred line population. Plant Omics Journal, 4(7): 418- 427.
15. De Vienne D. 2003. Molecular markers in plants genetics and biotechnology. Science Publishers Inc., Enfield, NH, USA, and Plymouth, UK.
16. Flores Berrios, E., Gentzbittel, L., Kayyal, H., Alibert, G. and Sarrafi, A. 2000. AFLP mapping of QTLs for in vitro organogenesis traits using recombinant inbred lines in sunflower (*Helianthus annuus L.*). Theoretical and Applied Genetics, 101: 1299-1306.
17. Gentzbittel, L., Vear, F., Zhang, Y.X., Berville, A. and Nicolas, P. 1995. Development of a consensus linkage RFLP map of cultivated sunflower (*Helianthus annuus L.*). Theoretical and Applied Genetics, 90: 1079-1086.
18. Gentzbittel, L., Mouzeyar, S., Badaoui, S., Mestries, E., Vear, F., Tourvieille de Labrouhe, D. and Nicolas, P. 1998. Cloning of molecular markers for disease resistance in sunflower (*Helianthus annuus L.*). Theoretical and Applied Genetics, 96: 519–525.
19. Haddadi, P., Ebrahimi, A., Langlade, N.B., Yazdi-Samadi, B., Berger, M., Calmon, A., Naghavi, M.R., Vincourt, P. and Sarrafi, A. 2012. Genetic dissection of tocopherol and phytosterol in recombinant inbred lines of sunflower through QTL analysis and the candidate gene approach. Molecular Breeding, 29: 717-729.
20. Hittalmani, S., Huang, N., Courtois, B., Venuprasad, R., Shashidhar, H.E., Zhuang, J.Y., Zheng, K.L., Liu, G.F., Wang, G.C., Sidhu, J.S., Srivan- taneeyakul, S., Singh, V.P., Bagali, P.G., Prasanna, H.C., McLaren, G. and Khush, G.S. 2003. Identification of QTL for growth and grain yield-related traits in rice across nine locations of Asia. Theoretical and Applied Genetics, 107: 679-90.
21. Holland, J.B. 2006. Estimating Genotypic Correlations and Their Standard Errors Using Multivariate Restricted Maximum Likelihood Estimation with SAS Proc MIXED. Crop Science, 46: 642-654.

22. Hu, J., Yue, B. and Vick, B. 2007. Integration of TRAP markers onto a sunflower SSR linkage map constructed from 92 recombinant inbred lines. Helia, 30: 25-36.
23. Hu, J., Seiler, G. and Kole, C. 2010. Genetics, genomics and breeding of sunflower. Routledge, USA, 342 pages.
24. Jan, C.C., Vick, B.A., Miller, J.F., Kahler, A.L. and Butler, E.T. 1998. construction of an RFLP linkage map for cultivated sunflower. Theoretical and Applied Genetics, 96: 15–22.
25. Kalia, R.K., Rai, M.K., Kalia, S., Singh, R. and Dhawan, A. K. 2011. Microsatellite markers: an overview of the recent progress in plants. Euphytica, 177: 309-334.
26. Kearsey, M.J. and Pooni H.S. 1996. The genetical analysis of quantitative traits. Chapman and Hall, London, UK.
27. Lai, Z., Livingstone, K., Zou, Y., Church, S.A., Knapp, S.J., Andrews, J. and Rieseberg, L.H. 2005. Identification and mapping of SNPs from ESTs in sunflower. Theoretical and Applied Genetics, 111: 1532–1544.
28. Langar, K., Lorieux, M., Desmarais, E., Griveau, Y., Gentzbittel, L. and Bervillé, A. 2003. A Combined mapping of DALP and AFLP markers in cultivated sunflower using F9 recombinant inbred line. Theoretical and Applied Genetics, 106: 1068-1074.
29. Leon, A.J., Lee, M., Rufener, G.K., Berry, S.T. and Mowers, R.P. 1995. Use of RFLP markers for genetic linkage analysis of oil percentage in sunflower. Crop Science, 35: 558-564.
30. Mestries, E., Gentzbittel, L., Tourvieille de Labrouhe, D., Nicolas, P. and Vear, F. 1998. Analysis of quantitative trait loci associated with resistance to *Sclerotinia sclerotiorum* in sunflower (*Helianthus annuus* L.) using molecular markers. Molecular Breeding 4: 215-226.
31. Mohler, V. and Singrun, C. 2004. General considerations: Marker-assisted selection. Pp. 305-318. In: Nagata, T., Lorz, H. and Wenzel, G. (ed), Biotecnology in Agriculture and Forestry: Molecular Marker System in Plant Breeding and Crop Improvement. Vol. 55, Springer, Berlin.
32. Mokrani, L., Gentzbittel, L., Azanza, E., Fitamant, L., Al-Chaarani, G. and Sarrafi, A. 2002. Mapping and analysis of quantitative trait loci for grain oil and agronomic traits using AFLP and SSR in sunflower (*Helianthus annuus* L.). Theoretical and Applied Genetics, 106: 149-156.
33. Nazir, S., Bashir, E. and Bantel, R. 1994. Crop production. National Book Foundation, Islamabad, Pakistan. Pp: 342-345.
34. Oliver, R.E., Lazo, G.R., Lutz, J.D., Rubenfield, M.J., Tinker, N.A., Anderson, J.M., Morehead, N.H.W., Adhikary, D., Jellen, E.N., Maughan, P.J., Guedira, G.L.B., Chao, S., Beattie, A.D., Carson, M.L., Rines, H.W., Obert, D.E., Bonman, J.M., Jackson, E.W. 2011. Model SNP development for complex genomes based on hexaploid oat using high-throughput 454 sequencing technology. BMC Genomics, 12: 77.
35. Poormohammad Kiani, S. and Sarrafi, A. 2010. Molecular mapping of complex traits. In Hu, J., Seiler, G. and Kole, C. (eds). Genetics, genomics and breeding of sunflower, pp. 135-172.
36. Poormohammad Kiani, S., Maury, P., Nouri, L., Ykhlef, N., Grieu, P. and Sarrafi, A. 2009. QTL analysis of yield-related traits in sunflower under different water treatments. Plant Breeding, 128: 363-373.
37. Poormohammad Kiani, S., Maury, P., Sarrafi, A. and Grieu, P. 2008. QTL analysis of chlorophyll fluorescence parameters in sunflower (*Helianthus annuus* L.) under well-watered and water-stressed conditions. Plant Science, 175: 565-573.
38. Poormohammad Kiani, S., Talia, P., Maury, P., Grieu, P., Heinz, R., Perrault, A., Nishinakamasu, V., Hopp, E., Gentzbittel, L., Paniego, N. and Sarrafi, A. 2007a. Genetic analysis of plant water status and osmotic adjustment in recombinant inbred lines of sunflower under two water treatments. Plant Science, 172: 773-787.
39. Poormohammad Kiani, S., Grieu, P., Maury, P., Hewezi, T., Gentzbittel, L., and Sarrafi, A. 2007b. Genetic variability for physiological traits under drought conditions and differential expression of water stress-associated genes in sunflower (*Helianthus annuus* L.). Theoretical and Applied Genetics, 114: 193-207.
40. Rachid Al-Chaarani, G., Gentzbittel, L., Huang, X. and Sarrafi, A. 2004. Genotypic variation and identification of QTLs for agronomic traits, using AFLP and SSR markers in RILs of sunflower (*Helianthus annuus* L.). Theoretical and Applied Genetics, 109: 1353-1360.
41. Rachid Al-Chaarani, G., Gentzbittel, L., Wedzony, M. and Sarrafi, A. 2005. Identification of QTLs for germination and seedling development in sunflower (*Helianthus annuus* L.). Plant Science, 169: 221-227.
42. Sarrafi, A. and Gentzbittel, L. 2004. Genomics as efficient tools: example sunflower breeding. Pp. 107-119. In: Nagata, T., Lorz, H. and Wenzel, G. (ed), Biotecnology in Agriculture and Forestry: Molecular Marker System in Plant Breeding and Crop Improvement. Vol. 55, Springer, Berlin.
43. Satyabrata, M., Hedge, M.R. and Chattopodhay, S.B. 1988. Hand Book of Annual Oilseed Crops. Oxford IBH Pub. Co. (Pvt.) Ltd. New Delhi, Pp.176.
44. Shi, J., Li, R., Qiu, D. Jiang, C., Long, Y., Morgan, C., Bancroft, I., Zhao, J. and Meng, J. 2009. Unraveling the complex trait of crop yield with quantitative trait loci mapping in Brassica napus. Genetics, 182: 851-861.

45. Skoric, D. and Marinkovic, R. 1986. Most recent results in sunflower breeding. Int. Symposium on sunflower, Budapest, Hungary, p: 118–119.

46. Tang, S., Yu, J.K., Slabaugh, M.B., Shintani, D.K. and Knapp, S.J. 2002. Simple sequence repeat map of the sunflower genome. Theoretical and Applied Genetics, 105: 1124- 1136.

47. Tuberosa, R., Sanguineti, M.C., Landi, P., Giuliani, M.M., Salvi, S. and Conti, S. 2002a. Identification of QTLs for root characteristics in maize grown in hydroponics and analysis of their overlap with QTLs for grain yield in the field at two water regimes. Plant Molecular Biology, 48: 697-712.

48. Veldboom, L.R., Lee, M. and Woodman, W.L. 1994. Molecular marker-facilitated studies of morphological traits in an elite maize population. 2. Determination of QTLs for grain yield and yield components. Theoretical and Applied Genetics, 89: 451-458.

49. Voorrips, R. E. 2002. MapChart: software for the graphical presentation of linkage maps and QTLs. Journal of Heredity, 93: 77-78.

50. Wang, S., Basten, C.J. and Zeng, Z.B. 2005. Windows QTL Cartographer V2.5. Department of Statistics, North Carolina State University, Raleigh NC, Available from http: // statgen .ncsu.edu /qtl-cart / WQTLCart.htm.

51. Yu, J.K., Tang, S., Slabaugh, M.B., Heesacker, A., Cole, G., Herring, M., Soper, J., Han, F., Chu, W.C., Webb, D.M., Thompson, L., Edwards, K.J., Berry, S., Leon, A.J., Olungu, C., Maes, N. and Knapp, S.J. 2003. Towards a saturated molecular genetic linkage map for cultivated sunflower. Crop Science, 43: 367-387.

52. Yue, B., Vick, B.A., Miller, J.F., Cai, X. and Hu, J. 2008a. Construction of a linkage map with TRAP markers and identification of QTL for four morphological traits in sunflower (*Helianthus annuus* L.). In: Proc 17th Int Sunflower Conf, 8–12 June, Cordoba, Spain, pp 655-660.

6

Estimation of genetic parameters for quantitative and qualitative traits in cotton cultivars (*Gossypium hirsutum* L. & *Gossypium barbadense* L.) and new scaling test of additive–dominance model

Gholamhossein Hosseini*

Cotton Research Institute of Iran
*Corresponding Author, Email: Hooshmandh2004@yahoo.com

Abstract

A complete diallel cross of nine cotton genotypes (*Gossypium hirsutum* L. & *Gossypium barbadense* L.) viz Delinter, Sindose-80, Omoumi, Bulgare-539, Termez-14, Red leaf (Native species), B-557, Brown fiber and Siokra-324 having diverse genetic origins was conducted over two years to determine the potential for the improvement of yield, its components, oil and fiber quality traits by means of genetic analysis, combining ability, heritability and heterotic effects. The detailed studies were based on F_1 generations where crossed seeds in the first year were used for F_1 generation in the second year. The successful hybrids were recognized and distinguished by morphological markers such as flower color, spot position and their colors in petal, fiber color, seed linter, leaf color and their shapes. Analysis of variance for Simple Square Lattice Design (SSLD) showed highly significant differences ($P \leq 0.01$) among various genotypes which allowed genetic analysis by Griffing, Hayman and Hayman-Jinks' method. Additive- dominance model and related correlation (Wr, Vr) were adequate for majority of the traits and partially adequate for some traits. Majority of the traits were influenced by non-additive gene action in F_1 generation. These results are encouraging for practical improvement through hybrid breeding programs and the contributions of additive genes through selection method. Significant variation for general combining ability (GCA) effects, specific combining ability (SCA) effects ($P \leq 0.05$) and high narrow sense heritability indicates the potential for improvement through selection. On the other hand, over-dominance gene action, low and moderate rate of narrow-sense heritability for some traits suggests that improvements should be made utilizing a combination and hybrid breeding approach.
Key words: Cotton, Hybrid, Genetics.

Introduction

Modern genetics can be traced to the rediscovery of Gregor Mendel's research in the early 1900s. Mendel recognized that organisms have two copies of each gene (alleles) and that one allele is contributed by each parent to the offspring. This phenomenon is observed in diploid organisms, those which have two sets of each chromosome in the genome. Mendel also concluded that alleles display dominance and recessiveness. However, today we recognize that other types of alleleic interaction can exist in which alleles are additive (the heterozygote value is the average of the two homozygotes), incomplete dominant (the heterozygote value lies closer to one of the two homozygotes), or overdominant (the heterozygote value exceeds either of the two homozygotes), as well as dominant. Linkage is a key genetic phenomenon impacting plant breeding. Linkage violates Mendelian independent assortment due to the arrangement of genes on chromosomes therefore every gene on a chromosome is inherited together. Many traits are said to be linked because the genes controlling them lie together on a chromosome and therefore have a higher random probability of being transmitted together to the progeny.

Many other genetic phenomena influence expression of traits. First, multiple alleles can exist for each gene in a population. Each individual may possess only two copies but those copies can differ among individuals (e.g., leaf shape in Upland cotton, *G. hirsutum* L.). Second, epistasis is a phenomenon in which the expression of one gene is affected by the genotype of a gene at a separate locus (e.g., expression of AA, Aa, and aa depends on the genotype at locus B). Third, pleiotropy is a phenomenon in which single gene can affect multiple traits. Fourth, heterosis is a phenomenon in which progeny between unrelated parents perform better than what would be expected based on the average performance of the parents; this is the phenomenon which has led to hybrid seed production for yield improvement in cotton and other crops and can be the result of combinations of the previous genetic phenomena. Finally, environment is a crucial modifier of gene expression (Ragsdale 2003).

Cotton, as a commercial crop, has played a vital role in agriculture, industrial development, and employment generation. This most important cash crop, besides providing raw material (fiber) for textile industry, also provides food in the form of oil and cotton seed cake for human and animal consumption. It also earns a huge amount of foreign exchange through the export of its raw materials as well as its finished products. Due to its undisputed importance, cotton has attracted maximum attention of geneticists and plant breeders and their sustained efforts have led to the evolution of high yielding cultivars for enhancing cotton production in the world. Evolution and utilization of high yielding, stresses tolerant cultivars to have primordial position in the crop production technology package. The research experience has repeatedly established uncontestable importance of transgressive hybridizations, the function of identification of genotypes and putting them to the point of the specific genotypic combinations (Hosseini, 2008).

In view of the pivotal importance of this type of research and its lasting impact upon the future cotton breeding strategies, a research programme is organized to study the genetic basis of different traits of cotton plant along with combining ability analysis and heterosis in a set of 9× 9 complete diallel cross experiment at Botany Department, University of Pune, India during 2004– 2007. Further, this research has three primary objectives: 1) To determine the potential of some quantitative and qualitative traits in screening for yield, oil content and fiber quality across upland and *barbadense* cotton genotypes; 2) To determine the potential to improve mentioned properties by a diallel analysis of nine upland cotton genotypes and 3) To determine the efficiency of correlation between Wr and Vr for epistasis testing instead and along with Wr-Vr ANOVA and b (Wr, Vr) regression test and to introduce new test scale for epistasis existence in diallel cross. The all tetraploids (2n = 52) genotypes which have been used in the present research belong to genomic group of $(AD)_1$ and $(AD)_2$ with large and small chromosomes.

Materials and Methods

The results reported in this study pertain to genetic analysis, combining ability and heterosis estimates in Upland cotton (*Gossypium hirsutum* L. & *Gossypium barbadense* L.), conducted at the Research Farm of Botany Department of Pune University (73°, 51'E longitude, 18°, 31' N latitude and altitude 559m) during 2004-2007. Breeding material comprised of nine different G. *hirsutum* L. & G. *barbadense* L. genotypes varied by origin, yield and its components and fiber as well as oil quality traits. The cultivars were Delinter, Sindose-80, Omoumi, Bulgare-539, Termez-14, Red leaf, B-557, Brown fiber and Siokra- 324.The brief description of cultivars is presented in Table 1.

Crossing Block

The seeds of the nine diverse genotypes were sown on 12th July 2005, in a non- replicated crossing block by dibbling on a well prepared seed bed. Plants were raised in nine rows, each of 36 meters length, at the spacing of 0.25 and 1.5 meters between plants and rows, respectively. NPK was applied at the rate of 60:60:60. The 1/3 dose of nitrogen and 2/3 dose of phosphorus and potash were applied at sowing time, the remaining nitrogen, phosphorus and potash in two split doses at four-leaf and 25cm of plant height stages. All cultural practices and plant protection were done regularly. The crop was ready for crossing on September 20, 2005. The genotypes were crossed in a complete diallel fashion by hand pollination. Crossing continued up to mid November 2005. All precautionary measures were observed to avoid undesirable contamination of genetic material while selfing and crossing in the crossing block of nine genotypes (Table 2). The ginning was performed with roll ginning machine and the seeds were kept safely for sowing F_1 experiment in thecoming year.

Table 1. The brief description of cultivars.

Characters	Cultivars				
	Delinter	**Sindose-80**	**Omoumi**	**Bulgare-539**	**Termez-14**
Origin	Iran	Greece	Iran	Bulgaria	Uzbekistan
Oil content (%)	18.7	17.4	21.7	17.13	20.3
Boll Weight (g)	3.7	2.45	1.7	2.7	2.2
Uniformity ratio (%)	47	46	47.5	48	46.5
Staple Length (mm)	24.3	27.5	29.3	24.5	33
Fiber bundle strength (g/tex)	18.4	22.3	25.7	20.1	26.9
Micronaire (μg/inch)	4.05	2.6	3.2	3.2	3.05
Earliness (days to flowering)	94.5	78	70	84.5	73

Table 1. Continued.

Characters	Cultivars			
	Red leaf	**B-557**	**Brown fiber**	**Siokra-324**
Origin	Iran	Bulgaria	Iran	Australia
Oil content (%)	18.5	15.9	18.3	16.4
Boll Weight (g)	2.9	2.2	2.5	2.8
Uniformity ratio (%)	43.5	45.5	47.5	48.5
Staple Length (mm)	17.6	26	23.8	27.8
Fiber bundle strength (g/tex)	21.5	19.55	16.95	22.2
Micronaire `(μg/inch)	2.75	2.75	3.85	3.15
Earliness (days to flowering)	98	93.5	94	81.5

F_1/Parents Experiment

The 9×9 F_1 complete diallel cross having seventy-two F_1 hybrids along with nine parental cotton were sown on 23th June, 2006, by dibbling on a well prepared seed bed. Each genotype was planted in four rows measuring 6 meters as hill method with conservation of four plants in one hill in a Simple Square Lattice Design (SSLD) with two replications. The row and plant spacing were 80 and 25 cm, respectively.

Cultural practices including fertilizer, hoeing, weeding, irrigation and plant protection measures were carried out as recommended for cotton production. The data were subjected to analysis of variance (ANOVA) on the basis of lattice design, using MSTATC, a computer software package. The data were analyzed using dial software(version 1.1) delivered by Mark Burow and James G.Coors and Dial 98 software that had been revised (September, 2006) and delivered by Yasuo Ukai.

Table 2. Crossing block of nine genotypes of cotton (*G.hirsutum* L. *& G.barbadense* L.) during 2005 -2006.

Cultivars	Delinter	Sindose-80	Omoumi	Bulgare-539	Termez-14	Red leaf	B-557	Brown fiber	Siokra-324
Delinter	X11	X12	X13	X14	X15	X16	X17	X18	X19
Sindose-80	X21	X22	X23	X24	X25	X26	X27	X28	X29
Omoumi	X31	X32	X33	X34	X35	X36	X37	X38	X39
Bulgare-539	X41	X42	X43	X44	X45	X46	X47	X48	X49
Termez-14	X51	X52	X53	X54	X55	X56	X57	X58	X59
Red leaf	X61	X62	X63	X64	X65	X66	X67	X68	X69
B-557	X71	X72	X73	X74	X75	X76	X77	X78	X79
Brown fiber	X81	X82	X83	X84	X85	X86	X87	X88	X89
Siokra-324	X91	X92	X93	X94	X95	X96	X97	X98	X99

Xij = X♂♀

Morphological Markers

There were more and enough morphological markers for recognition of all successful hybrids such as 1) Petal spot that inherited from parents of Omoumi and Termez-14 and the expression of this marker in crossing with non-petal spot parents was demonstrated from light red petal spot (smaller in size) to dark red petal spots (bigger in size) in the related hybrids (Row & Column 3 and 5 on Fig. 1) and absence

Fig 1. Morphological markers of cotton flower in 9×9 diallel cross.

of petal spot in non-successful hybrids. 2) Yellowness of petals as marker varies from less yellowness to more yellowness with more yellowness also originated from parents of Omoumi and Termez-14 and their successful hybrids demonstrated a moderate yellowness petals (Row & Column 3 and 5 on Fig. 1) in hybrids. 3) Red color petal marker that originated from Red leaf parent and its hybrids varies from less red petals to more red petals in its related hybrids (Row & Column 6 on Fig. 1) and absence of red color in non-successful hybrids. 4) Brown fiber marker that was converting from

Brown cotton and its crossing with white color parents had light, intermediate and dark brown color fiber (Row & Column 8 on Fig. 4) and absence of brown color in non-successful hybrids. 5) Lint less seed marker that originated from Delinter parent and in its hybrids removing the fiber from seed was easier than non-successful hybrids (Row & Column 1 on Fig. 3). 6) Red leaf marker that originated from genes of Red leaf and its hybrid had intermediate color between green and red color in its successful hybrids (Row & Column 6 on Fig. 2) and green color leaf in

Fig 2. Morphological markers of cotton leaf in 9×9 diallel cross.

non-successful hybrids.7) Leaf lobbing originated from Siokra-324 parent and those hybrids that had Siokra-324 as one of their parents had leaf lobbing variation from less deeper, intermediate and deeper leaves and consequently without leaf lobbing in non-successful

hybrids (Row & Column 8 on Fig. 2).

Results

Analysis of variance in a Simple Square Lattice Design (SSLD) showed highly significant diversity (P ≤ 0.01) among genotypes of the studied traits and those which allowed genetic analysis by Hayman (1954) and Griffing,s (1956) methods (Table 3).

In F1 generation, the analysis of variance of arrays indicated epistasis effects due to the significance of Wr-Vr for uniformityratio, fiber bundle strength, seed index, seed cotton yield and boll weight; dominance effects due to the significance of Wr+Vr for all traits except boll weight and non-epistasis effects due to the significance of b value for all traits except uniformity. Such results confirmed additive-dominance model for mentioned traits. It was also found that the assumptions of the Hayman-Jinks model are not fulfilled for some traits such as fiber bundle strength, seed index,

seed cotton yield and boll weight which makes the model partially adequate for them and non adequate only foruniformity.

Additive- dominance model was adequate for the remaining traits that adequacy of additive-dominance model is with no nonalleleic interaction and independence of gene action for random gene recombination. These results are confirmed with testing of additive- dominance model by means of significant correlation between Wr and Vr that is presented for the first time in this study.

In F_1 generation of uniformity ratio, the regression analysis indicated that regression coefficient (b) differed non-significantly from zero but significantly from unity.

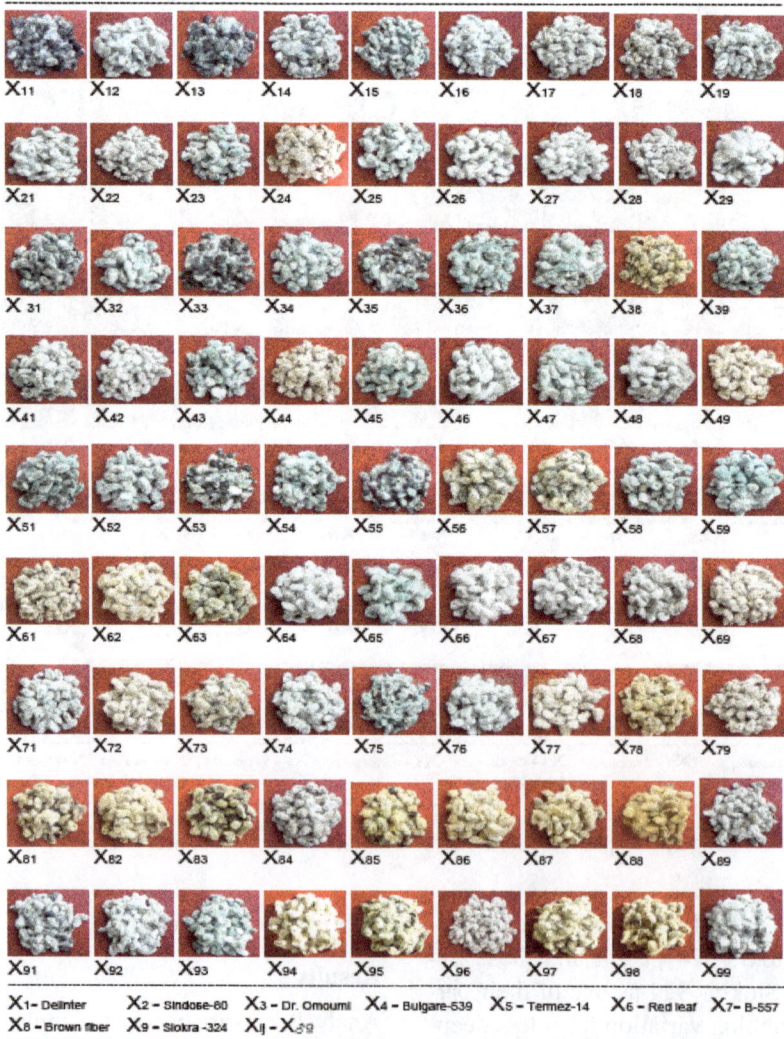

X_1- Delinter X_2 – Sindose-80 X_3 – Dr. Omoumi X_4 – Bulgare-539 X_5 – Termez-14 X_6 – Red leaf X_7– B-557
X_8 – Brown fiber X_9 – Siokra -324 X_{ij} – X_{99}

Fig 3. Morphological markers of cotton seed in 9×9 diallel cross.

The analysis of variance of arrays revealed that Wr+Vr and Wr-Vr were significant, showing existence of dominance with nonallelic interaction and the dependence of genes on random associations in their actions. Also there is non-adequacy of additive-dominance model with nonalleleic interaction and dependent

gene action for random gene recombination. It got confirmed by "r" test due to its nonsignificant value indicating non-adequacy of model with nonalleleic interaction (Table 4) thus the assumptions of the Hayman-Jinks model are not fulfilled which makes the model partially adequate (Jinks, 1954).

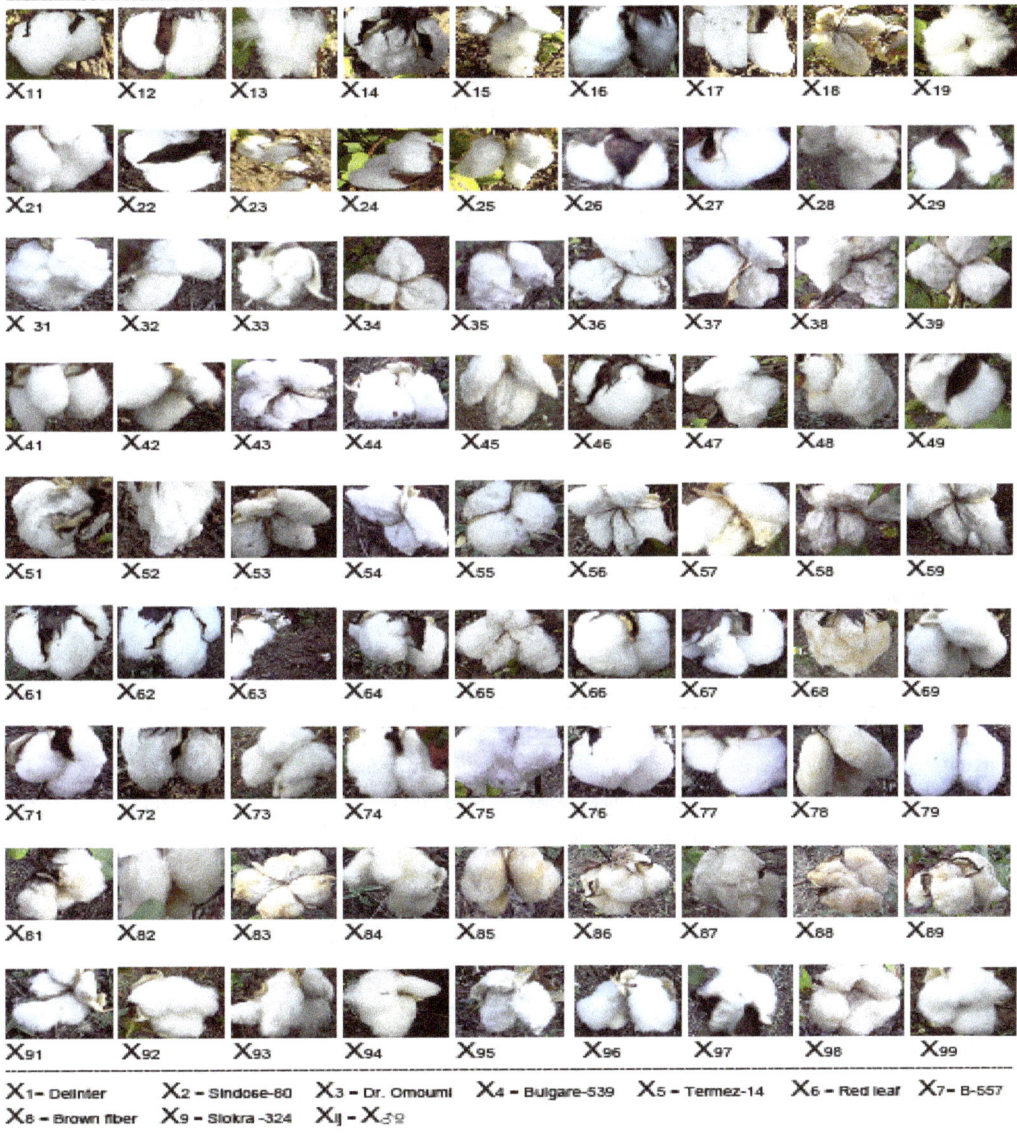

Fig 4. Morphological markers of cotton fiber in 9×9 diallel cross.

All the genetic components of variance, the additive (D), dominance (H_1, H_2) and F were significant and h^2 was positive and nonsignificant.

The additive component (D) was smaller than dominance components (H_1, H_2) and the mean degree of dominance ($\sqrt{H1/D} = 1.59$) was more than 1 indicating non-additive type of gene action and is in increasing position as confirmed by positive and non-significance of $h^2(0.028)$ as well as by the value of Kd/Kd+Kr (0.653). Unequal values of H1 and H2 indicating dis-similar distribution of positive and negative genes was also confirmed by the ratio $H_2/4H_1$ (0.203) which has been showed on Table 5 (Mather1971).

Table 3: Estimation of mean squares and F ratio's along with CV% at 80 D.F for analysis of variance for different yield and quality traits of cotton (*G.hirsutum* L. *G.barbadense* L.) in F_1 generation during 2006-2007.

Source of Variance	DF	Mean Square					
		Oil Content (%)	Staple Length (mm)	Uniformity Ratio (%)	Micronaire (µg/inch)	Fiber Bundle Strength (g/tex)	Earliness (day)
Replications	1	0.025	0.831	4.840	0.005	4.173	33.802
Treatments							
Unadjusted	80	6.705**	21.255**	5.123**	0.206**	24.587**	124.863**
Adjusted	80	6.705**					
Blocks within Reps (adj.)	16	0.696	0.125	0.353	0.005	0.089	3.677
Error							
Effective	64	0.648					
RCB Design	80	0.649	0.135	0.490	0.006	0.250	5.215
Intra block	64	0.637	0.138	0.524	0.006	0.291	5.599
Relative Efficiency (RCB)		100.5	Less than RCB	Less than RCB	Less than RCB	Less than RCB	Less than RCB
CV%		4.105	1.266	1.493	2.396	2.148	2.907

**. Significant at 0.01 level (2-tailed).
*. Significant at 0.05 level (2-tailed).

Table 3: Continued.

Source of Variance	DF	Mean Square					
		Lint% (G.O.T)	Seed Index (g)	Seed cotton yield(g)	Boll weight (g)	Bolls/Plant	Plant Height (cm)
Replications	1	5.111	0.155	11.239	0.007	0.747	34.722
Treatments							
Unadjusted	80	44.172**	7.558**	56.518**	0.464**	1.156**	454.878**
Adjusted	80			56.518**	0.464**	1.156**	454.878**
Blocks within Reps (adj.)	16	1.477	0.007	8.156	0.004	0.741	4.056
Error							
Effective	64			3.050	0.002	0.366	3.674
RCB Design	80	4.178	0.009	3.783	0.002	0.412	3.685
Intra block	64	4.853	0.009	2.69	0.002	0.330	3.592
Relative Efficiency (RCB)		Less than RCB	Less than RCB	124.02	106.5	112.46	100.29
CV%		4.749	1.04	7.197	1.584	7.077	1.973

**. Significant at 0.01 level (2-tailed).
*. Significant at 0.05 level (2-tailed).

Table 4: Scaling test of additive-dominance model "*b*" regression analysis, array analysis of variance and correlation (Wr,Vr) for a 9×9 diallel cross experiment of cotton (*G.hirsutum* & *G. barbadense*) in F_1 generation.

Traits	*b* value ± SE	Correlation (Wr, Vr)	Source of variance		D.F	MS	CV%
Oil Content %	(1.016 ± 0.16)**	0.923**	Wr+Vr	Between Arrays	8	8.977**	22.22
				Within Arrays	9	0.7474	
			Wr-Vr	Between Arrays	8	0.356	-75.80
				Within Arrays	9	0.300	
Staple Length (mm)	(0.978 ± 0.093)**	0.97**	Wr+Vr	Between Arrays	8	93.24**	5.48
				Within Arrays	9	0.479	
			Wr-Vr	Between Arrays	8	1.402	-32.42
				Within Arrays	9	0.285	
Uniformity Ratio (%)	(0.053 ± 0.245)	0.081	Wr+Vr	Between Arrays	8	1.833**	29.93
				Within Arrays	9	0.42	
			Wr-Vr	Between Arrays	8	1.532**	-54.17
				Within Arrays	9	0.275	
Micronaire (μg/inch)	(0.925 ± 0.114)**	0.951**	Wr+Vr	Between Arrays	8	0.014**	10.34
				Within Arrays	9	0.0001	
			Wr-Vr	Between Arrays	8	0.0001**	44.83
				Within Arrays	9	0.00001	
Fiber Bundle Strength(g/tex)	(0.786 ± 0.105)**	0.943**	Wr+Vr	Between Arrays	8	67.245**	8.99
				Within Arrays	9	1.681	
			Wr-Vr	Between Arrays	8	2.5**	-229.05
				Within Arrays	9	0.372	
Earliness **(day)**	(0.879 ± 0.091)**	0.965**	Wr+Vr	Between Arrays	8	8408.3**	8.67
				Within Arrays	9	62.35	
			Wr-Vr	Between Arrays	8	193.136	-226.29
				Within Arrays	9	116.802	
Lint% (G.O.T)	(0.781 ± 0.187)**	0.845**	Wr+Vr	Between Arrays	8	546.77**	29.11
				Within Arrays	9	46.908	
			Wr-Vr	Between Arrays	8	48.95	-49.6
				Within Arrays	9	22.823	
Seed index (g)	(0.851± 0.203)**	0.846**	Wr+Vr	Between Arrays	8	7.072**	2.12
				Within Arrays	9	0.007	
			Wr-Vr	Between Arrays	8	0.591**	-4.62
				Within Arrays	9	0.005	
Seed cotton yield(g)	(0.403 ± 0.130)*	0.76**	Wr+Vr	Between Arrays	8	309.62**	16.56
				Within Arrays	9	37.342	
			Wr-Vr	Between Arrays	8	90.161**	-80.52
				Within Arrays	9	16.266	
Boll weight (g)	(0.579 ± 0.227)*	0.692**	Wr+Vr	Between Arrays	8	0.013	40.98
				Within Arrays	9	0.011	
			Wr-Vr	Between Arrays	8	0.004*	-583.66
				Within Arrays	9	0.001	
Bolls/Plant	(0.900 ± 0.151)**	0.914**	Wr+Vr	Between Arrays	8	0.172**	17.69
				Within Arrays	9	0.025	
			Wr-Vr	Between Arrays	8	0.014	-100.33
				Within Arrays	9	0.049	
Plant Height (cm)	(0.873 ± 0.073)**	0.976**	Wr+Vr	Between Arrays	8	49319.2**	11.76
				Within Arrays	9	980.4	
			Wr-Vr	Between Arrays	8	743.82	-49.47
				Within Arrays	9	366.447	

**. Significant at 0.01 level (2-tailed).
*. Significant at 0.05 level (2-tailed).

Table 5. Analysis of variance for various traits in a 9×9 diallel cross of cotton (*G.hirsutum* L. & *G.barbadense* L.) in F_1 generation based on Griffing method I, model mixed-B (due to GCA, SCA and reciprocal effects), Hayman (due to SCA and reciprocal components) and Hayman-Jinks method (estimation of genetic components of variance in F_1 generation).

Source of Variance	DF	Mean Square					
		Oil Content (%)	Staple Length(mm)	Uniformity Ratio (%)	Micronaire (μg/inch)	Fiber Bundle Strength(g/tex)	Earliness (day)
Replications	1	0.025	0.831*	4.840**	0.005	4.173**	33.802*
Treatments	80	6.700**	21.255**	5.123**	0.206**	24.586**	124.863**
GCA(a)	8	40.976**	155.782**	10.326**	0.936**	183.85**	754.694**
SCA(b)	36	4.156**	11.136**	4.712**	0.11**	8.644**	102.638**
b1	1			0.60			
b2	8			3.03**			
b3	27			5.31**			
RECIP	36	1.628**	1.48**	4.354**	0.138**	5.138**	7.126
c	8			5.24**			
d	28			4.27**			
Error(Me)	80	0.324	0.68	0.245	0.003	0.125	2.6
MSGCA/MSSCA		9.86	13.99	2.19	8.51	21.27	7.35
Degree of Dominance(Griffing)		0.88	0.72	4.36	1.13	0.46	1.32
$2\sigma^2$gca/2 σ^2gca+ σ^2sca		0.53	0.58	0.19	0.47	0.63	0.43
Heritability(ns)(Griffing)		0.49	0.58	0.17	0.45	0.68	0.42
D				2.068**			
H_1				5.174**			
H_2				4.191**			
F				1.999*			
h^2				0.028			
Kd/(kd+kr)				0.653**			
h				0.346			
uv				0.203**			
$\sqrt{H_1/D}$				1.59**			
h^2/H_2				0.0076			
D/D+E)				0.894**			
Heritability(bs)		0.906*	0.992**	0.865**	0.965**	0.99**	0.962**
Heritability (ns)		0.651*	0.754**	0.289**	0.641**	0.82**	0.607**

**. Significant at 0.01 level (2-tailed).
 *. Significant at 0.05 level (2-tailed).

Table 5: Continued.

Source of Variance	DF	Mean Square					
		Lint% (G.O.T)	Seed Index (g)	Seed cotton yield (g)	Boll weight (g)	Bolls/Plant	Plant Height (cm)
Replications	1	5.111	0.155**	11.239	0.007	0.747	34.722
Treatments	80	44.172**	7.558**	56.518**	0.464**	1.156**	454.53**
GCA(a)	8	185.31**	42.354**	360.456**	2.69**	8.396**	3637.98**
SCA(b)	36	41.54**	6.034**	38.204**	0.32**	0.68*	199.906**

Table 5: Continued.

RECIP	36	15.436**	1.88**	7.29**	0.114**	0.022	1.722
Error (Me)	80	2.84	0.005	1.89	0.001	0.206	1.842
MSGCA/MSSCA		4.46	7.02	9.43	8.41	12.34	18.20
Degree of Dominance(Griffing)		2.09	1.44	0.98	1.2	0.34	0.55
$2\sigma^2 gca/2 \sigma^2 gca + \sigma^2 sca$		0.32	0.41	0.50	0.45	0.75	0.65
Heritability(ns) (Griffing)		0.30	0.41	0.48	0.45	0.54	0.64
Heritability(bs)		0.902**	0.999**	0.938**	0.912**	0.714**	0.993*
Heritability (ns)		0.466**	0.609**	0.653**	0.647**	0.619**	0.799*

**. Significant at 0.01 level (2-tailed).
*. Significant at 0.05 level (2-tailed).

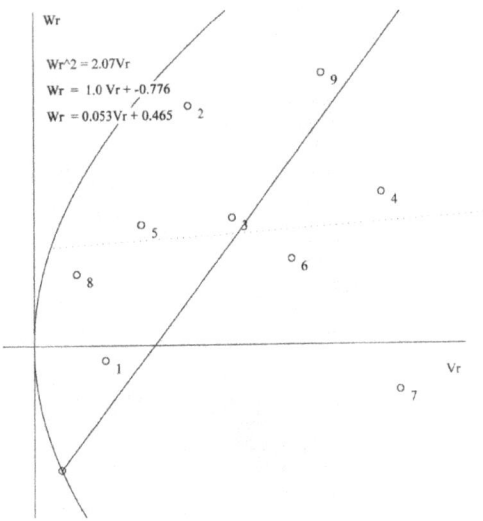

No	Pr	Wr	Vr	Fr
1	Delinter	-0.074	0.458	0.876
2	Sindose-80	1.145	0.989	0.609
3	Omoumi	0.608	1.272	0.647
4	Bulgare-539	0.723	2.239	0.482
5	Termez-14	0.577	0.689	0.741
6	Red leaf	0.410	1.661	0.618
7	B-557	-0.220	2.361	0.608
8	Brown fiber	0.343	0.272	0.841
9	Siokra-324	1.296	1.855	0.453

Fig 5. Scattering of Parents on Wr/Vr Regression Line and Limitting Parabola for Uniformity Ratio and Their F_1 Fr Values (F_1 Generation).

Low estimates of narrow (0.289) and moderate broad (0.865) and true (0.894) sense heritability were also recorded in F_1 (Table 5). Non- significant positive correlation coefficient ($r = 0.108$) with b value of 0.07 obtained between Wr+Vr and parental means enunciated that parents containing recessive genes were responsible for increased uniformity ratio, while dominance genes were responsible for decreased uniformity in F_1 generation. The Wr/Vr graph plotted in Fig. 5 and Fr values for F_1 uniformity ratio, show that the regression line (b=1) intercepted Wr axis below the origin on negative side which suggested an over-dominance type of gene action, while regression line (b= 0.053) cuts Wr axis above the origin on positive side which suggested a partial domi-

nance type of gene action. The distribution of array points along the regression line conceive that Delinter (1) and Brown color (8) bearing maximum and positive Fr values and being nearest to the point of origin, had large number of dominant genes, in contrast to the cultivar Bulgare-439 (4) and Siokra (9) being farther and possessing minimum Fr values, had maximum recessive genes. Mean squares due to GCA and SCA were highly significant for all the traits in F_1 generation by Griffing method indicating the importance of additive and nonadditive effects of genes for genetic controlling of traits. High estimations of MSGCA/MSSCA for all traits and also high narrow-sense heritability for all traits except uniformity ratio (non-adquate trait for additive-dominance model) and lint% by Griffing and Hayman method (differences of herita-bility between two methods belong to inter-action of environment and genetic parameters in Griffing method) displaythe importance of additive effects of genes in genetic control of more traits. This is also confirmed by the de-gree of dominance estimated near to 1 or less than 1 for some traits. For further improvement and breeding of these traits, selection method should be more efficient. For example, the hy-brids of parents manifested the highest oil con-tent because the mean of parents for oil content was 18.282% while the mean of hybrids was 19.77% and the best crosses with over domi-nance gene effect for this trait were Omoumi × Brown fiber (22.95%), Delinter × Omoumi (22.78%) and Delinter × Termez-14 (22.495%) indicating 26% increase in oil content (22.95-18.25/18.25×100) which is commercially valu-able. For oil content Delinter, Omoumi and

Termez-14 were the best general combiner and Sindose-80 ×Siokra-324, Termeze-14 × B-557 and Omoumi × Brown fiber were the best spe-cific combiner. For other traits including staple length, uniformity, micronaire, fiber bundle strength, earliness, Lint%, seed index, seed cotton yield, boll weight, bolls/plant and plant height the value of their top F1 hybrids has been increased 7.22mm (34.4-27.18), 3.4%, 1.08µg/ inch, 8.08g/tex, -16.72days, 11.69%, 5.62g, 20.6g, 1.96g, 2.15 and 36.06cm in comparison with the mean of their parents respectively. Heterosis of varying magnitude was found in F_1 generation. Plant height, seed cotton yield and earliness components showed maximum hetero-sis, while uniformity and micronaire showed low heterosis and the remaining traits showed mod-erate heterosis. This indicates the higher perfor-mance of F_1 hybrids for related traits which is normal from physiological point of view.

Table 6. Top means of parents, hybrids, crosses, GCA, SCA, interaction effects and average heterosis and increased coefficient in F_1 generation.

Traits	Oil Content (%)	Staple Length (mm) (2.5 SL)	Uniformity Ratio (%)	Micronaire (µg/inch)	Fiber Bundle Strength (g/tex)	Earliness (day)
Mean of three top parents	3 = 21.7	5 = 33.1	4 = 48	1 = 4.05	5 = 26.95	3 = 70
	5 = 20.3	3 = 29.35	3 = 47.5	8 = 3.85	3 = 25.7	5 = 73
	1 = 18.7	9 = 27.85	8 = 47.5	9 = 3.15	9 = 22.2	2 = 78
Mean of three top crosses (TC)	8×3 = 22.95	9×5 = 34.4	2×4 = 50	1×8 = 4.25	8×5 = 29.6	3×9 = 68.5
	1×3 = 22.78	6×5 = 34.3 .	1×9 = 49.5	1×9 = 3.9	7×5 = 29.6	9×3 = 68.5
	1×5 = 22.49	9×3 = 33.9	9×2 = 49.5	8×9 = 3.55	3×2 = 29.5	6×3 = 69
Mean of parents (MP)	18.28	27.18	46.6	3.17	21.52	85.22
Mean of hybrids	19.77	29.26	46.8	3.12	23.51	77.7
LSD (α = 0.05)	1.608	0.731	1.392	0.149	0.995	4.544
LSD (α = 0.01)	2.137	0.970	1.846	0.197	1.320	6.025
Three top GCA	3 = 2	5 = 3.9	5 = 0.61	8 = 0.26	5 = 4.17	3 = -7.6
	5 = 1.46	3 = 3.16	1 = 0.33	1 = 0.20	3 = 3.31	5 = -5.08
	1 = 0.09	9 = -0.44	3 = 0.28	9 = 0.058	9 = -0.31	9 = -3.72
Three top SCA	2×9 = 1.73	1×3 = 2.05	6×7 = 1.8	1×9 = 0.59	5×7 = 2.6	1×2 = -9.69
	5×7 = 1.43	3×7 = 1.9	2×4 = 1.4	2×8 = 0.28	2×3 = 2.1	5×6 = -8.88
	3×8 = 1.26	2×3 = 1.76	3×5 = 1.2	3×5 = 0.21	1×9 = 1.78	3×6 = -8.11
Increased coefficient (TC/MP) %	26	27	7.2	34	38	20
Average heterosis	1.4	2.07	0.187	-0.53	1.98	-7.5
Two top interaction effects	4×8 = -1.27	3×5 = -1.30	2×8 = -3	1×8 = 0.55	5×8 = -3.62	
	8×9 = 1.1	2×7 = 1.17	2×6 = -2.25	5×8 = 0.4	1×5 = -3.02	

Table 6. Continued.

Traits	Lint % (G.O.T)	Seed Index (g)	Seed cotton yield (g)	Boll weight (g)	Bolls/Plant	Plant Height (cm)
Mean of three top parents	8 = 48.98	5 = 9.53	6 = 26.88	1 = 3.72	6 = 9.6	3 = 114
	2 = 45.99	3 = 8.63	1 = 26.84	6 = 2.94	4 = 8.6	5 = 113
	4 = 44.74	6 = 8.42	4 = 22.35	9 = 2.7	7 = 8.1	9 = 91
Mean of three top crosses (TC)	7×1 = 53.2	6×3 = 13.13	4×1 = 40.1	4×1 = 4.55	6×4 = 10.1	3×5 = 121.5
	2×6 = 51.53	7×3 = 12.76	1×4 = 38.34	1×4 = 4.41	6×7 = 10.1	5×3 = 119.5
	7×2 = 50.24	5×6 = 12.62	6×1 = 38.08	7×8 = 4.01	6×1 = 10.1	3×6 = 119.5
Mean of parents (MP)	41.51	7.51	19.50	2.59	7.95	85.44
Mean of hybrids	43.22	9.12	24.86	3.02	8.62	98.61
LSD (α = 0.05)	4.067	0.185	3.489	0.094	1.209	3.829
LSD (α = 0.01)	5.39	0.245	4.636	0.125	1.606	5.088
Three top GCA	1 = 1.66	5 = 1.86	6 = 5.03	1 = 0.53	6 = 1.13	3 = 17.57
	8 = 1.64	3 = 1.78	1 = 4.14	6 = 0.19	4 = 0.37	5 = 16.71
	9 = 1.58	6 = 0.09	4 = 1.9	4 = 0.1	7 = 0.048	9 = 0.24

Table 6. Continued.

	Lint % (G.O.T)	Seed Index (g)	Seed cotton yield (g)	Boll weight (g)	Bolls/Plant	Plant Height (cm)
Three top SCA	1×7 = 4.97	5×6 = 1.52	1×4 = 8.8	1×4 = 0.87	3×8 = 0.64	5×8 = 10.56
	3×6 = 4.69	3×6 = 1.50	5×9 = 4.2	5×9 = 0.50	3×6 = 0.55	2×5 = 9.45
	3×5 = 4.46	3×7 =1.42	5×7 = 4.2	7×8 = 0.44	1×6 = 0.53	3×4 = 8.34
Increased coefficient (TC/MP) %	28	74	105	76	27	42
Average heterosis	1.7	1.6	5.3	0.43	0.67	13.16
Two top interaction effects	1×6 = -5.07	4×8 = -1.11	7×8 = 6.104	7×8 =0.790		
	3×7 = 3.155	3×4 = -1.02	2×3 = 2.354	2×3 =0.320		

1- Delinter	2- Sindose-80	3 - Omoumi	4-Bulgare-539	5-Termez -14
6 -Red leaf	7-B-557	8- Brown fiber	9- Siokra-324	

Discussion

The lint%(G.O.T) was controlled by nonadditive genes in F1 generation, and the results are not in agreement with those reported by Bhatade and Bhale (1983) and McCarty et al.(1996) who reported additive type of gene action with partial dominance for inheritance of lint%. The results are in great resemblance with the findings of Kohel (1980) and Avtonomov et al. (1981) who determined significant heterosis over mid and also better parents for oil content in F_1 generation. The obtained results are authenticated by the findings of Percy and Turcotte (1992) for improvement of fiber properties in F_1 generation. Related results to micronaire were largely in agreement with the findings of Percy and Turcotte (1992), as they did not notice heterosis for fiber fineness in intra-*hirsutum* and intrabarbadense hybrids, although interspecific crosses of G.*hirsutum* × G.*barbadense* displayed a varying level of heterosis in some studies.

It is concluded that the additive-dominance model was adequate for majority of the traits and partially adequate for some traits. Majority of the traits were influenced by additive genes in F_1 generation. These results are encouraging for practical improvement through hybridization and selection method. Significant variation for genotypic, general combining ability (GCA) effects, and specific combining ability (SCA) effects ($P \le 0.05$) was identified for all the studies traits and indicates the potential for improvement through selection. In addition for other agronomic traits, it is suggested that improvements should be made through utilizing a backcross approach. We can also produce and use new hybrids that were the best crosses on the basis of our purpose and 12 studied traits commercially. Plant breeders will be able to use data of mean performance, estimation of heterosis, heterobeltiosis, combining ability (GCA and SCA) and interaction effects of all traits while producing new cultivars depending on the annual demand for fiber quality, oil content and other characteristics in F_1 generation. For example, traits including oil content, staple length, uniformity ratio, micronaire, fiber bundle strength, earliness, Lint%, seed index, seed cotton yield, boll weight, bolls/plant and plant height value of their top F_1 hybrids have been increased by 25%, 7.22mm, 3.4%, 1.08µg/inch, 8.08g/tex, -16.72days, 11.69%, 5.62g, 20.6g, 1.96g, 2.15 and 36.06cm respectively in comparison with means of their parents. In seed production programme we can use the best general combiner and the best specific crosses in the view of their interaction effects.

References

1. Avtonomov, V., T. Sokolova., S. Rakhmankulov, and L. Sydykhodzhaeva, 1981: Seed oil content in distant interspecific hybrids. Plant .Breeding. 52(4): 3151.
2. Bhatade, S. S, and N. L. Bhale, 1983: Combinig ability for seed and fiber characters and its interaction with location in *G.arboreum*. Ind.J.Agric. Sci. 53(6):
3. 418-422. Basal, H., and I. Turqut, 2003: Heterosis and combining ability for yield components in half diallel cotton. (G.hirsutum L.). Turk. J. Agric. 27: 207-212.
4. Breese, E., 1972: Biometrical genetics and its application. Eucarpia Congress, Cambridge.1971:135-146.
5. Echekwu, C. A., and S. O. Alabi, 1994: A diallel analysis of earliness in interspecific crosses of cotton. Discovery & Innovation, 6(3): 241-243.
6. Fehr, W. R., 1993: Principles of cultivar development: Theory and technique. Macmillan, New York.
7. Griffing, B., 1956: Concept of general and specific combining ability in relation to diallel crossing systems. Aust. J. Biol. Sci. 9: 463-493.
8. Hayman, B. I., 1954 a: The theory and analysis of diallel crosses. Genetics, 39: 789-809.
9. Hayman, B. I., 1954b: The analysis of variance of diallel cross. Biometrics, 10: 235-245.
10. Hosseini, GH., 15.12.2008. Crisis of white gold. Kargozaran newspaper. Page7.
11. Hosseini, GH., 2012: Applied parametric statistics. Rahnama press. ISBN: 978-964- 367-503-5.
12. Hosseini, GH., and R. J. Thengane, 2007a: Estimation of genetic parameters for salinity tolerance

in early growth stages of cotton (Gossypium hirsutum L.) genotypes. International Journal of Botany. 3(1): 103-108.
13. Hosseini, GH., and R. J. Thengane, 2007b: Salinity tolerance in cotton (*Gossypium hirsutum* L.) genotypes. International Journal of Botany. 3(1): 48-55.
14. Hosseini, GH., and R. J. Thengane, 2007c: Gene action and earliness in progenies of colored fiber (Brown) cotton (*Gossypium hirsutum* L.). NATOINAL SEMINAR ON CURRENT TRENDS IN PLANT BIORESOURCE UTILIZATION. Department of Botany, University of Pune 26 and 27, Feb. 2007.
15. Hosseini, GH., and R. J. Thengane, 2007d: Estimation of the number of requirement crosses between two-parents for approaching all of the F_2 expected segregated genotypes. NATOINAL SEMINAR ON CURRENT TRENDS IN PLANT BIORESOURCE UTILIZATION. Department of Botany, University of Pune . 26 and 27, Feb. 2007.
16. Innes, N. L., 1975: Estimates of genetic parameters for lint quality in Upland cotton (*Gossypium hirsutum* L.). Theoretical and Applied Genetics.. 46(5): 249-256.
17. Jinks, J. L., 1954: The analysis of continuous variation in diallel crosses of *Nicotiana rustica* varieties. Genetics, 39: 767-788.
18. Jinks, J. L., 1955: A survey of genetical bases of heterosis in a vaeiety of diallel crosses. Heredity, 9: 223-238.
19. Jinks, J. L., 1956: The F_2 and back cross generation from a set of diallel crosses. Heredity. 10: 1-30.
20. Kohel, R. J., 1980: Genetic studies of seed oil in cotton. Crop Sci., 20(6): 684-686.
21. Kohel, R. J., and C. F. Lewis, 1984: Cotton.Madison, Wisconsin, USA. pp. 605.
22. Mather, K., and J. L. Jinks, 1971: Biometrical Genetics. Chapman & Hall Ltd. London., 2nd ed. pp. 382.
23. Mather, K., and J. L. Jinks, 1977: Introduction to Biometrical Genetics. Chapman & Hall Ltd. London., 1st .ed.pp. 73-80.
24. McCarty, J. C., J. N. Jenkins., B. Tang, and C. E. Watson, 1996: Genetic analysis of primitive cotton germplasm accessions. Crop Sci., 36 (3): 581-585.
25. Nasisri, P., M, Ebrahimi., and GH, Hosseini, 2012. Advanced procedures of statistics. Yadvareh. ISBN: 978-600-6221-01-4.
26. Percy, R. G., and E. L. Turcotte, 1991: Early maturing, short stature American Pima cotton parents improve agronomic traits of interspecific hybrids. Crop Sci., 31: 709- 712.
27. Ragsdale, P.I., 2003. Diallel analysis of within-boll seed yield components and fiber properties in upland cotton (*Gossypium hirsutum* L.) and breeding potential for heat tolerance. Texas A&M University.
28. Roy, D., 2000: Plant breeding (Analysis & exploitation of variation). Narosa
29. publishing house. pp, 701.

7

Study of genetic diversities and relatedness of Iranian citrus genotypes using morphological and molecular markers

Hajar Abedinpour[1*], Nad Ali Babaeian Jelodar[1],
Gholam Ali Ranjbar[1] & Behrouz Golein[2]

1. Department of Plant Breeding, Sari Agricultural Sciences and Natural Resources University, Sari, Iran.
2. Iran Citrus Research Institute, Ramsar, Iran.
   *Corresponding Author, Email: h_abedinpour@ymail.com

Abstract
Having knowledge about genetic relationships among accessions is necessary for developing breeding strategies to produce improved cultivars. In present study, genetic diversity and inter-relationship among 29 genotypes of citrus were comparatively analyzed using morphological and RAPD markers. Significant variability was observed among citrus genotypes for 61 quantitative and qualitative morphological characters of leaves, fruits and seeds. Furthermore, the RAPD markers revealed a high polymorphism rate (91.82 %). A pair-wise similarity value between genotypes ranged from 0.14 to 0.97 with average of 0.62. Both morphological and molecular analysis indicated a high degree of variation among studied genotypes. In current research, genotypes "pummelo" and "mandarin" were confirmed as true species of citrus in distinct cluster. Results of present study proved that both of morphological and molecular markers are potential tools for determining genetic diversities and genetic relationships of citrus genotypes and can be used in citrus breeding programs.
Keywords: Citrus, Cluster analysis, Genetic variability, Molecular markers.

Introduction

Citrus is one of the most economically important fruit crops in worldwide (20). *Citrus* and its closed relatives are represented by 28 genera from tribe Citreae of subfamily Aurantioideae in family Rutaceae (32). Citrus taxonomy and phylogeny are very complicated, controversial and confusing, mainly due to sexual compatibility between *Citrus* and its related genera, the high frequency of bud mutations and the long history of cultivation and wide distribution (24).
Elucidating relationships, taxonomy, and diversity are important for developing breeding strategies, conserving biodiversity, and improving breeding efficiency. Understanding the genetic diversity in citrus is also critical for characterizing germplasm, controlling genetic erosion and registration of new cultivars (12, 4).

Morphological characterization in combination with molecular markers would be more rewarding in terms of accurate identification and characterization of most closely related cultivars at intra-specific level. Molecular marker techniques are routinely used for proper characterization, management and conservation of germplasm collections of horticultural species (16). Among molecular markers, random amplified polymorphic DNA (RAPD) markers have been employed most widely for characterization of plant species. RAPD markers are simple, fast, and sensitive. They require no prior knowledge about DNA sequence and can amplify a large number of DNA fragments for reaction (35). RAPD markers are routinely used for proper characterization, management and conservation of germplasm collections of *Citrus* species. For example, RAPD has been used to generate linkage map for citrus (6). Federici *et al.* (11)

examined the phylogenetic relations of 88 accessions representing 45 *Citrus* species and six related genera by utilizing RFLP and RAPD markers. Overall, these previous studies demonstrated that molecular markers are powerful tools for elucidating genetic diversity, determining parentage, and revealing phylogenetic relationships among various citrus species. Nicolosi *et al.* (24) used RAPD, SCAR, and cpDNA markers to elucidate phylogenetic relationships and genetic origins of hybrids in 36 accessions of *Citrus* and one accession from each of four related genera and indicated that *Fortunella* is phylogenetically close to *Citrus* while the other three related genera are distant from *Citrus* and from each other. Dehestani *et al.* (9) evaluated the genetic diversity in 52 genotypes of Navel orange in Mazandaran province (Iran) using RAPD marker and reported high polymorphism (70.13%). Malik *et al.* (20) investigated genetic diversity and inter-relationship among 22 cultivars of *C. sinensis* based on morphological and RAPD markers. In their study, RAPD markers proved to be useful for germplasm characterization and diversity analysis in *C. sinensis* cultivars. Pal *et al.* (27) studied genetic variability and relationships of mandarins using morphological and molecular markers. Their study revealed that both mor-

phological and molecular markers can be successfully utilized for inferring genetic diversity and genetic relationship of mandarin group. Tripolitsiotis *et al.* (33) evaluated genetic similarity among 36 accessions of the Greek *Citrus* germplasm using RAPD and ISSR markers and indicated that both techniques were proven to be equally analytical with an average discrimination power above 0.9. The RAPD and ISSR markers were highly correlated and clustering based on their results are highly correspondence. *Citrus* accessions formed separate clusters according to their species, even though sweet orange and mandarin cultivars revealed high affinity, while lemons were more divergent.

Little is known about the genetic variability of the Iranian citrus accessions. The objective of the present study was to assess genetic diversity and relationship of some important *Citrus* genotypes using morphological and RAPD markers.

Materials and methods

Plant material and sample collection

A total of 29 genotypes of Citrus were collected from Iran Citrus Research Institute, located at Tonekabon, Iran. These genotypes were used for morphological and molecular studies (Table 1). Flower, leaf and fruit sam-

Table 1. Plant materials utilized for morphological and RAPD analysis.

Plant code	Common name	Scientific name	Plant code	common name	Scientific name
G1	Sour orange	*Citrus aurantium*	G61	Unknown natural type	*Citrus* sp.
G2	Marss orange	*Citrus sinensis*	G63	Unknown natural type	*Citrus* sp.
G3	Thomson navel orange	*Citrus sinensis*	G65	Unknown natural type	*Citrus* sp.
G4	Local orange (Siavaraz#1)	*Citrus sinensis*	G67	Unknown natural type	*Citrus* sp.
G5	Local orange (Siavaraz#2)	*Citrus sinensis*	G70	Unknown natural type	*Citrus* sp.
G6	Local orange (Siavaraz#3)	*Citrus sinensis*	G71	Unknown natural type	*Citrus* sp.
G7	Local orange (Siavaraz#4)	*Citrus sinensis*	G72	Unknown natural type	*Citrus* sp.
G8	Moallemkoh (Natural type)	*Citrus* sp.	G73	Unknown natural type	*Citrus* sp.
G9	Shelmohalleh (Natural type)	*Citrus* sp.	G74	Unknown natural type	*Citrus* sp.
G10	Atabaki mandarin	*Citrus reticulata*	G76	Unknown natural type	*Citrus* sp.
G11	Unshiu mandarin	*Citrus unshiu*	G78	Unknown natural type	*Citrus* sp.
G12	Dancy mandarin	*Citrus reticulata*	G79	Unknown natural type	*Citrus* sp.
G13	Bami mandarin	*Citrus reticulata*	G80	Unknown natural type	*Citrus* sp.
G14	Local mandarin	*Citrus reticulata*			
G15	Clementine mandarin	*Citrus clementina*			
G16	Pummelo	*Citrus grandis*			

ples of each genotypes were collected for confirmation of taxonomic identity, characterization and DNA extraction.

Morphological characters

For achieving uniformity in current study only genotypes from Iran Citrus Research Institute were used. 15 leaves, 10 flowers and 10 fruits were randomly collected from each plant with three replications. 61 morphological characters (qualitative and quantitative) of flower, leaf, fruit and seed were determined according to the International Plant Genetic Resources Institute (IPGRI) protocols (13).

Samplings were done by randomly collection of 15 leaves, 10 flowers and 10 fruits from each plant in three replications. According to the criteria provided by protocols of International Plant Genetic Resources Institute (IPGRI) 61 morphological characters (qualitative or quantitative) of flowers, leaves, fruits and seeds were determined (13). All of the 61 morphological characters were converted to bi- and multistate code. A pair-wise similarity matrix was generated based on simple matching coefficient method using software NTSYS ver. 2.10e (29). A cluster analysis was performed using the unweighted pair group method with arithmetic

average (UPGMA) based on simple matching coefficient using XLSTAT software version 2012.3.01 (2). Principal coordinate analysis (PCo) was also carried out for studying correlations among the variables and establishing relationships among genotypes using the Genalex ver 6.5 software (28). The two-way Mantel test (21) for goodness of fit for the UPGMA cluster was also performed using the NTSYS ver. 2.10e software.

DNA isolation

From each genotyepe, five young leaves were taken and total genomic DNA was isolated from leaves using the CTAB (hexadecyltrimethylammonium- bromide) method (22). The DNA concentration was determined spectrophotometrically (Nano Drop 1000) at 260 nm and its quality was checked by electrophoresis on 0.8 % agarose gel. The extracted DNA was diluted to 20ng/μl and stored at -20°C for PCR amplification.

PCR amplification

Thirty RAPD primers were initially screened and finally 19 primers that produced scorable polymorphic bands were selected for further analysis (Table 2).

DNA amplification was carried out in 25 μL

Table 2. Statistical analysis and results of genetic diversity of 29 citrus genotypes.

Row	Primer Name	Primer Sequence 5' → 3'	Annealing temperature	Total number of Bands	Number of Polymorphic Bands	% polymorphism	PIC
1	OPB-12	CCTTGACGCA	37	15	13	86/66	0.216
2	OPE-09	CTTCACCCGA	37	16	16	100	0.120
3	OPA-04	AATCGGGCTG	37	15	14	93/33	0.263
4	OPA-07	GAAACGGGTG	37	19	19	100	0.235
5	OPA-08	GTGACGTAGG	37	10	9	90	0.282
6	OPA-19	CAAACGTCGG	37	12	11	91/66	0.267
7	OPG-05	CTGAGACGGA	37	11	9	81/81	0.247
8	OPG-06	GTGCCTAACC	37	16	14	87/5	0.226
9	OPB-08	GTCCACACGG	35	14	12	85/71	0.215
10	OPA-12	TCGGCGATAG	35	13	12	92/30	0.249
11	OPA-05	AGGGGTCTTG	37	9	8	88/88	0.224
12	OPA-18	AGGTGACCGT	37	19	18	94/73	0.197
13	OPM-11	GTCCACTGTG	37	17	15	88/23	0.265
14	OPM-14	AGGGTCGTTC	35	12	12	100	0.309
15	OPM-18	CACCATCCGT	37	11	10	90/90	0.189
16	OPG-04	AGCGTGTCTG	37	16	15	93/75	0.267
17	OPC-07	GTCCCGACGA	37	16	15	93/75	0.223
18	OPA-10	GTGATCGCAG	37	16	15	93/75	0.213
19	OPA-09	GGGTAACGCC	37	12	11	91/66	0.266
Mean	-	-		14/15	13/05	91/82	0.230

Table 3. Cophenetic coefficients obtained from algorithms with similarity coefficient.

	UPGMA algorithm	Simple connection algorithm	Complete connection algorithm
Dice similarity coefficient	0.972	0.979	0.986
Jaccard similarity coefficient	0.975	0.984	0.989*
Simple matching similarity coefficient	0.979	0.977	0.986

reactions containing 20 ng of template DNA, 0.2 mM dNTPs, 10μM primer, 2.5 μL of 10× PCR Buffer (CinnaGen, Iran), 3 mM of magnesium chloride, 17.2 μL ddH$_2$O and 1.5 unit of Taq polymerase (CinnaGen, Iran). PCR amplification was carried out in a PTC- 10096V Thermocycler (MJ Research, Inc, USA). The thermal cycler conditions for PCR reactions were an initial denaturation of 1 min at 94°C followed by 40 cycles comprising 1 min at 94°C, 1 min at 35-37°C (annealing temperature was optimized for each primer) (Table 2), and 1 min and 30 s at 72°C. An additional step of 7 min at 72°C was used for final extension. Amplification products were separated by electrophoresis (8.3 V.cm^{-1}) in 1.5% agarose gel and stained by ethidium bromide (10μg.ml^{-1}). A photographic record was taken under UV illumination.

Data analysis

Only clear and repeatable amplification products were scored as 1 for present bands and 0 for absent ones. Data were analyzed with the NTSYS-pc software package version 2.10 (29). A cluster analysis was performed using the unweighted pair group method with arithmetic average (UPGMA) based on simple matching coefficient using XLSTAT software version 2012.3.01 (2). The representativeness of dendrograms was evaluated by estimating cophenetic correlation for the dendrogram and comparing it with the similarity matrix, using Mantel's matrix correspondence test (21). The result of this test is a cophenetic correlation coefficient, r, indicating how well the dendrogram represents similarity data. The percent of polymorphism was calculated using the formula (number of polymorphic bands/ total bands). Polymorphism information content (PIC) was calculated for dominant markers that the allelic relationship between their bands was unclear with the formula PIC=Σ [2fi (1-fi)]. The principal coordinate analysis (PCo) of the original binary data matrix was also per-

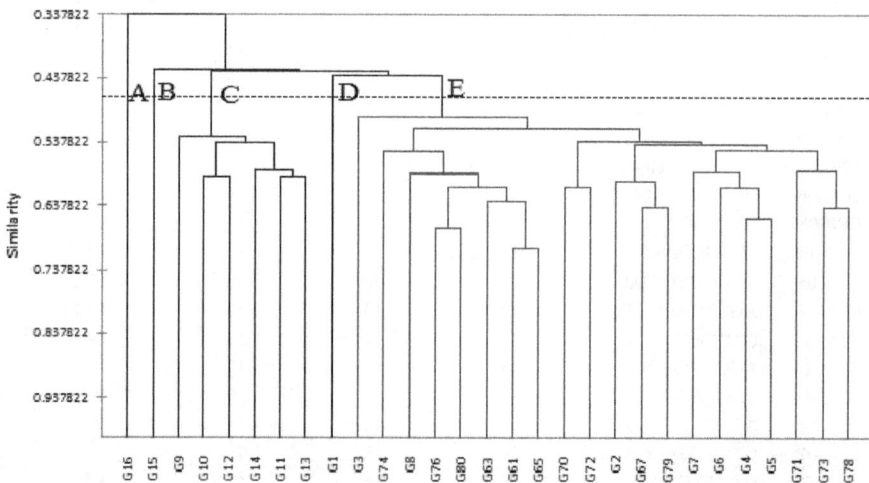

Figure 1. Dendrogram generated using UPGMA, showing relationships between 29 citrus genotypes based on morphological data.

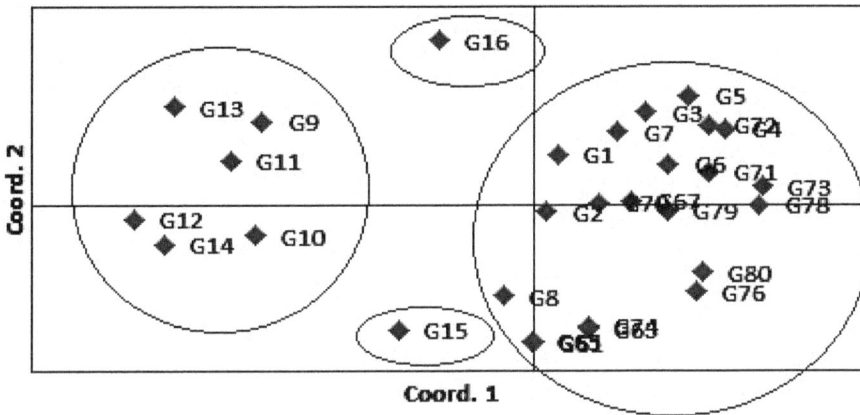

Figure 2. Principle Coordinate Analysis (PCo) ordination based on morphological data.

formed using Genalex ver 6.5 software (28).

Results and Discussion

Morphological analysis

The average genetic similarity among citrus genotypes was 0.47, with values ranging from 0.22 to 0.70. Genotypes G61 and G65 showed the highest degree of similarity (0.70), indicating that these pairs are closely related genotypes. On the other hand, the pummelo (G16) and G74 genotypes indicated the lowest similarity values (0.22).

The dendrogram obtained with 29 genotypes based on 61 quantitative and qualitative morphologic characteristics, separated citrus genotypes into five clusters (A, B, C, D and E), which diverged at a similarity index of 0.47 (Figure 1). Cluster A comprised pummelo (G16). Cluster B comprised Clementine mandarin (G15). Cluster C was divided into two sub-clusters, so that, sub-cluster C1 consisted of Shelmohalleh (G9) and sub-cluster C2 contained Dancy (G12), Local (G14), Bami (G13), Unshiu (G11) and Atabaki (G10) mandarins. In current study, Clementine (G15) and Unshiu (G11) were classified into two distinct clusters. In view of morphological characteristics of mandarins, similar results have been reported in previous studies (18, 7). Cluster D, contained sour orange (G01). Cluster E, the largest group, consisting genetically unknown local genotypes (G8, G61, G63, G65, G67, G70, G71, G72, G73, G74, G76, G78, G79 and G80), Siavaraz (G4, G5, G6 and G7), Thomson Na-

vel (G3) and Marss (G2) oranges. Within this cluster, the genetically unknown local genotypes G61 and G65 showed 0.70 genetic similarities. The Siavaraz oranges (G4, G5, G6 and G7) were very similar to Thomson navel (G3) (Figure 1). Rouhi Ghorabaie et al. (30) reported high similarity among oranges using morphological traits which is in agreement with the present study.

The cophenetic analysis comparing the UPGMA cluster analysis and the simple matching similarity matrix demonstrated a correlation r= 0.86, indicating that data in matrix was well represented by the dendrogram.

Principal coordinate analysis:

Principal Coordinate Analysis was drawn with two dimensional graph using 61 quantitative and qualitative morphologic characteristics (Figure 2). A two-dimensional plot generated from PCA showed four groups which supporting the clustering pattern of the UPGMA dendrogram, except for genotype sour orange (G01) which was included in oranges group, while it was present in cluster D in UPGMA clustering.

The analysis oriented the first three principal components, which contributed 62.27 % of the total variability of collected genotypes. Maximum variability was contributed by the first coordinate (27.07 %) followed by the second coordinate (20.47 %), and the third coordinate (14.72 %).

RAPD analysis

From total of 19 screened primers, 269 bands

with high intensity were scored. The number of bands scored per primer combination ranged from 9 (OPA-05) to 19 (OPA-07 and OPA-18), with a mean of 14. Overall, the polymorphic band number varied between 8 (OPA-05) and 19 (OPA-07), with a mean of 13. The PIC values for the 19 primers ranged from 0.120 to 0.309, with an average of 0.230 (Table 2).

In order to classify genotypes based on RAPD data, Dice, Jaccard and the Simple matching similarity coefficient were calculated. Based on comparison of the correlations in matrices of similarity, each matrix of similarity was used to draw clusters using UPGMA algorithms, simple and complete connection. Cophenetic coefficient was calculated for every cluster. This coefficient shows the severity of similarity between matrix and cluster. Therefore, greater number resulted from the comparison between coefficient matrix and cophenetic matrix, indicating goodness of fit of the cluster analysis to the similarity matrix (23). Accordingly, Jaccard similarity coefficient and UPGMA algorithms were chosen as the most compatible similarity coefficient and clustering algorithm (Table 3).

A similarity matrix was calculated using RAPD data according to Jaccard coefficient (14). dendrogram was constructed using the UPGMA method (Figure 3). Cophenetic correlation between ultrametric similarities of tree and similarity matrix was found to be high (r= 0.98, P < 0.01), suggesting that the cluster analysis strongly represents the similarity matrix. The genotypes studied had similarity values ranging from 0.14 to 0.97.

Results of similarity matrix showed that the highest genetic similarity (0.97) was existed between genotypes G74 and G73 and the lowest genetic similarity (0.14) was observed between genotypes of pummelo and Local mandarin. UPGMA dendrogram was generated by RAPD data and average similarity (0.49) for all genotype pairs was used as a cut off value for defining the clusters (Figure 3). From this dendrogram, 29 genotypes could be classified into five classes (A, B, C, D and E).

Considering the dendrogram (Figure 3), cluster A, included sour orange (G1). Sour orange showed similarity values of 0.26 and 0.42 with pummelo and mandarin, respectively. According to previous works, have suggested that sour orange is a natural hybrid between mandarin and pummelo (5, 4, 1) which was consistent with this study. Unshiu mandarin (G11) is placed into cluster B. The cluster C, the largest group, consist of unknown local genotypes (G61, G63, G65, G67, G70, G71, G72, G73, G74, G76, G78, G79 and G80), Siavaraz (G4, G5, G6 and

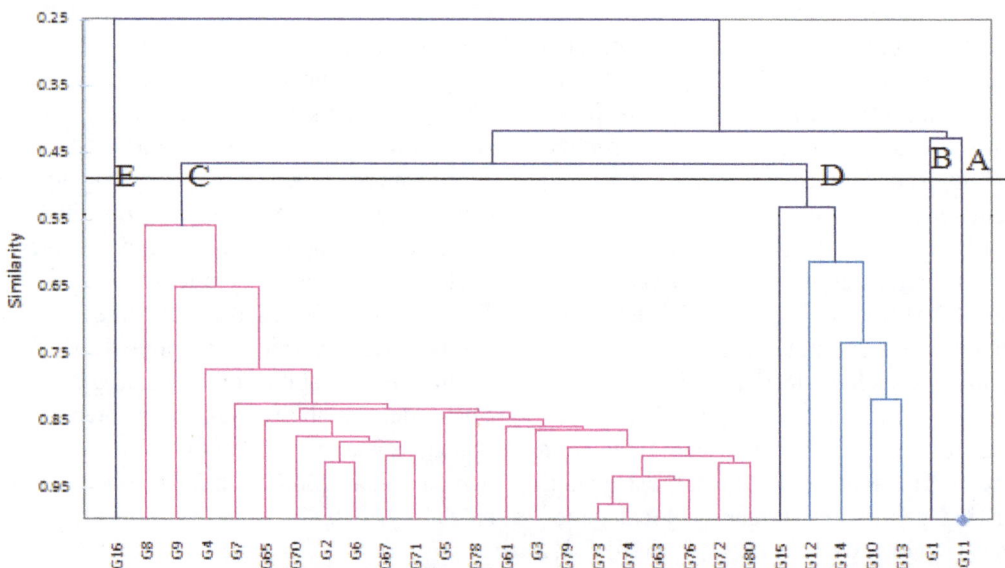

Figure 3. Dendrogram generated using UPGMA, showing relationships between 29 citrus genotypes, using RAPD data.

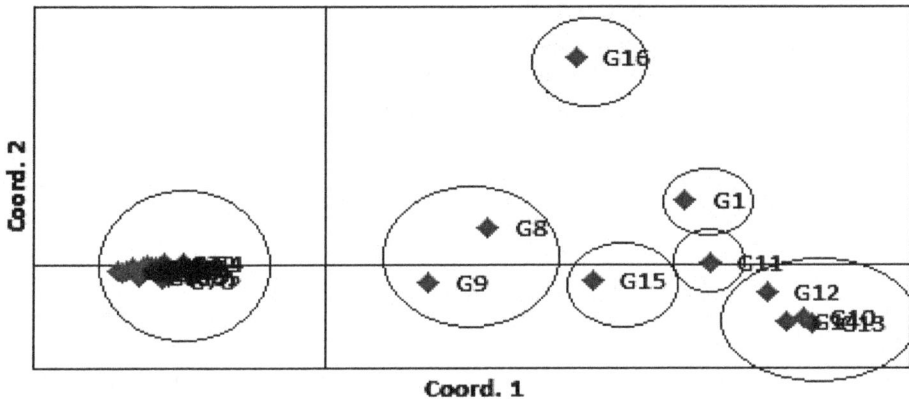

Figure 4. Principle Coordinate Analysis (PCo) ordination based on RAPD data

G7), Thomson navel (G3) and Marss (G2) oranges. Within this cluster, the genotypes G74 and G73 showed 0.97 genetic similarity. Sweet orange showed low level of genetic diversity according to lots of previous studies (19, 25, 26, 34, 20). It is notified that, most of sweet oranges were mutations of unique ancestor tree. However, despite differences in morphological characters, genetic variation of sweet orange was low (10). According to a recent study using a large number of oranges, there is a high level of genetic similarity among oranges (34). Furthermore, accessions arising from spontaneous mutations are also often difficult to be distinguished (4). According to our data, Siavaraz genotypes were high similar to Thomson navel orange (G3), indicating that probably originated from bud mutation, and the idea was supported by SSR analysis (17, 15, 30).

Shelmohalleh (G9) had high similarity (0.68) to Siavaraz 3, whereas based on morphological data it was clustered in mandarin group. Also, Moallemkoh (G8) showed high similarity to Siavaraz 2 and Thomson navel. The present findings were consistent with the results of Jannati et al. (15). Using SSR markers, they reported that Moallemkoh (G8) had similarity to Thomson and Siavaraz and they concluded that this genotype probably obtained through the hybridization between them or a bud mutation. Cluster D comprised with two subclusters, D1 including Clementine mandarin (G15) and D2 consisting of Dancy (G12), Local (G14), Bami (G13) and Atabaki (G10) mandarins. Within subcluster D2, high genetic similarity (0.82) was observed between the genotypes of Bami

(G13) and Atabaki (G10) in spite of morphological differences. In both RAPD and morphological analysis, Clementine (G15) and Unshiu (G11) were classified into two distinct clusters. Coletta Filho et al. (8) reported that they belong to two different groups. A high level of polymorphism (86%) was also reported by Campos et al. (7) in mandarins based on morphological and AFLP markers. Although Coletta Filho et al. (8) reported very narrow genetic base of mandarin group using RAPD marker and proposed mandarin group as a single species mandarin. Mandarins are one out of three citrus types that Barrett and Rhodes (5) proposed as true species and a number of researches (24, 4, 34) supported this idea.

Using both of morphological and RAPD markers, Pummelo (G16) was placed into group E separately and showed a little similarity in comparison with other genotypes. Pummelo was reported as one of the three true citrus species by Barrett and Rhodes (5) and most of subsequent studies were followed by this statement (11, 24, 4, 34). Thus, Pummelo has played an important role as the parent of many citrus fruits, such as lemons, oranges and grapefruits.

Principal coordinate analysis:

The principal Coordinate analysis was performed for better visualization of relationship among studied accessions. The classical principal Coordinate analysis (PCo) is likely an example of dimensionality reduction. The results of PCo are demonstrated in Figure 4.

A two-dimensional plot generated from PCo showed seven groups which was found to be almost similar to the clustering pattern of UPG-MA dendrogram. The reason behind observing low differences is that two to three of the first components cannot represent the diversity of primary variables (total number of bands). In a 2D plot analysis, genotypes Shelmohalleh (G9) and Moallemkoh (G8) were stayed together at the same group, whereas, in UPGMA clustering they were presented at two different clusters. Clemantine mandarin (G15) also formed a separate group in 2D plot but in dendrogram, it was placed at cluster D.

The analysis demonstrated that the first three principal components have contributed 77.74 % of total variability of collected genotypes. Maximum variability was created by the first component (46.17 %) followed by the second (20.44%), and the third components (11.13 %). In molecular set of data, two or three of the first components can defined about 10-20 percent of changes, in which, are not statistically suitable for graphic display, but represents genetically desirable sample of total genome (31). In present study, 77.74% of total changes were determined mainly because of measuring only a few numbers of components. The applied primers have covered only a little chromosomal regions and have poor dispersion in various parts of genome. Hence, further investigation using more primers is necessary for covering whole plant genome.

Comparison between RAPD and morphological data

Comparing matrices of RAPD and morphological data showed a weak correlation between dendrograms (r= 0.47, P=0.99) following 500 random permutations with the Mxcomp procedure from NTSYS program. Despite the weak correlation between morphological and molecular analysis, similar groups were placed at respective dendrograms. Formation of five clusters was consistently found in both analyses, however, some discrepancies could be found between two dendrograms. For example, Clementine mandarin (G15) was clearly separated in morphological analysis, however, it was grouped in subgroup D1 based on RAPD analysis. Another discrepancy concerning the Shelmohalleh (G9) that clustered into mandarin genotypes within subgroup C1 in morphologi-

cal dendrogram, but it was clustered closely to orange genotype within cluster C based on molecular analysis. Similar results were found in mandarins by Koehler-Santos et al. (18), who detected differences between dendrograms generated from morphological and SSR data, and suggested that morphological and molecular differences were apparently independent, due to different selection and evolutionary factors. The reasons for the non-correlation or weak correlation between morphological and molecular markers can be fallowing by these acts: 1- Low number of primers that are probably not cover as well genomic level and resulting in a weak correlation between the genes those controllers the molecular and morphological traits. 2- Morphological traits are affected by the environment or the effects of genotype and environment interaction or maybe have involved the effects of dominance, epistatic and pleiotropic or different allelic combinations may be lead to similar phenotypes and the resulting morphological differences that is not consistent with genetic differences. 3- Morphological characteristics are compared by RAPD sequences that may be have various levels of changes in evolutionary, as a nucleotide change that can alter RAPD phenotype but morphological trait may be preserve du 87e to compatibility despite random mutations contingency. 4- Most of the genome of eukaryotic organisms are comprises non-coding regions that during evolution were exposed to mutation. RAPD sequences existed in the coding and non-coding sequences. Therefore many of the molecular markers are created in non-coding regions that have not linkage with coding genes (3).

Conclusion

Comparison of morphological and molecular characterization data is of immense importance to conclude the extent of genetic diversity present in the set of cultivars (20). Generally, the applied primers in present study showed high PIC that indicating polymorphism and high efficiency. Furthermore, the morphological data showed highly variation among the selected citrus genotypes. However, since morphological variation influences by environmental conditions, more accuracy will be achieved by application of molecular markers for grouping the genotypes. In both RAPD and morphological analysis, 29 genotypes were classified into five groups. however, some difference could be found between two dendrograms. Although the correlation between morphological and RAPD data is low, both techniques can be used complementarily in citus genetic diversity. This study represented the first attempt to use morphological trait with RAPD markers to study genetic diversity of Iranian citrus genotypes. Present study revealed that morphological and molecular markers could be successfully utilized for inferring genetic diversity and genetic relationship of citrus. Results derived from present study are useful for citrus breeding programs and enhancing citrus industry.

Acknowledgments

The help and cooperation received from Citrus Research Institute, I.R. Iran is fully acknowledged.

References

1. Abkenar, A.A., Isshiki, S., Matsumoto, R., and Tashiro, Y. 2007. Comparative analysis of RUJD 2014. Genetic diversity of diploid wheat (Triticum urartu) using morphological traits and RAPD markers. J Plant Prod Res, 21(1): 149-166.
2. Barkley, N.A., Roose, M.L., Krueger, R.R., and Federici, C.T. 2006. Assessing genetic diversity and population structure in a citrus germplasm collection utilizing simple sequence repeat markers (SSRs). Theor Appl Genet, 112(8): 1519-1531.
3. Barrett, H.C., and Rhodes, A.M. 1976. A numerical taxonomic study of affinity relationships in cultivated Citrus and its close relatives. Syst Bot, 105-136.
4. Cai, Q.G.C.L., Guy, C.L., and Moore, G.A. 1994. Extension of the linkage map in Citrus using random amplified polymorphic DNA (RAPD) markers and RFLP mapping of cold-acclimation-responsive loci. Theor and Appl Genet, 89(5): 606-614.
5. Campos, T.E., Gutiérrez Espinosa, M.A., Warburton, M.L., Santacruz Varela, A., and Villegas Monter, Á., 2005. Characterization of madarin (Citrus spp.) using morphological and AFLP markers. Interciencia, 30(11): 687-693.
6. Coletta Filho, H.D., Machado, M.A., Targon, M.L.P.N., Moreira, M.C.P.Q.D.G., and Pompeu Jr, J. 1998. Analysis of the genetic diversity among mandarins (Citrus spp.) using RAPD markers. Euphytica, 102(1): 133-139.
7. Dehestani, A., Kazemitabar, S.K., and Rahimian, H. 2007. Assessment of genetic diversity of na-

vel sweet orange cultivars grown in Mazandaran province using RAPD markers. Asian J Plant Sci, 6:1119–1124.
8. Fang, D.Q., and Roose, M.L. 1997. Identification of closely related citrus cultivars with inter-simple sequence repeat markers. Theor Appl Genet, 95(3): 408-417.
9. Federici, C.T., Fang, D.Q., Scora, R.W., and Roose, M.L. 1998. Phylogenetic relationships within the genus Citrus (Rutaceae) and related genera as revealed by RFLP and RAPD analysis. Theor Appl Genet, 96(6-7): 812-822.
10. Herrero, R., Asins, M.J., Carbonell, E.A., and Navarro,L. 1996. Genetic diversity in the orange subfamily Aurantioideae. Intraspecies and intragenus genetic variability. Theor Appl Genet, 92(5): 599-609.
11. IPGRI (International Plant Genetic Resource Institute), 2000. Descriptors of Citrus. International Plant Genetic Resources Institute, Rome, Italy. p75.
12. Jaccard, P. 1908. Nouvelles recherches sur la distribution florale. Bull Soc Vaud Sci Nat, 44: 223- 270.
13. Jannati, M., Fotouhi, R., Pourjanabad, A., and Salehi, Z. 2009. Genetic diversity analysis of Iranian citrus varieties using micro satellite (SSR) based markers. J Hortic For, 1(7): 120-125.
14. Karp, A,. Kresovich, S,. Bhat, K.V., Ayad, W.G., and Hodgkin, T. 1997. Molecular tools in plant genetic resources conservation: a guide to the technology. IPGRI Bull, 2-47.
15. Kianoush, S., Babaeian Jelodar, N., and Asadi Abkenar, A. 2009. Evaluation of genetic diversity in citrus germplasm using microsatellite (SSR) Markers. J Agri Sci Nat Resour, 15(6): 109-117.
16. Koehler-Santos, P., Dornelles, A.L.C., and Freitas, L.B.D. 2003. Characterization of mandarin citrus germplasm from Southern Brazil by morphological and molecular analyses. Pesqu Agropecu Bras, 38(7): 797-806.
17. Luro, F., Laigret, F., Bové, J.M., and Ollitrault, P. 1995. DNA amplified fingerprinting, a useful tool for determination of genetic origin and diversity analysis in Citrus. Hortscience, 30(5): 1063-1067.
18. Malik, S.K., Rohini, M.R., Kumar, S., Choudhary, R., Pal, D., and Chaudhury, R. 2012. Assessment of genetic diversity in Sweet Orange [Citrus sinensis (L.) Osbeck] cultivars of India using morphological and RAPD markers. Agric Res, 1(4): 317-324.
19. Mantel, N. 1967. The detection of disease clustering and a generalized regression approach. Cancer Res, 27: 209-220.
20. Murray, M. G., and Thompson, W.F., 1980. Rapid isolation of high molecular weight plant DNA. Nucleic Acids Res, 8(19): 4321-4326.
21. Nei, M., and Feldman, M. W. 1972. Identity of genes by descent within and between populations under mutation and migration pressures. Theor

Popul Biol, 3(4): 460-465.

22. Nicolosi, E., Deng, Z.N., Gentile, A., La Malfa, S., Continella, G., and Tribulato, E., 2000. Citrus phylogeny and genetic origin of important species as investigated by molecular markers. Theore Appl Genet, 100(8): 1155-1166.

23. Novelli,V.M., Cristofani, M., Souza, A.A., and Machado, M.A., 2006. Development and characterization of polymorphic microsatellite markers for the sweet orange (Citrus sinensis L. Osbeck). Geneti Mol Biol, 29(1): 90-96.

24. Novelli, V.M., Machado, M.A., and Lopes, C.R. 2000. Isoenzymatic polymorphism in Citrus spp. and Poncirus trifoliata (L.) Raf.(Rutaceae). Genet Mol Biol, 23(1): 163-168.

25. Pal, D., Malik, S.K., Kumar, S., Choudhary, R., and Sharma, K. C. 2013. Genetic variability and relationship studies of mandarin (Citrus reticulata Blanco) using morphological and molecular markers. Agric Res, 2(3): 236-245.

26. Peakall, R.O.D., and Smouse, P.E. 2006. GENAL-EX 6: genetic analysis in Excel. Population genetic software for teaching and research. Mol Ecol Notes, 6(1): 288-295.

27. Rohlf, F.J. 2000. NTSYS-pc: numerical taxonomy and multivariate analysis system, ver. 2.10e. Exeter Ltd., Setauket.

28. Rouhi Ghorabaie, H.R.R., Ghazvini, R.F., Golein, B., and Nabipour, A.R. 2010. Identification of some citrus accessions in a citrus germplasm utilizing simple sequence repeat markers (SSRs). Hortic Environ Biotechnol, 51(4): 343-347.

29. Siahsar, B.A., Allahdoo, M., and Shahsavand, H. 2010. Evaluation of genetic diversity of ttiitipyrum, triticale and wheat lines through RAPD and ISJ markers. Iran J Agric Sci, 41(3): 555-568.

30. Swingle, W.T., and Reece, P.C. 1967. The botany of citrus and its wild relatives in the orange subfamily. In: Reuther, W., Webber, H.J., Batchelor, L.D. (Eds.), The Citrus Industry. University of California Press, Berkeley. pp:190– 430.

31. Tripolitsiotis, C., Nikoloudakis, N., Linos, A., and Hagidimitriou, M. 2013. Molecular characterization and analysis of the greek citrus germplasm. Not Bot Horti Agrobotanici Cluj- Napoca, 41(2): 463-471.

32. Uzun, A., Yesiloglu, T., Tuzcu, O., and Gulsen, O. 2009. Genetic diversity and relationships within Citrus and related genera based on sequence related amplified polymorphism markers (SRAPs). Sci Hortic, 121(3): 306-312.

33. Williams, J.G., Kubelik, A.R., Livak, K.J., Rafalski, J.A., and Tingey, S.V. 1990. DNA polymorphisms amplified by arbitrary primers are useful as genetic markers. Nucleic Acids Res, 18(22): 6531-6535.

8

Involvement of Cytosine DNA methylation in different developmental stages of *Aeluropus littoralis*

S. H. Hashemi-Petroudi[1*], Gh. A. Nematzadeh[1], H. Askari[2] and S. Ghahary[3]

1. Genetic and Agricultural Biotechnology Institute of Tabarestan, Sari University of Agricultural Sciences and Natural Resources, Iran.
2. Department of Biotechnology, Faculty of New Technologies and Energy Engineering, Shahid Beheshti University, G. C., Tehran, Iran.
3. Department of Agronomy, Tarbiat Modares University, Iran.
   *Corresponding Author, Email: irahamidreza@yahoo.com

Abstract

DNA methylation as epigenetic mark plays a key role in normal differential and developmental processes as well as in dynamic gene regulation at the genomic level. To assess DNA methylation pattern in different developmental stages of *Aeluropus littoralis,* methylation sensitive amplified polymorphism (MSAP) was used. Methylation and demethylation status at the CCGG recognition site were tracked by two sets of cytosine methylation-sensitive enzymes (*Msp*I and *Hpa*II), which were classified into three types. The percentage of total bands per type I (non methylation), type II (CpG methylation) and type III (CpCpG methylation) fragments were 75.7, 19.4 and 4.9, respectively. The most frequent methylation events (19.4%) were observed in type II fragment in which full methylation pattern occurred. Out of 480 bands, 33 bands showed methylation alterations between differential developmental stages in all three types of detectable methylation levels. In this study, polymorphic bands had two main directions associated with methylation or demethylation patterns in which methylation level increased during plant development. The methylation and demethylation events at CG sites could be related to developmental stage-specific gene regulation.

Additional keywords: Epigenetic, DNA methylation, *Aeluropus littoralis*, Halophyte, Developmental stages.

Introduction

Genetic and epigenetic information are essential determinants of structural and functional states of organisms (Loidl 2004). Epigenetic regulation governs expression of the genome by processes often associated with chromatin structure. The regulation may be coupled to histone variants, histone post-translational modifications, and DNA methylation, which are involved in a broad spectrum of biological behaviors (Chinnusamy and Zhu 2009). Methylation is a universal DNA modification (Hendrich and Tweedie 2003) which is raised from the conversion of cytosine into 5-methylcytosine at nuclear DNA. The direct addition of a methyl group can provide docking sites for proteins to alter the chromatin state or affect the covalent modification of resident histones (Alis *et al.* 2007). In plants, cytosine methylation is shared among CpG, CpHpG, and CpHpH contexts, where H stands for adenine, cytosine, or thymine (Feng *et al.* 2010; Chinnusamy and Zhu 2009). The highly frequent occurrence of the genome methylation can be considered on the basis of *Arabidopsis* genome methylation which is placed around 24% of CG dinucleotides in the collection of methylated contexts (Cokus *et al.* 2008). Methylation plays a crucial role in the regulation of gene expression, genome plasticity and gene silencing (Choi and Sano 2007; Loidl 2004). Transcriptional and post-transcriptional gene silencing are linked with hypermethylation of promoter sequences and transcribed or coding sequences, respectively (Paszkowski and Whitham 2001). It seems that resetting of epigenetic status across development has a significant role in gene expression patterns. In comparison with animals, much less is known about epigenetic modifications in plants (Zhang *et al.* 2010). The epigenetic status in mammals is systematically reprogrammed during developmental stages, featuring erasure and reestablishment of epigenetic marks, and spatially and temporally redefined cytosine methylation patterns (Reik *et al.* 2001). DNA methylation profiling has become an important technology in many areas of epigenetic researches. The approaches principally investigate methylation patterns either across the entire genome or at specific areas of interest like promoters and other regulatory elements (Rauch and Pfeifer 2010). In general, two techniques are used to assess methylation: (i) the sequencing after modification of genomic DNA with sodium bisulfite and (ii) PCR amplification after digestion of genomic DNA with methylation-sensitive endonuclease (s) (Fuso *et al.* 2006; Tollefsbol 2004). Amplified fragment length polymerphism (AFLP) technique (Vos *et al.* 1995) was modified by Substitution of *Mse* I with two isoschizomer enzymes *Hpa* II and *Msp*I, which has been entitled to methylation sensitive amplification polymorphism (MSAP). In the MSAP, the genome-wide DNA methylation profiling can be investigated based on CCGG site (Xu *et al.* 2000). This technique has been successfully used for plant epigenome studies in a number of plant species viz. maize (Zhao *et al.* 2007, Lu *et al.* 2008; Tan 2010), rice (Sha *et al.* 2005), pepper (Portis *et al.* 2004), hops (Peredo *et al.* 2008), *Brassica napus* L. (Labra *et al.* 2004), *Brassica oleracea* (Salmon *et al.* 2008), and Azalea (Meijon *et al.* 2009).

Halophytic plants are very important genetic resources to study the relationships among genome, epigenome and abiotic stresses. The poaceae halophyte *A. littoralis* as a perennial, rhizomatous monocotyledonous is a diploid (2n-=2X-=14) plant with a relatively small haploid genome of 349Mb and C4 photosynthesis (Barhoumi *et al.* 2007; Wang 2004). Thus, *A.*

littoralis has the potential to become a precious genetic resource for understanding the molecular mechanisms of stress-responses in important monocot crops (Zouari *et al.* 2007; Ben Saad *et al.* 2010; Hashemi et al. 2013). On the other hand, seedling development and plant establishment under natural environmental conditions are among the major concerns in the field of crop research. Therefore, understanding the pattern and extent of cytosine methylation during plant growth and development could provide valuable information about chromatin status and its regulatory role in biological process. The present study employed methylation-sensitive amplified polymorphism technique to analyze genome-wide DNA methylation pattern during *A.littoralis* seedling development.

Materials and methods

Plant materials and growth conditions
The surface-sterilized *Aeluropus litto-ralis* seeds were transferred on 1/2MS medium (Murashige and Skoog 1962) supplemented with 3% sucrose and 0.7% agar in glass plates. The cultures were incubated at 25 ± 2oC with 16h light/8h dark photoperiod at 100 μmol m-2 s-l photon flux density using cool-white fluorescent light. After four days, seedlings were sampled at three sequential growth stages including: the last stage of coleoptile elongation, the complete expansion of seed leaf, and the appearance of the third leaf. At each sampling time, 150 seedlings were harvested in three replications and were immediately placed in liquid nitrogen.

DNA isolation and quantification

One hundred mg of each sample was powdered and genomic DNA was extracted using DNeasy Plant Mini Kit (QIAGEN, Germany). Quantification and qualification of isolated DNA were determined by agarose electrophoresis and spectrophotometer.

Methylation sensitive amplified polymorphism (MSAP)

MSAP analysis was carried out according to Vos *et al.* (1995) with some modifications. The isoschizomers of *Hpa*II and *Msp*I (Roche Applied Science, Germany) were employed as 'frequent-cutter' enzymes instead of *Mse* I. Both *Hpa*II and *Msp*I recognize the same tetra-nucleotide sequence (5'-CCGG-3'), but they display different sensitivity to DNA methylation (Xu *et al.* 2000; Fang and Chao 2007; Peng and Zhang 2009). Genomic DNA was digested overnight using the *Msp*I/*Eco*RI and *Hpa*II/*Eco*RI enzyme combinations and then linked to the adapters. Sequences of the adapters and primers used in this study are listed in Table 1. Pre-amplification was performed in a 20μl reaction mixture, containing 50ng ligated DNA, 250μM of each dNTPs, 1x Taq buffer PCR (10mM Tric-HCl; 50mM KCl, pH 8.8; 0.08% Nonidet P40), 2mM MgCl2, 1U of *Taq* polymerase (Fermentas, Lithuania) and 0.25μM of each non-selective primer pairs (Alpha DNA, Canada) using a MJ Mini thermal cycler (Bio-

Rad, USA). The PCR program consisted of 20 cycles: 94°C for 1min, 55°C for 1min, 72°C for 2min. The pre-amplification product was diluted 20-fold and used as a template for selective amplification.

The PCR program consisted of 12 cycles at 94°C for 30s, 65°C for 30s, and 72°C for 60s with annealing temperature decreasing 0.7°C per cycle; followed by 25 cycles at 94°C for 30s, 56°C for 30s and 72°C for 60s and final elongation at 72°C for 10min. Selective amplified products were separated by a 6% denaturing PAGE using Sequi-Gen GT Cell system (Bio-Rad, USA) and visualized by silver staining. The gels were dried at room temperature for 12h and were scanned with an Imaging Densitometer model GS-800 (Bio-Rad, USA).

Table 1. The list of primer and linker sequences.

Sequence ID	Sequence	Sequence ID	Sequence
*Eco*RI linker 1	CTC GTA GAC TGC GTA CC	HM Linker 1	GAT CAT GAG TCC TGC T
*Eco*RI linker 2	AAT TGG TAC GCA GTC TAC	HM Linker 2	CGA GCA GGA CTC ATG A
*Eco*RI (A)	GAC TGC GTA CCA ATT CA	HM	ATC ATG AGT CCT GCT CGG
*Eco*RI (N)	GAC TGC GTA CCA ATT CN	HM (TCAA)	ATC ATG AGT CCT GCT CGG AA
HM (TCAC)	ATC ATG AGT CCT GCT CGG TCA C	*Eco*RI (AGG)	GAC TGC GTA CCA ATT CAG G
*Eco*RI (AAC)	GAC TGC GTA CCA ATT CAA C	*Eco*RI (GGA)	GAC TGC GTA CCA ATT CGG A
*Eco*RI (AAG)	GAC TGC GTA CCA ATT CAA G	*Eco*RI (AAA)	GAC TGC GTA CCA ATT CAA A
*Eco*RI (ACA)	GAC TGC GTA CCA ATT CAC A	*Eco*RI (AGT)	GAC TGC GTA CCA ATT CAG T
*Eco*RI (ACC)	GAC TGC GTA CCA ATT CAC C	*Eco*RI (ACT)	GAC TGC GTA CCA ATT CAC T
*Eco*RI (ACG)	GAC TGC GTA CCA ATT CAC G	*Eco*RI (AGC)	GAC TGC GTA CCA ATT CAG C

Data analysis

Differences between *Eco*RI/*Msp*I and *Eco*RI/*Hpa*II patterns were considered as epigenetic changes. Each pair of isoschizomers was classified into three types for the analysis of variation in methylation status (Table 2). In type I, *Hpa*II and *Msp*I recognition sites were not methylated and showed the same band patterns. In type II, internal methylation occurred in mCpG site at both strands which could be detected by *Msp*I (full-methylated sequence) while in Type III, methylation happened in mCmCpG or mCpG sites at one strand (hemi-methylated sequence) and was capable of being visualized with *Hpa*II enzyme (Peredo *et al.* 2008).

Results

As shown in Table 3, differential plant development was completely evident during the sampling time increments (four days). It seems that plant seedlings at the second developmental stage (DS2) showed a sharp increase in length and fresh weight where they increased up to 3-and 6-fold in comparison with the first sampling time (DS1), respectively. The DS2 seems to follow growth rather than to follow organ development while the final developmental stage (DS3) tends to generate more plant organs.

The assessment of DNA methylation involvement in regulation of plant growth and development was established on the basis of MSAP

Table 2. Schematic presentation of cytosine methylation patterns in MSAP analysis (Salmon *et al.* 2008).

Description	Band type	*Hpa*II	*Msp*I	Enzyme restriction site			
Non-methylated	I	+	+	C	C	G	G
				G	G	C	C
CpG methylation	II	-	+	C	mC	G	G
				G	G	mC	C
CpCpG methylation	III	+	-	mC	mC	G	G
				G	G	C	C
		+	-	C	mC	G	G
				G	G	C	C
Hypermethylation	IV	-	-	mC	mC	G	G
				G	G	mC	mC

profiling technique where different primer combinations were tested. Primer pairs of HM and *Eco*RI (A) used for pre-selective amplification. The 13 primer combinations for *Eco*RI (ANN) and HM (TCAW) were used as selective primers. PCR products between 100 and 800 base pairs with high frequent occurrence in different reactions were considered while fragments with weak intensity and low reproducibility were not evaluated for analysis.

Table 3. Plant characteristics at three different developmental stages.

Developmental stage	Plant age (days after sowing)	Number of leaves	Plant length (mm)		Plant fresh weight (µg)	
			Shoot	Root	Shoot	Root
DS1	5	0	2.8	7.4	75	27
DS2	9	1	7.1	22.2	487	178
DS3	13	2	14.3	27.9	1650	632

Out of the 13 primer combinations, 11 primer pairs produced polymorphic bands at different stages of seedling development (Table 4) while two combinations detected no variation in methylation sites. In total, 480 repro-ducible amplification products were observed in CCGG sites with an average of 37 bands per each enzymatic combination. The highest and the lowest numbers of bands were 26 and 51 for *Eco*RI (CTA) /HM (TCAC) and *Eco*RI (AAG) /HM (TCAA) primer combinations, respectively. Epigenetic marks as methylation and demethylation at the CCGG recognition site were tracked by the differences between *Eco*RI /*Msp*I and *Eco*RI / *Hpa*II patterns. The percentage of total bands in type I, II and III were 75.7, 19.4 and 4.9, respectively (Table 4). The most frequent methylation events (19.4%) were observed in type II fragment in which full methylation pattern occurred in CCGG recognition site.
Out of 480 bands, 33 bands showed methyl-

ation alterations among differential developmental stages which belonged to all three types of methylation. The number of polymorphic bands (NPB) produced by each primer combination ranged from one (*Eco*RI (AGT) / HM (TCAA)) to seven (*Eco*RI (ACA) /HM (TCAA); *Eco*RI (ACT)/HM (TCAA); *Eco*RI (AGG) / HM (TCAA)) (Table 4).
Twenty one percent of total detected methylation (21%) was identical among different developmental stage. The level of polymorphism for the three types of fragments decreased from type I to III which were 4.7 %, 1.9% and 0.4 % in type I, II, and III, respectively. In Figure. 1, a sample of DNA methylation pattern in primer combination of *Eco*RI ACT/HM TCAA is presented. Different developmental stages were compared using polymorphism frequency of 33 enzymatic recognition sites. *Msp*I recognized 27, 27, and 8 and *Hpa*II restricted 21, 19, and 7 sites at different developmental stages of DS1,

Table 4. Data of primer combinations were used in methylation sensitive amplification polymorphism. M: monomorphic bands; P: polymorphic bands.

Type of bands	Type I		Type II		Type III	
Number of bands	M	P	M	P	M	P
EcoRI (AAA) / HM (TCAC)	24	2	1	1	-	-
EcoRI (AGT) / HM (TCAA)	32	1	4	-	-	-
EcoRI (AAC) / HM (TCAA)	29	3	8	1	1	1
EcoRI (CTA) / HM (TCAC)	20	-	6	-	-	-
EcoRI (AAG) / HM (TCAA)	39	2	9	-	1	-
EcoRI (ACA) / HM (TCAA)	26	3	6	-	2	-
EcoRI (ACC) / HM (TCAA)	27	-	7	-	2	-
EcoRI (AAC) / HM (TCAC)	18	1	6	1	4	-
EcoRI (ACG) / HM (TCAA)	21	-	6	1	8	-
EcoRI (ACT) / HM (TCAA)	26	3	9	2	1	-
EcoRI (AGC) / HM (TCAA)	18	3	8	2	1	1
EcoRI (AAA) / HM (TCAA)	26	2	4	-	1	-
EcoRI (AGG) / HM (TCAA)	30	2	4	1	-	-
Total	336	22	78	9	21	2
Percent	72	4.7	16.6	1.9	4.4	0.4

(Primer combination)

DS2, and DS3, respectively. It appears that the variation of DNA cytosine methylation followed an increasing trend along with seedling differentiation (Figure. 2) owing to increase in polymorphism amount. The lowest level of methylation was observed in the developmental stage 1 (DS1) with the cumulative polymorphism percent value equal to 63 and 81 for

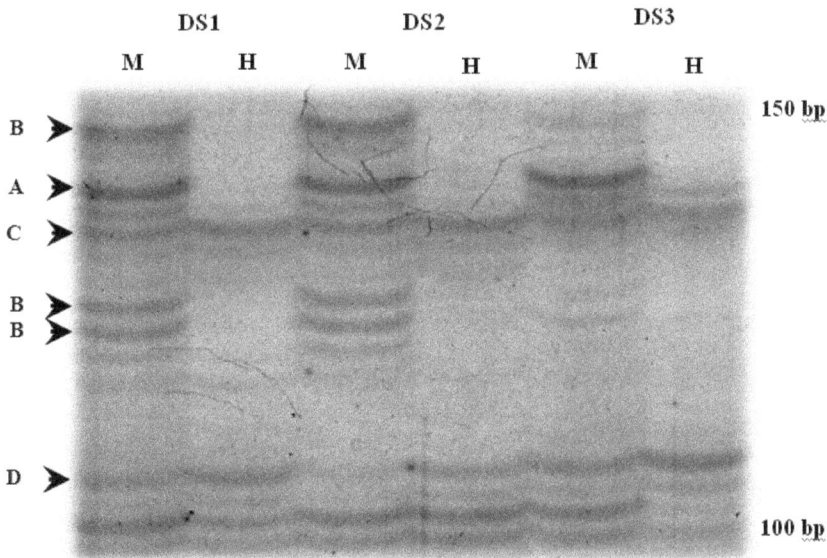

Figure. 1. MSAP profile of Aeluropus littoralis across different developmental stages (DS1, DS2 and DS3) detected by EcoRI (ACT) /HM (TCAA) primer combination. Each sample digested by EcoRI+Hpa II (H) or EcoRI+MspI (M). The arrows indicate different types of bands. The arrow represents those digested by EcoRI+MspI but not by EcoRI+HpaII as type II fragments (CpG methylation). In the B arrow indicated polymorphism in type II fragment that MspI site disappear in DS3. The C and D arrow showed types I fragment that in D arrow, methylation in MspI site were detected.

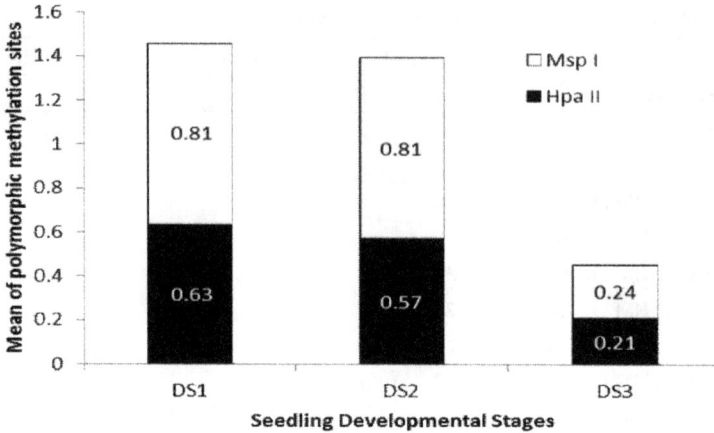

Figure. 2. The mean of 33 polymorphic methylation sites across three developmental stages (DS1, DS2 and DS3) detected by MSAP.

*Msp*I and *Hpa*II sites. Consequently, the polymorphism values were reduced in DS2 (57 and 81) and DS3 (21 and 24).

Discussion

AFLP-based MSAP technique is a robust method to make a global overview of methylation pattern that is able to amplify a large number of methylated and demethylated fragments simultaneously with no need for sequence information. It should be noted that MSAP can only analyze two states of methylation at CCGG sites including full methylation of both the external and internal cytosine and hemi-methylation of the internal cytosine (Zhao *et al.* 2007) and therefore the methylation events could be underestimated. Besides, the isochimeric restriction enzymes, *Msp*I and *Hpa*II, theoretically generate four distinct patterns of which type IV as a hypermethylation status cannot be practically detectable.

Plant growth cycle runs across a series of spatial and temporal developmental events so that successful performance of each phase is a precursor for the proper possibility of other successive immediate events. Seed-seed cycle is a major concept in economic productivity of crop plants in different environmental conditions. Seed germination and early seedling development are the critical points in the growth cycle and plant establishment on the basis of wide reprogramming of cell differentiation. DNA cytosine methylation constructs an integral part of the epigenetic con-

trolling network to tune tissue, organ and developmental stage-specific gene expression across plant development. Regulation of cytosine methylation across the plant development is an essential dimension for gene expression remodeling and consequent differentiation.

In plants, the occurrence of different types of DNA methylation is dependent on its own maintenance regulation pathway and mode of inheritance (Salmon *et al.* 2008). The present results indicated that methylation rate increased along with progression in plant development. Total occurrence of methylation in the internal cytosines on both strands (type II) was always more frequent than hemi-methylation on the external cytosine (type III) with 18.5% and 4.8 %, respectively. Despite the different values for the methylation types, the percentage of polymorphic fragments in each type were similar and estimated 11.5% and 9.5%, respectively. In contrast, type I fragments displayed higher levels of methylation alternation compared to types II and III fragments.

In this study, polymorphic bands had two main directions associated with methylation or demethylation patterns. Out of 22 bands in type I, methylation value was higher than demethylation (18 versus 3 bands) and was mainly limited to DS3 stage while no changes were observed in other stages. Demethylation status occurs in both *Msp*I and *Hpa*II recognition sites. Of 18 methylation fragments, 15 bands showed hyper-methylation while the remaining devoted to

CpG (one band) and CpCpG (two bands) sites. Similarly, in type II, 7 methylation sites were detected in DS3 and two demethylation events were identified only in DS1 and DS2. Most likely, these methylation and demethylation play main role for switching on or off the key genes in developmental processes. As shown in Figure. 2, *Msp*I and *Hpa*II are strikingly able to detect methylation patterns, which are linked to the developmental stages. In other word, different developmental stages may be explained using differential sensitivity of genome recognition sites to *Msp*I and *Hpa* II enzymes. These methylation and demethylation events at CG sites are probably related to developmental stage-specific gene regulation. Methylation in specific regions of genes or in their vicinity can inhibit the expression of these genes, while demethylation of genes has been shown to result in reactivation (Grunau *et al.* 2001). The evidence indicates that, an increase or a decrease in methylation level is temporary and will alter accordingly along with the growth process and developmental stage-dependent manners (Lu *et al.* 2008; Zhang *et al.* 2010). CG full methylated sites (as recognized by *Msp*I) are established during cell divisions by methyltransferase 1 (MET1), whereas CG hypomethylated sites (as recognized by *Hpa* II) are maintained by chromodomain-containing methyltransferase (CM T3) and domains rearranged methyltransferases (DRMs) (Salmon 2008). It seems that dynamic interaction between the methylation of a genome and the activity of the enzymes, DNA methyltransferases and demethylases, determine the final genome-wide pattern of DNA methylation (Rival *et al.* 2008).

Based on the data, along with the progression in plant development and the increase in specialized cell populations, the methylation statuses of genome altered. It means that cells probably turn on (or off) specific of genes by DNA methylation mechanism during differen-tiation and development. At a given developmental stage, plant response to the environmental variations will strik-ingly depend on accessible parts of genome and its gene contents. It may partly explain why plant response to the environment alterations can be related to the developmental

stage at that condition. The results clearly demon-strated that the MSAP is a highly efficient technique for genome-wide DNA methylation analysis in the *Aeluropus littoralis*.

References

1. Allis, C.D., Jenuwein, T. and Reinberg, D. 2007. Overview and Concepts, In: Allis, C.D., Jenuwein, T., Reinberg, D., Caparros, M.L. (eds.) Epigenetic, Cold Spring Harbor Laboratory Press pp, 23-40.
2. Barhoumi, Z., Djebali, W., Chaibi, W., Abdelly, C. and Smaoui, A. 2007. Salt impact on photosynthesis and leaf ultrastructure of *Aeluropus littoralis*. *J Plant Res*, 120: 529-537.
3. Ben-Saad, R., Zouari, N., Ben-Ramdhan, W., Azaza, J., Meynard, D., Guiderdoni, E. and Hassairi, A. 2010. Improved drought and salt stress tolerance in transgenic tobacco overexpressing a novel A20/AN1 zinc-finger "AlSAP" gene isolated from the halophyte grass *Aeluropus littoralis*. *Plant Mol Biol*, 72: 171-190.
4. Chinnusamy, V. and Zhu, J.K. 2009. RNA-directed DNA methylation and demethylation in plants. *Sci China Ser C-Life Sci*, 52: 331-343.
5. Choi, C.S. and Sano, H. 2007. Abiotic-stress induces demethylation and transcriptional activation of a gene encoding a glycerophosphodiesterase-like protein in tobacco plants. *Mol Genet Genomics*, 277: 589-600.
6. Cokus, S.J., Feng, S., Zhang, X., Chen, Z., Merriman, B., Haudenschild, C.D., Pradhan, S., Nelson, S.F., Pellegrini, M. and Jacobsen, S.E. 2008. Shotgun bisulphite sequencing of the Arabidopsis genome reveals DNA methylation patterning. *Nature*, 452: 215-219.
7. Fang, J. and Chao, C.T. 2007. Methylation-sensitive amplification polymorphism in date palms (*Phoenix dactylifera* L.) and their off shoots. *Plant Biol*, 9: 526-533.
8. Feng, S., Cokus, S.J., Zhangd, X., Chen, P.Y., Bostick, M., Golle, M.G., Hetzel, J., Jaine, J., Straussf, S.H., Halperne, M.E., Ukomadug, C., Sadlerh, K.C., Pradhani, S., Pellegrini, M. and Jacobsen, S.E. 2010. Conservation and divergence of methylation patterning in plants and animals. *PNAS*, 107: 8689-8694.
9. Fuso, A., Scarpa, S., Grandoni, F., Strom, R. and Lucarelli, M. 2006. A reassessment of semi quantitative analytical procedures for DNA methylation: Comparison of bisulfite and *Hpa* II polymerase chain reaction based methods. *Anal Biochem*, 350: 24-31.
10. Grunau, C., Renault, E., Rosenthal, A. and Roizes, G. 2001. MethDB-a public database for DNA methylation data. *Nucleic Acid Res*, 29: 270-274.
11. Hashemi, S.H., Nematzadeh, Gh., Askari, H. and Ghasemi, Y. 2013. Pattern of DNA cytosine

methylation in *Aeluropus littoralis* during temperature stress. *J Plant Mol Breed*, 1: 16-24.

12. Hendrich, B. and Tweedie, S. (2003) The methyl-CpG binding domain and the evolving role of DNA methylation in animals. *Trends Genet*, 19: 269-277.

13. Labra, M., Grassi, F., Imazio, S., Fabio, T.D., Citterio, S., Sgorbati, S. and Agradi, E. 2004. Genetic and DNA-methylation changes induced by potassium dichromate in *Brassica napus* L. *Chemosphere*, 54: 1049-1058.

14. Loidl, P. 2004. A plant dialect of the histone language. *Trends Plant Sci*, 9: 85-90.

15. Lu, Y., Rong, T. and Cao, M. 2008. Analysis of DNA methylation in different maize tissues. *J Genet Genomics*, 35: 41-48.

16. Meijon, M., Valledor, L., Santamarıa, E., Testillano, P.S., Risueno, M.C., Rodrıguez, R., Feito, I. and Canal, M.J. 2009. Epigenetic characterization of the vegetative floral stages of azalea buds: Dynamics of DNA methylation and histone H4 acetylation. *J Plant Physiol*, 166: 1624-1636.

17. Murashige, T. and Skoog, F. 1962. A revised medium for rapid growth and bio assays with tobacco tissue cultures. *Physiol Plant*, 15: 473-497.

18. Paszkowski, J. and Whitham, S.A. 2001. Gene silencing and DNA methylation processes. *Curr Opin Plant Biol*, 4: 123-129.

19. Peng, H. and Zhang, J. 2009. Plant genomic DNA methylation in response to stresses: Potential applications and challenges in plant breeding. *Prog Nat Sci*, 19: 1037-1045.

20. Peredo, E.L., Arroyo-Garcia, R., Reed, B.M. and Revilla, M.A. 2008. Genetic and epigenetic stability of cryopreserved and cold-stored hops (*Humulus lupulus* L.). *Cryobiology*, 57: 234-241.

21. Portis, E., Acquadro, A., Comino, C. and Lanteri, S. 2004. Analysis of DNA methylation during germination of pepper (*Capsicum annuum* L.) seeds using methylation-sensitive amplification polymorphism (MSAP). *Plant Sci*, 166: 169-178.

22. Rauch, T.A. and Pfeifer, G.P. 2010. DNA methylation profiling using the methylated-CpG island recovery assay (MIRA). *Methods*, 52: 213-217.

23. Reik, W., Dean, W. and Walter, J. 2001. Epigenetic reprogramming in mammalian development. *Science*, 293: 1089-1093.

24. Rival, A., Jaligot, E., Beule, T. and Finnegan, E.J. 2008. Isolation and expression analysis of genes encoding MET, CMT, and DRM methyltransferases in oil palm (*Elaeis guineensis* Jacq.) in relation to the 'mantled' somaclonal variation. *J Exp Bot*, 59: 3271-3281.

25. Salmon, A., Clotault, J., Jenczewski, E., Chable, V.R. and Manzanares-Dauleux, M.J. 2008. *Brassica oleracea* displays a high level of DNA methylation polymorphism. *Plant Sci*, 174: 61-70.

26. Sha, A.H., Lin, X.H., Huang, J.B. and Zhang, D.P. 2005. Analysis of DNA methylation related to rice adult plant resistance to bacterial blight based on methylation-sensitive AFLP (MSAP) analysis. *Mol Gen Genomics*, 273: 484-490.

27. Tan, M. 2010. Analysis of DNA methylation of maize in response to osmotic and salt stress based on methylation-sensitive amplified polymorphism. *Plant Physiol Biochem*, 48: 21-26.

28. Tollefsbol, T.O. 2004. Methods of epigenetic analysis. *Methods Mol Biol*, 287: 1-8.

29. Vos, P., Hogers, R., Bleeker, M., Reijand, M., Vande-Lee, T., Hornes, M., Fritjers, A., Pot, J., Paleman, J., Kuiper, M. and Zabeau, M. 1995. AFLP: a new technique for DNA fingerprinting. *Nucleic Acids Res*, 23: 4407-4414.

30. Wang, R.Z. 2004. Plant functional types and their ecological responses to salinization in saline grasslands, Northeastern China. *Photosynthetica*, 42: 511-519.

31. Xu, M.L., Li, X.Q. and Korban, S.S. 2000. AFLP-based detection of DNA methylation. *Plant Mol Biol Rep*, 18: 361-368.

32. Zhang, M., Kimatu, J.N., Xu, K. and Liu, B. 2010. DNA cytosine methylation in plant development. *J Genet Genomics*, 37: 1-12.

33. Zhao, X., Chai, Y. and Liu, B. 2007. Epigenetic inheritance and variation of DNA methylation level and pattern in maize intra-specific hybrids. *Plant Sci*, 172: 930-938.

34. Zouari, N., Ben-Saad, R., Legavre, T., Azaza, J., Sabau, X., Jaoua, M., Masmoudi, K. and Hassairi, A. 2007. Identification and sequencing of ESTs from the halophyte grass *Aeluropus littoralis*. *Gene*, 404: 61-69.

9

The effect of medium and plant growth regulators on micropropagation of Dog rose (*Rosa canina* L.)

Mahboubeh Davoudi Pahnekolayi[1], Leila Samiei[2*], Ali Tehranifar[1] & Mahmoud Shoor[1]

1. Horticultural Department, College of Agriculture, Ferdowsi University of Mashhad, Mashhad, Iran.
2. Research Center for Plant Sciences, Ferdowsi University of Mashhad, Mashhad, Iran.
   *Corresponding Author, Email: samiei@um.ac.ir

Abstract
Dog rose (*Rosa canina* L.) is one of the most important ornamental and medicinal plants which are used as a rootstock for ornamental roses such as *Rosa hybrid* and *Rosa floribunda*. *In vitro* propagation of rose has a very important role in rapid multiplication of species with desirable traits and in production of healthy and disease-free plants. Micropropagation of *Rosa canina* L. was revised, using its nodal segments under different combinations of BAP (0, 0.5, 1, 1.5 and 2 mgl^{-1}), GA3 (0 and 0.5 mgl^{-1}) and NAA (0 and 0.5 mgl^{-1}) on Murashige and Skoog medium (MS) and Van der Salm medium (VS) in proliferation stage and using different combinations of NAA and IBA (0, 0.3, 0.6 and 0.9 mgl^{-1}) and ½ VS medium in rooting stage. The highest shoot proliferation was obtained on VS medium containing 2 mgl^{-1} BAP. Furthermore, the highest root induction obtained in ½ VS containing 0.6– 0.9 mgl^{-1} of NAA or IBA. The present study presents an *in vitro* protocol for *R. canina*.
Key words: Micropropagation, Native plants, Plant growth regulators (PGRs), *Rosa canina* L.

Introduction

Rosa canina L. (belongs to *Rosaceae* family) is a medicinal plant that its fruits have many essential medicinal properties unknown for many people especially in Iran. However, this plant can be produced commercially and its orchards can be established like other fruit trees (12, 18, 19). Fruits of rose species are rich in minerals, vitamins (A, B1, B2, B3, C and K), sugars, phenolic compounds, carotenoids, tocopherol, bioflavonoid, tannins, organic acids, amino acids, volatile oils, vanillin and other photochemicals such as antioxidant and antimicrobials. Medicinal properties and benefits of roses are: nutrient, mild laxative, mild diuretic, mild astringent, carminative, ophthalmic, tonic and verminfuge (18, 19). Dog rose has also been used as a rootstock for ornamental roses (8, 9, 10). More than 200 species are present in the genus *Rosa* (21) from which 14 wild species are found in Iran. Traditionally, most roses are

heterozygous and do not breed true to type. Commercial propagation of roses is usually done by cutting, although they can be propagated by budding and grafting, which are difficult and undesirable processes (4). Conventional rose propagating methods are very slow and time consuming. Tissue culture is becoming increasingly popular as an alternative to conventional rose propagation methods (16). Micropropagation of Rosa species is carried out through the meristem-tip, shoot-tip and axillary buds culture. In recent years, several reports have been published on *in vitro* proliferation of roses, and have shown that tissue culture provides an alternative method for rapid multiplication (13). A successful micropropagation protocol proceeds through a series of stages, each with a specific set of requirements. These are (i) initiation of aseptic cultures, (ii) shoot multiplication, (iii) rooting of microshoots, and (iv) hardening off and field transfer of tissue culture raised plants (11). The presence of cytokinin in the culture medium helped in the year round multiplication of shoots in hybrid roses (17). Although the rate of growth was found to be very slow. They observed a high percentage of bud break in a hormone-free medium within 10-12 days. However, Media supplemented with BAP or BAP+GA$_3$ resulted in early bud break within 6-8 days of culture with enhanced rates of shoot multiplication. The addition of BAP (2.0-3.0 mgl^{-1}) as the only growth regulator in the culture medium resulted in feeble callusing at the cut ends of the explants and the shoot elongation was considerably slow in a 60-days culture period and the explants response varied from 63 to 80%. Incorporation of GA$_3$ at low concentrations (0.1-0.25 mgl^{-1}) in the BAP supplemented medium improved explants response up to 95%. BAP was the most effective growth regulator in stimulating shoot proliferation. Further, the use of 3 types of auxins in combination with BAP showed that NAA was more effective than IAA or IBA in the production of multiple shoots (15, 20). The objective of this study was to investigate the best combinations and media for *in vitro* micropropagation of *Rosa canina* L. as rootstock for grafting.

Materials and methods

Sterilization of explants
When the buds obtained chilling requirement in February 2012, axillary buds of *R. canina* (grown at the Botanical Garden of Ferdowsi University of Mashhad, Iran) were cut and washed with running tap water (1h), then decontaminated with 70% (w/v) ethanol (30s) and sodium hypochlorite (2.5%) (15min). All explants were washed three times with sterile distilled water.

Shoot proliferation
In this experiment, shoot explants with 3 axillary buds and 2 cm in length were transferred to 2 different media: MS and VS medium. 20 treatments with 8 replications (one explant in each glass vials) were considered for each medium (MS and VS) in this stage. Different combinations of BAP (0, 0.5, 1, 1.5 and 2 mgl$^-$

^{1}), GA3 (0 and 0.5 mgl^{-1}) and NAA (0 and 0.1 mgl^{-1}) were considered for MS and VS media. The shoot number, shoot length, shooting percentage, leaf number and percentage of green leaves were recorded after 60 days. The pH of media was adjusted to 5.8 before adding 8 grl^{-1} agar. Media were autoclaved for 15min at 121 C° and 1.2 kgf/cm pressure. All the *in vitro* cultures were placed under 16/8h light/dark cycle in a growth chamber and were maintained at 24±1 C°.

Rooting stage
New shoots (1-2 cm in height) were transferred VS and ½ VS media containing different hormonal concentrations of IBA and NAA (0, 0.3, 0.6 and 0.9 mgl^{-1}) for rooting. Root length, root number and the percentage of root induction were recorded after 30 days.

Acclimatization of plantlets
The plantlets were cultivated for 5 weeks in plastic glasses containing sterile mixture of Peat/Perlite (3/1). The plantlets were put in growth chamber under a 16/8-hours photoperiod, at a temperature of 23/25 C0 (night/day) and 80% relative humidity. The shoot survival percentage was recorded after 30 days.

Experimental design and statistical analysis
The shoot proliferation experiment was performed in completely randomized design (CRD) with 40 treatments (composition of BAP (0, 0.5, 1, 1.5 and 2 mgl^{-1}), GA3 (0, 0.5 mgl^{-1}) and NAA (0 and 0.1 mgl^{-1}) in MS and VS medium) and 8 replications (one explant in each replication). Rooting experiment was carried out in a completely randomized factorial base on design with 16 treatments (composition of IBA and NAA (0, 0.3, 0.6 and 0.9 mgl^{-1}) in VS and ½ VS medium) and 15 replications (one explant in each replication). Data were analyzed with SPSS software (version 19) and the comparison of means was performed by Duncan's new multiple range test at $P \leq 0.05$. Before carrying out ANOVA the percentage data were transformed using angular transformation (Arc Sin$\sqrt{}$%).

Results

Shoot proliferation

The results showed that the effect of treatments (type and concentrations of PGRs) on vegetative traits of *R. canina* in proliferation stage was significant ($P \leq 0.05$) (Table 1). The maximum number of new leaves (19.96), shoot per explant (4.21) and shoot percentage (41.10%) were produced on the medium containing 2 mgl^{-1} BAP (Figure 1A, 1B, 1C), whereas the maximum percentage of green leaves (84.84%) and the highest increase in plantlet height (1.27 cm) were obtained on the medium containing 1.5 mgl^{-1} BAP (Figure 1D, 1E). Also, the effect of medium on vegetative traits in proliferation stage was signif-

Table 1. Analysis of Variance (mean of square) of *R.canina* in proliferation stage ($P \leq 0.05$).

Source of Variation	Degree of Freedom (df)	Shoot number	Leaf number	Shoot length	Percentage of green leaves	Shoot percentage
Medium	1	4.82 ns	319.8 **	4.05 **	8318.1 **	308.40 ns
Hormone	19	9.56 **	267.84 **	0.723 **	4395.11 **	512.98 **
Error	280	1.7	35.86	0.197	407.62	97.32

ns: non- significant

*,**: significant at P< 0.05 and 0.01, respectively.

icant ($P \leq 0.05$) (Table 1). The lowest shoot multiplication was observed on MS medium while the highest shoots were formed on VS medium (Table 2).

Kim *et al*. (6) indicated that *in vitro* shoot proliferation and multiplication of roses are largely based on medium formulations containing cytokinins as a major plant growth regulators, although low concentrations of auxins or GA_3 are also essential. The results of the present study demonstrated that inclu-sion of 0.5 $mgl^{-1}GA_3$ to the culture media didn't increase the number of axillary shoots and stem height in all of the BAP concentrations. The results showed that in some of the traits the differences between the treatments were not significant, however, the maximum number of axillary shoots were significantly higher in the VS medium containing 2 mgl^{-1} BAP. As the concentration of BAP was lower than 2 mgl^{-1}, a reduced growth rate was noted. 2 types of MS and VS medium were

Table 2. Effect of different media on some vegetative traits of *R.canina* in proliferation stage.

medium	shoot number	leaf number	shoot length (cm)	green leaves (%)	Shooting (%)
MS	2.15 a	8.37 b	0.768 b	46.29 b	26.25 a
VS	2.4 a	10.37 a	0.993 a	56.49 a	28.21 a

Different letters indicate significant difference according to Duncan Test (*P*<0.05)

also used in this experiment and the results showed that the best plantlets (in proliferation stage) obtained in VS medium because of the alleviating effect of iron deficiency in leaves.

Root initiation and acclimatization

ANOVA of the effect of two media on root number, root length and rooting percentage revealed significant differences (P≤0.05) (Table 3). The results showed that the effect of concentrations and the types of PGRs on root num-ber and rooting percentage were significant but there was no difference in root length (P≤0.05) (Table 3). The highest root length (1.85 cm), root number (2.28) and rooting percentage (24.75%) were recorded in ½ Vs medium (Table 4). The maximum root number (2.14) and rooting percentage (23.08%) were also observed in 0.6 and 0.9 mgl^{-1} IBA or NAA, respectively (Table 5). The results showed that there are differences between IBA and NAA treatments in root length in comparison with control, while significant differences in root number and rooting percentage was recorded. Therefore, the PGRs concentrations had more importance than the PGRs types in rooting stage.

The regenerated shoots cultured on VS medium containing ½ and full strength of VS (macro, micro elements and vitamins) showed different responses to rooting after 4 weeks of culture. The best rooting were obtained on medium containing ½ strength of VS macro, micro salts and vitamins.

Table 3. Analysis of variance (mean of square) on some traits of *R. canina* in rooting stage ($P \leq 0.05$).

Source of variation	Degree of freedom	Root number	Root length	Rooting percentage (%)
PGRs type (A)	1	13.43 [*]	0.74 [ns]	882.25 [*]
PGRs concentration (B)	3	14.46 [*]	0.48 [ns]	871.28 [*]
Culture Medium (C)	1	85.12 [**]	45.88 [**]	6036.00 [**]
A x B	3	11.69 [*]	0.85 [ns]	704.061 [*]
A x C	1	36.85 [**]	18.07 [**]	2523.17 [**]
B x C	3	19.03 [**]	9.15 [*]	1312.36 [**]
A x B x C	3	5.23 [ns]	3.33 [ns]	324.209 [ns]
Error	224	4.54	3.44	295.96

ns: non- significant

*,**: significant at P< 0.05 and 0.01, respectively.

Table 4. Effect of different media on some traits of *R.canina* in rooting stage.

Medium Culture	Root number	Root length (cm)	Rooting percentage (%)
Vs	1.01 b	0.922 b	14.03 b
½ Vs	2.28 a	1.85 a	24.75 a

Different letters indicate significant difference according to Duncan Test (P<0.05)

Acclimatization of the rooted plantlets were easily carried out at ±24 C° and the relative humidity of 80% during initial stages of development and after that it was gradually reduced to 40%. After 4 weeks of culture acclimatized plants were transferred to greenhouse for flowering.

Discussion

Tissue culture techniques are used extensively for growing plants commercially. In this study, according to the responses of explants cultured on two media, VS and ½ VS were relatively the best and the most appropriate treatments for shoot and root formation. For initiation of aseptic cultures, a thorough knowledge of the physiological status and the susceptibility of the plant species to different pathological contaminants are required. So, utilization of 70% (w/v) ethanol (for 30s), and sodium hypochlorite (2.5%) (for 15min) will be effective. Other researchers have reported a high percentage of breaking bud dormancy on hormone-free medium within 10-12 days, but the growth rate was very low in roses (17). In our study the effects of different concentrations of BAP (0 to 2 mgl⁻¹), GA₃ (0 and 0.5 mgl⁻¹) and NAA (0 and 0.1 mgl⁻¹) were obvious on bud break and growth in VS medium, however, the growth rate was lower on MS and VS medium without growth regulators.

The VS medium with additive Fe (FeEDDHA chelat) was also better than MS medium (with EDTA chelat) in all stages of micropropagation of this plant.

Table 5. Effect of concentrations of PGRs on rooting traits of *R.canina*.

Concentration (mg/L)	Root number	Root length (cm)	Rooting percentage
0	0.77 b	1.42 a	12.29 b
0.3	1.4 ab	1.27 a	17.78 ab
0.6	2.14 a	1.38 a	23.08 a
0.9	1.83 a	1.48 a	20.86 a

Different letters show significant differenced according Duncan Test (p<0.05).

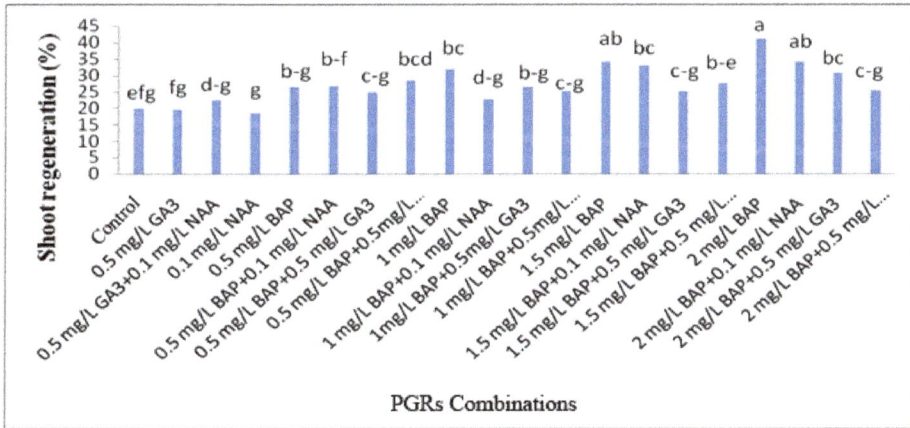

(A)

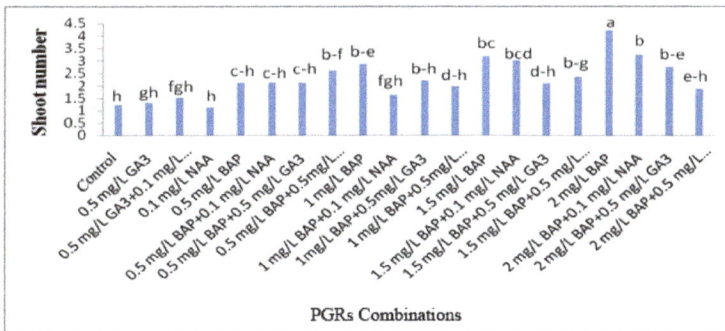

(B)

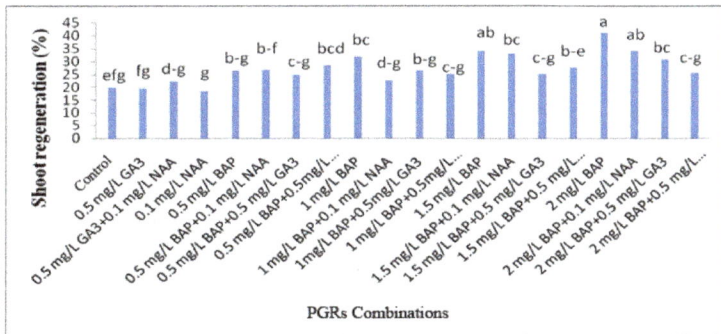

(C)

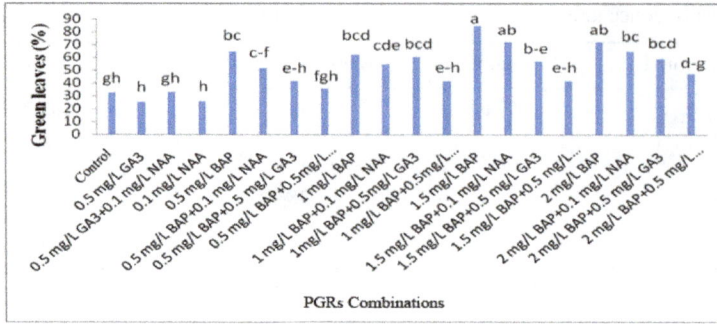

(D)

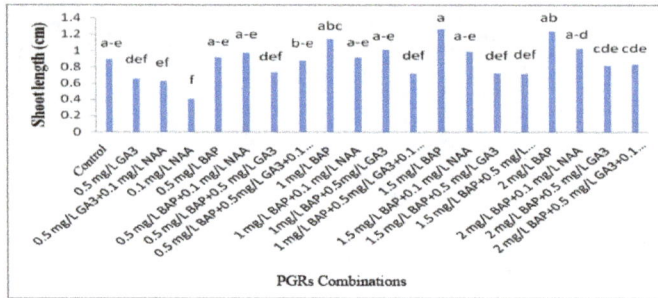

(E)

Figure 1. Effect of Hormones (type and concentrations of PGRs) on vegetative traits of *R.canina* in proliferation stage, showing; A) Average number of new leaves B) Average number of shoot per explant C) Average Shoot percentage (%) D) Average percentage of green leaves (%)E) Average Shoot length (cm).

Figure 2. Different stages of micropropagation of *Rosa canina* L: A) Proliferation stage B) Rooting stage C) Acclimatization stage.

In vitro shoot multiplication relies largely on medium formulations containing BAP as the major PGRs in combination with a low concentration of NAA (22). In the present study, 2 mgl-1BAP was the optimum treatment for *in vitro* multiplication of *Rosa canina* L. The result is different from the findings on optimal BAP concentration (4.4– 13.2 µM) (2). BAP is needed for proliferation of *Rosa canina* L. plants, but a high concentration is undesirable, At a concentration higher than 2.2 µM, it will lead to multiplication of shoots, which is not beneficial to shoot elongation and this is not

in line with our results. When the culture medium contains NAA at higher concentration, the bud can form more callus from the base section, which will greatly affect the young seedlings for absorption of water and nutrition, and thus inhibits their growth. The inhibition is especially obvious when NAA concentration is increased to 0.5 mgl^{-1}. Auxin is a rooting hormone and the application of synthetic auxin i.e. IBA might have increased the biosynthesis of indole acetic acid (IAA) or could act as synergistic to IAA. (3). Another possible reason for higher rooting and early root initiation by IBA might be involvement of IBA in ethylene bio-synthesis (1). It has been suggested that auxin induced ethylene may induce adventitious root formation instead of action of auxin itself (14). A high concentration of GA_3 (0.5 mgl-1) in present study caused the chrolosis and the death of young shoots, which appeared water-logged and that is in line with Xing *et al* (22). The best results in rooting stage achieved with 0.6 or 0.9 mgl^{-1} IBA/NAA in ½ VS medium. Decreasing medium salt strength of MS medium generally increases rooting in rose micropropagation. (5) The same results were reported in *R.hybrida* cv. Peace (7) and *R.rugosa* (22). Our study yielded a practical protocol for efficient axillary bud multiplication from *Rosa canina* L. explants. Here we demonstrate high-efficiency micro- propagation of *R. canina*. In conclusion, the present study provides an *in vitro* propagation protocol for *R. canina*. At the proliferation stage, maximum number of axillary shoots was achieved in the medium containing 2 mgl^{-1} BAP.

At the rooting stage, 0.6 and 0.9 mgl^{-1} IBA/NAA were effective.

References

1. Arteca, R. 1990. Hormonal stimulation of ethylene biosynthesis. In Polyamines and Ethylene: Biochemistry, Phisiology and Interactions, American Society of Plant Phisiologists, Rockville, MD, 216–223.

2. Carelli, BP. and Echeverrigaray, S. 2002. An improves system for the *in vitro* propagation of rose cultivars. Sci Hortic, 92: 69-74.

3. Dixon, RA. and Gonzales, RA. 1993. Plant Cell Culture. A practical approach. 2nd Edn. Plant Biology division. The Samuel Roberts Noble Foundation. P.O.Box 2180. Ardmore, Oklahoma 73402, USA (At Oxford University Press. Oxford, New York Tokyo).

4. Horn, W. 1992. Micropropagation of rose (*Rosa* L.). Biotechnol Agric Springer, (20): 320- 342.

5. Hyndman, S., Hasegawa, P.M. and Bressan, R.A. 1982. The role of sucrose and nitrogen in adventitious root formation on cultured rose shoots. Plant Cell Tissue Organ Cult, (1): 229–238.

6. Kim Chkou, J.U., Jee, S.O. and Chung, J.D. 2003. *In vitro* Micropropagation of *Rosa hybrid* L.J. Plant Biotechnol, 5: 115-119.

7. Kirichenko, E.B., Kuz-Mina, T.A. and Kataeva, N.V. 1991. Factors in optimizing the multiplication of ornamental and essential oil roses *in vitro*. Byullenten Glavnogo Bot kogo Sada, 159: 61-67.

8. Khosh-Khui, M. and Sink, K.C. 1982a. Micropropagation of new and old world Rosa species. Am J Hortic Sci, 57: 315-9.

9. Khosh-Khui, M, and Sink, K.C. 1982b. Rooting enhancement of *Rosa hybrid* for tissue culture propagation. Sci Hortic, 17: 371-6.

10. Khosh-Khui, M, and Sink, K.C. 1982c. Callus induction and culture of Rosa. Sci Hortic, 17: 361-70

11. Kumar Pati, P., Prasad Rath, S., Sharma, M., Sood, A. and Singh Ahuja, P. 2006. *In vitro* propagation of rose– a review, Biotechnol Adv, 24: 94–114.

12. Lunca, A. 2005. A guide to medicinal plants in North Africa. Center for Mediterranean cooperation, Malaga (Spain), p. 256.

13. Motallebi-Azar, A., Shirdel, M., Masiha, S., Mortazavi, N., Maloobi, M. and Sharafi, Y. 2011. Effects of inorganic nitrogen source and NH4:NO3 ratio on proliferation of dog rose (*Rosa canina* L.). J Med Plants Res, 5(18): 4605-4609.

14. Mudge, M.W. 1989. Effect of ethylene on rooting. In adventitious root formation in cuttings, eds., TD Davis; BE Haissing and N Sankhla. Dioscorides Press. Portland, DR, pp. 150-161.

15. Pati, PK., Sharma, M., Sood, A., and Ahuja, PS, 2005. Micropropagation of *Rosa damascena* and R. bourboniana in liquid cultures. In: Hvoslef Eide AK, Preil W, editors. Liquid systems for in vitro mass propagation of plants. Neth Kluwer Academic Publishers.

16. 16. Roberts, A.V. and Schum, A. 2003. Cell Tissue & Organ Culture. Encyclopedia of rose science. Oxford: Elsevier Academic Press, 57-110.

17. Rout, G.R. and Jain, S.M. 2004. Micropropagation of ornamental plants– cut flowers. Propagation ornam Plants, 4: 3-28.

18. Sharafi, Y. 2010a. Biological characteristics of pollens in some genotypes of *Rosa canina* L. as main factors affecting fruit set. Afr J Med Plants Res, 2 (20): 2173-2175.

19. Sharafi, Y. 2010b. Suitable *in vitro* medium for studying pollen viability in some of the Iranina hawthorn genotypes. J Med Plants Res, 4(19): 1967-1970.

20. Vijaya, N., Satyanarayana, G., Prakash, J., and Pierik, RLM. (1991). Effect of culture media and Growth Regulators on *in vitro* propagation of rose. Curr Plant Sci Biotechnol Agric, 12: 209-214.

21. Wissemann, V. 2003. Classification. Encyclopedia of rose science. Oxford: Elsevier Academic Press, 111-117.

22. Xing, W. 2010. Micropropagation of Rosa *Rougosa through* axillary shoot proliferation, Acta Biologica Cracoviensia. Ser Bot, 69–75.

10

The transient expression of coat protein of Foot and Mouth Disease Virus (FMDV) in spinach (*Spinacia oleracea*) using Agroinfiltration

M. Habibi[*1]**, S. Malekzadeh-Shafaroudi**[1]**, H. Marashi**[1]**, N. Moshtaghi**[1]**, M. Nassiri**[1]
& S. Zibaee[2]

1. Department of Biotechnology and plant breeding, Faculty of Agriculture, Ferdowsi University of Mashhad, Mashhad, Iran.
2. Razi Vaccine and Serum Research Institute, Mashhad, Iran.
   *Corresponding Author, Email: maziar.habibi.p@gmail.com

Abstract

Foot and Mouth Disease (FMD) is a very dangerous livestock disease which causes a serious loss in the production of milk and meat. Therefore, producing an effective recombinant subunit vaccine virus this disease is of great importance. Transient gene expression is a valuable tool to reach rapid and acceptable recombinant vaccine. An Agrobacterium-mediated transient gene expression assay was carried out in spinach (*Spinacia oleracea*) leaves for expression of a chimeric gene encoding a part of capsid protein of Foot and Mouth Disease virus called VP1. The plant leaves were transformed via agroinfiltration procedure. The presence of foreign gene and its expression in transformed plants were confirmed through polymerase chain reaction (PCR), real time PCR, protein Dot blot and ELISA. The results obtained in this examination showed quite a high level of gene expression in spinach leaves, showing that transient gene expression can be applied as an effective and time-saving procedure for the production of recombinant proteins. The procedures for transformation, detection of recombinant protein and its application for molecular experiments are described in the study.

Keywords: Agroinfiltration, FMDV, Recombinant Vaccine, Spinach, VP1.

Introduction

Green plants are appropriate systems for the expression of foreign pharmaceutical proteins including recombinant vaccines; however, the long time required for producing transgenic plants together with the high cost and low protein yield are the major obstacles to the comercialization of plant-based molecular farming (Doran 2006). An ideal solution for this shortcoming is application of transient gene expression using *Agro-bacterium tumefaciens*. In this method, the suspension of *A.tumefaciens* containing the gene of interest is transferred to plant leaves either with a needle-free syringe or a vacuum infiltration and the expression of foreign genes on TDNA usually reaches to its maxim-um at 2–3 days post-infiltration (Sohn *et al.* 2011). Transient gene expression has been reported as a simple, cost-effective, fast and reliable method for a wide range of experiments including gene function (Wroblewski *et al.* 2005), protein production (Vaquero *et al.* 1999), host– pathogen interaction (Tang *et al.* 1996), protein–protein interaction (Ihara *et al.* 2007) and protein localization (Bhat *et al.* 2006) .

So far many plant species have been used for the production of recombinant vaccine (Walmsley *et al.* 2000); the most notably examples are tobacco, potato, tomato, banana, corn, lupine and lettuce (Carter and Langridge 2002). Choosing the plant species for expression of recombinant vaccine is an important task which is mainly determined by considering how the vaccine is going to be used. Edible plant species such as vegetables are appropriate candidates if the vaccine is planned for raw consumption (Sala *et al.* 2003).

Foot and Mouth Disease (FMD) is a highly contiguous animal disease with harmful effects on milk and meat-producing animals (Wang *et al.* 2002). There have been many efforts to produce recombinant vaccines against this disease in plant systems (Habibi and Zibaee 2013). However, to the best of our knowledge, no investigation has been reported on the production of FMD recombinant vaccine via transient gene expression in plant host has been reported. The capsid of Foot and Mouth Disease virus (FMDV) is composed of four structural polypeptides designated VP1, VP2, VP3 and VP4 (Bachrach *et al.* 1975). The prominent G–H loop of the VP1 capsid protein of FMDV, spanning residues 134–158, has been identified as the major immunogenic site for neutralizing antibodies (Rodriguez and Grubman 2009). Moreover, G-H loop flanking regions have been shown to boost its immunogenicity by inducing B cells and T-helper cells (Wang *et al.* 2002).

This paper reports on the production of a novel recombinant vaccine against FMD in spinach leaves through *Agrobacterium*-mediated transient gene expression. The synthetic gene designed for this study included a DNA fragment encoding 129 to 169 amino acids of VP1 capsid protein. This involved both G-H loop and its flanking regions, so was expected to be an effective tool for inducing immune response in animal host. The gene construct was further elaborated by the inclusion of eukaryotic ribosome binding site (Kozak sequence) and an endoplasmic reticulum signal peptide (SEKDEL) as was described in materials and methods section. Spinach was adopted as a host plant in this study since it is an edible vegetable and a suitable candidate for the production of recombinant vaccines especially those produced for veterinary use.

Materials and methods

Construction of synthetic VP1 gene

A 120 bp long fragment of VP1 coding 129-169 amino acids of VP1 capsid protein was designed as the main part of expression construct. A eukaryotic ribosome binding site called Kozak sequence, GCCACC, was introduced prior to the start codon and an endoplasmic reticulum signal peptide called SEKDEL consisting of six amino acids was attached to 31' end just before stop codon. Start codon (AUG) and stop codon (UAA) were also added into the 5' and 3' ends of the construct, respectively. Recognition sites of *Bam*HI and *Sac*I restriction enzymes were introduced into the 5' and 3' ends of the synthetic gene, respectively (Figure 1). The construct was synthesized and cloned into the pGem T-Easy vector (Bioneer, South Korea).

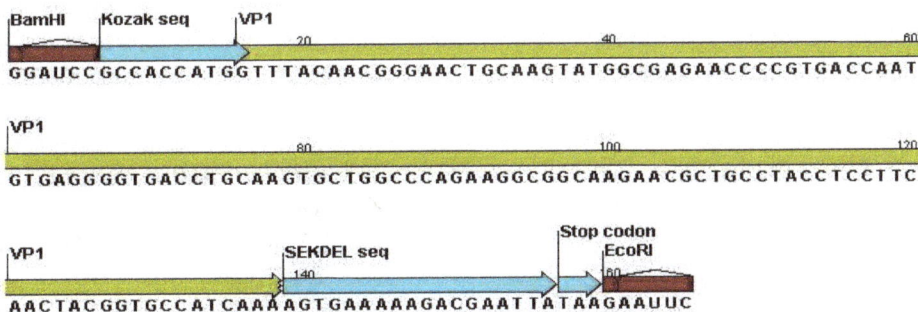

Figure 1. Schematic presentation of the synthetic VP1 gene.

Construction of a Binary Plant Expression Vector

The synthetic VP1 gene fragment was digested from pGem T-Easy vector by *Bam*HI and *Sac*I and was inserted into the plant expression vector pBI121 downstream of the CaMV 35S promoter and upstream of the nopaline synthase (NOS) terminator, yielding pBI121- VP1 vector. The ligation reaction mixture was used to transform *E. coli* strain DH5-α and kanamycin-resistant colonies were isolated after overnight incubation at 37°C. After amplification, the plasmid was extracted from bacterial cells using alkaline lysis method. The plasmid was introduced into *Agrobacterium tumef- aciens* strain GV3101 by heat shock method. In summary, a suspension of bacterium with OD=0.6 (600nm) was placed on ice for 15min. 1.5ml of the suspension was centrifuged at 4000g for 10min. Supernatant was removed and 1ml of cold $CaCl_2$ (20 mM) was added to bacterial pellet. The pellet was solved by vortex. One μgr of recombinant plasmid was added and mixed. The reaction tube was frozen in liquid nitrogen for 2min and then placed at 37°C for 5min. One milliliter of LB medium was added and the solution was placed in shaker incubator at 28°C. The suspension was again centrifuged; the supernatant was removed so that only 100μl of the suspension along with the bacterial pellet remained in the tube. The pellet was mixed with the culture medium and spread on solid culture containing LB agar medium supplemented with 50mg/l Kanamycin, 50mg/l Rifampicin and 20mg/l gentamicin. The recombinant colonies appeared after 48 hours. The putative transformed cells were further evaluated by PCR assay (Yasmin *et al* 2013).

Plant transformation

Single colony of *Agrobacterium* containing pBI121-VP1 plasmid was cultured for 48h on LB medium (NaCl 10 g/L, yeast extract 5 g/L, tryptone 5 g/L) supplemented with gentamicin 10mg/l, rifampicin 50mg/l and kanamycin 50mg/L. after reaching density of OD_{600}=1.5, the cultures were centrifuged, the supernatant was discarded and the pellet was resuspended in infiltration medium (10 mM $MgCl_2$, 10 mM MES pH 5.6, and 150 μM acetosyringone) and density was adjusted to OD_{600}=0.5. The suspension was incubated for 2h at room temperature. Agroinfiltration was then carried out to transform spinach leaves (Wroblewski 2005). The bacterial suspension was transferred to spinach leaves with a needle-free syringe as described by Sparkes *et al* [12]. Spinach plants were placed in growth chamber for three days under 25°C, 16h light/8h darkness photoperiod and 75% humidity and then they were analyzed.

Detection of VP1 gene in transgenic plants

Detection of VP1 gene in transgenic lines and other molecular analyses were conducted three days after agroinfiltration. PCR analysis was performed to evaluate presence of the expression cassette in the leaf tissue of transformed spinach plants. Genomic DNA was extracted from leaves of transgenic plants using modified Dellaporta method and used as template for PCR analysis using specific primers. The sequence of forward and reverse primers were 5' ATGGAAATTG-TAA-GTATGGAGA 3' and 5' GAAGAAA-GC-GAAAGGAGC 3' respectively. The forward primer matches a sequence within VP1 and reverse primers matches NOS terminator. Genomic DNA of wild type plants was used as negative control. PCR was carried out by 30 cycles of 94°C for 45 s, 58°C for 45 s and 72°C for 45s, followed with a final extension step at 72°C for 10min.

Real Time PCR assay

Real Time PCR assay was performed to analyze gene expression at transcription level. Total RNA was extracted from leaf tissue and complementary DNA (cDNA) was synthesized via reverse transcription using oligo (dT) 20 primer (Wroblewski *et al* 2005). The cDNA mixtures were used as templates for real-time PCR. Expression of the synthetic gene was quantitatively analyzed using a Real-Time PCR system (BioRad). Real-Time PCR was carried out in a 20 μL reaction volume containing 0.5μM of each primer and 10μl of SYBR Green Real time PCR master mix (Genet Bio, South Korea). Quantitative Real-Time PCR experiments were performed in duplicate for each sample. Forward and reverse primers for Real-Time PCR were 5' ATGGAAA- TTGTAAG-TATGGAGA 3' and 5' ATT-AAAAGAAGTTG-GAAGAGTT 3', respectively.

Protein dot blot assay

Expression of VP1 gene in spinach leaves was evaluated using protein dot blot assay. Briefly, total protein was extracted using Tris-HCL method. Small samples of the protein (3 μl) were dotted on nitrocellulose membrane and allowed to dry. BSA (Bovine Serum Albumin) was used to prevent non-specific antibody re-actions. The membrane was then incubated for 60min at 37 °C with primary antibody (1:2000 dilution), washed three times with PBS (Phosphate Buffer Saline) and PBST and finally incubated with secondary conjugated antibody (1:1500). Color was developed by adding OPD (Ortho- Phenylenediamine). Protein sample of non-transformed plant was used as negative control and a small volume of FMDV vaccine serotype O (about 3 µl) was used as positive control.

ELISA assay

Expression of the foreign gene was further evaluated using enzyme-linked immunosorbent assay (ELISA). ELISA plate was coated with total soluble proteins from the wild type and transformed plants and known FMDV VP1 antigen at 37 °C for one hour; followed by incubation with 1% bovine serum albumin (BSA) in PBS for 2h at 37 °C to prevent non-specific binding. The well was washed with PBST/PBS, incubated with antiserum reactive against FMDV (1:1000 dilutions) and then alkaline phosphatase conjugated with anti rabbit IgG (1:1500). Wells were developed with TMB (Tetramethyl benzidine) substrate; the color reaction was stopped by 2N H2SO4 and read at 405 nm of wavelength.

Results

The presence of expression cassette in *A. tume-*

faciens and transformed plants was evaluated using PCR analysis. PCR products were separated on 1% agarose gel by electrophoresis. The 587 bp band of foreign gene was observed in transgenic plant and *A. tumefaciens* colony. No band was amplified from non-transformed plant (Figure 2).

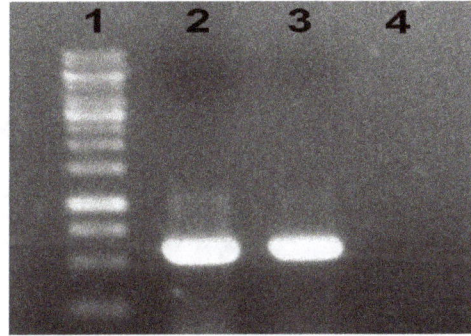

Figure 2. PCR analysis for detection of VP1 gene in transformed leaves of spinach. 1) 1 kb ladder; 2) plasmid pBI121VP1 (positive control); 3) transformed spinach plant; 4) wild type plant (negative control).

Expression of foreign gene was measured at transcription level using Real Time PCR. The results of Real Time PCR confirmed VP1 gene expression in all transformed samples but no signal was detected for control line (Figure 3). As can be inferred from Figure 3, transcription rate was quite high in transformed leaves. Expression of VP1 was further evaluated in translational level by dot blot and ELISA assays.

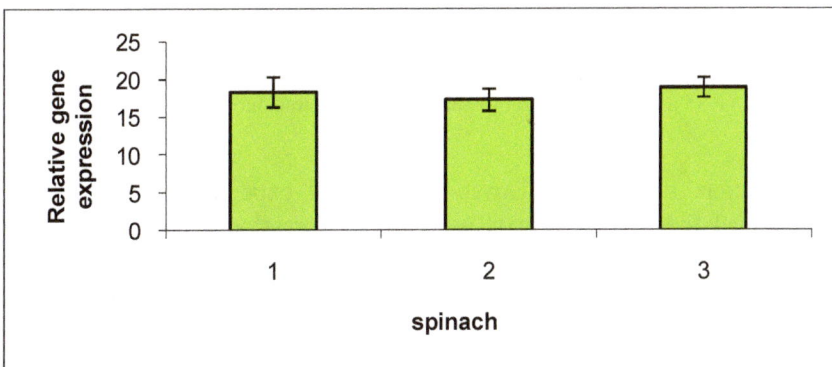

Figure 3. Quantitative measurement of VP1 gene transcription in transformed leaves of spinach via Real Time PCR. Data presented in this graph are obtained from three samples of transformed plants.

The production of recombinant VP1 protein was evaluated by dot blot assay. Positive signal showing specific antigen/antibody reaction was observed for protein samples obtained from transformed spinach plants and for those samples from positive control as well. As expected, no signal was detected for protein sample of non- transformed plant (Figure 4).

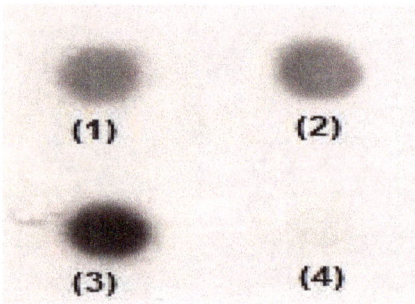

Figure 4. Dot blot assay for detection of recombinant protein in transiently transformed leaves of spinach. (1) and (2): protein samples of transformed plants, (3): commercial FMD vaccine as positive control, (4): protein sample of non-transformed plant.

Expression of VP1 recombinant protein was quantitatively assessed using ELISA assay. ELISA results showed that the recombinant protein was produced in all the three samples obtained from transformed plants, whereas no detectable signal was observed for that of non-transformed plant (Figure 5).

Discussion

In the present study, spinach leaves were transiently transformed with a chimeric construct of VP1 gene via agroinfiltration method. The method has been reported as an efficient and rapid procedure for transient gene expression in plants (Sohn *et al.* 2011).

The results of this study demonstrated that agroinfiltration can be a fast and efficient tool for production of recombinant vaccines in intact plants. As confirmed by Real Time PCR assay, transient expression level of the transgene was fairly high which was in agreement with the results obtained by Leckie and Stewart (2011) who reported high level of gene expression in leaves of *Nicotiana benthamiana*

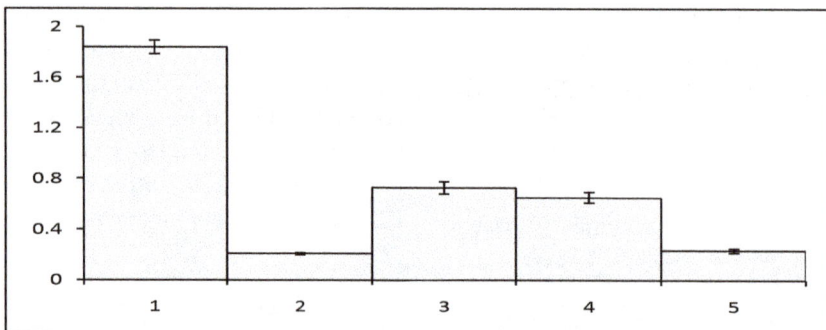

Figure 5. Quantification of recombinant VP1 expression in two transgenic spinach plants by ELISA. (1): positive control, (2) PBS as negative control (blank), (3) and (4): transformed plants, (5): wild type plant.

through agroinfiltration. Indeed, some investigators have claimed that transgene expression level in transient expression assays can be up to 1000 fold higher than that of stable transformation (Janssen and Gardner 1989). Although such a high expression level was not observed in the present study, the expression of VP1 was of great magnitude as shown by both Real Time PCR and ELISA assays. It is quite surprising that in spite of the wide

range of experimental purposes of transient gene expression, there have been few studies on the application of this transformation approach for producing recombinant vaccines in plant systems. Most of the works in the field of transient antigen expression in plant hosts have been conducted by means of plant viruses as vehicle for gene delivery and expression, in which the epitope of interest is usually inserted within the coat protein gene (Koprowski and

Yusibov 2001). This method has proved to be an efficient and rapid way for the production of recombinant protein in plants but is limited by the fact that construction of viral vector for expression of foreign protein is much laborious and time-consuming. Moreover, when the size of foreign gene exceeds a certain threshold, efficiency of the viral vector is reduced (Sala *et al*. 2003). In contrary, genes with large size can be efficiently expressed in plants via *Agrobacterium*- mediated genetic transformation. In the other word, agroinfiltration (and other types of *Agrobacterium*-mediated transient gene expression) combines advantages of both viral-based transient gene expression, that is the production of recombinant protein in a short time, and *Agrobacterium* mediated transformation, the ability to transfer large foreign genes. This makes agroinfiltration a promising alternative for the production of recombinant vaccines in plant-based systems.

The choice of plant species for production of recombinant vaccine is of great importance at least for two reasons. First, it is well documented that plant genetic background has a significant effect on transient gene expression level (Santos-Rosa *et al*. 2008). Yasmin *et al* (2013), for instance, reported that two out of three genotypes of rose showed higher susceptibility to agro-infection whereas one genotype never allowed for agro- infection. Spinach plants used in the present study showed high level of VP1 expression when transiently transformed with *A. tumefaciens*. This high level of gene expression was evident in both transcription (Figure 3) and translation levels (Figure 5). Although it is more reasonable to perform a stable genetic transformation program for permanent production of recombinant vaccine, transient gene expression can be regarded as a complementary process for achieving large amount of antibody for detection methods such as ELISA, western blotting, etc. The antigen can be quickly expressed in plant system through transient gene expression and the expressed recombinant protein can be parenteraly injected to animal models. This will trigger antibody production in immune system of the recipient animal. Based on the results, spinach is an appropriate platform for production of recombinant antigen

of FMDV. The transformed spinach lines can be parenteraly injected or orally administered to animals, because the crop is a palatable vegetable that can be easily incorporated in animal diet.

Conclusion

In this investigation, the efficacy of agroinfiltration for transient expression of VP1 protein in spinach plants was demonstrated. The expression level of the foreign gene was quite high in transformed plants. We believe that this method can be used as an effective and quick way for the production of recombinant antigens. The expressed antigen can be used as a recombinant vaccine or, more realistically, as a valuable source for production of specific antibody in veterinary diagnosis or molecular detection processes.

References

1. Bachrach, H.L., Moore, D.M., Mckercher, P.D. and Polatnick, J. 1975. Immune and antibody response to an isolated capsid protein of foot and mouth disease virus. Journal of Immunology, 115: 1636-1641.
2. Bhat, R., Lahaye, T., and Panstruga, R. 2006. The visible touch: in planta visualization of protein-protein interactions by fluorophorebased methods. Plant Methods, 2: 12-26.
3. Carter, J.E. and Langridge, W.H.R. 2002. Plant-based vaccines for protection against infectious and autoimmune diseases. Crit Rev Plant Sci, 21:93–109.
4. Doran, P.M. 2006. Foreign protein degradation and instability in plants and plant tissue cultures. Trends Biotechnol, 24:426– 432.
5. Habibi, M., and Zibaee, S. 2013. Plant- based recombinant vaccines. International Journal of Agriculture and Crop Sciences, 6: 27-30.
6. Ihara, Y., Nagano, M., Muto, S., Uchimiya, H., and Kawai, M. 2007. Cell death suppressor Arabidopsis Bax inhibitor- 1 is associated with calmodulin binding and ion homeostasis. Plant Physiol, 143: 650–660.
7. Janssen, B. and Gardner, R. 1989. Localized transient expression of GUS in leaf discs following cocultivation with Agrobacterium. Plant Mol Biol, 14: 61–72.
8. Koprowski, H. and Yusibov, V. 2001. The green revolution: plants as heterologous expression vectors. Vaccine, 19:2735– 41.
9. Leckie, B.M., and Stewart, C.N. 2011. Agroinfiltration as a technique for rapid assays for evaluating candidate insect resistance transgenes in plants. Plant Cell Rep, 30:325–334.

10. Rodriguez, L.L. and Grubman, M.J. 2009. Foot and mouth disease virus vaccines. Vaccine, 27: 90-94.

11. Sala, F., Rigano, M., Barbante, A., Basso, B., Walmsley, A.M. and Castiglione, S. 2003. Vaccine antigen production in transgenic plants: strategies, gene constructs and perspectives. Vaccine, 21:803–808.

12. Santos-Rosa, M.A., Poutaraud, D., Merdinoglu, D. and Mestre, P. 2008. Development of a transient expression system in grapevine via agro-infiltration. Plant Cell Rep, 27:1053-1063.

13. Sohn, S., Huh, S.M., Kim, K.H., Park, J.W. and Lomonossoff, G. 2011. Effect of Rice stripe virus NS3 on transient gene expression and transgene co-Silencing. Plant Pathol. J, 27(4) : 310-314.

14. Tang, X., Frederick, R.D., Zhou, J., Halterman, D.A., Jia, Y. and Martin, G.B. 1996. Initiation of plant disease resistance by physical interaction of AvrPto and Pto kinase. Science, 274: 2060–2063.

15. Vaquero, C., Sack, M., Chandler, J., Jürgen, D. Schuster, F., Monecke, M., Schillberg, S. and Fischer, R. 1999. Transient expression of a tumor-specific single-chain fragment and a chimeric antibody in tobacco leaves. Proc. Natl. Acad. Sci USA, 96: 11128–11133.

16. Walmsley, A.M., Arntzen, C.J. 2000. Plants for delivery of edible vaccines. Curr Opin Biotechnol, 11:126–9

17. Wang, C.Y., Chang, T.Y., Walfield, A.M., Ye, J. and Shen, M. 2002. Effective synthetic peptide vaccine for foot-and- mouth disease in swine. Vaccine, 20: 2603–2610.

18. Wroblewski, T., Tomczak, A. and Michelmore, R. 2005. Optimization of Agrobacterium-mediated transient assays of gene expression in lettuce, tomato and Arabidopsis. Plant Biotech. J, 3: 259–273.

19. Yasmin, A., Jalbani, A., Ali, M., Nasreen, A. and Debener, T. 2013. Development of Agrobacterium-based transient gene expression assay in rose leaves. Pak. J. Bot, 45(3): 1005-1009.

11

Assessment of seed storage protein composition of six Iranian adopted soybean cultivars [*Glycine max* (L.) Merrill.]

M. Arefrad[1], N. Babaian Jelodar[2*], Gh. Nematzadeh[1], M. Karimi[1] and S.k. Kazemitabar[2]

1. Genetics and Agricultural Biotechnology Institute of Tabarestan (GABIT). Sari, Mazandaran, Iran. P.O. Box: 578.
2. Department of Biotechnology, Sari Agricultural Sciences and Natural Resources University (SANRU). Sari, Mazandaran, Iran. P.O.Box: 578.
   *Corresponding Author, Email: Nbabaeian@yahoo.com

Abstract

Seed protein quality is an important topic in the production of soybean. The quality of soybean proteins is limited by anti-nutrient proteins and low levels of essential sulfur amino acids. In this study, protein content and solubility of six cultivars were evaluated and seed storage proteins were analyzed using SDS-PAGE and scanning densitometry. The results showed that seed storage protein bands were similar among soybean cultivars. However, concentration of β-conglycinin (7S), glycinin (11S) proteins and related subunits were statistically different among the soybean cultivars. According to the results of this study, 033 and DPX cultivars were characterized by high levels of protein content (42.45 %) and protein solubility (76.58 mg g^{-1}) respectively. Two cultivars DPX and JK were also identified by high 11S/7S ratio (1.39 and 1.43 % respectively). Besides, the JK was considered by the lowest concentration of 7S protein (20.35 %). The results showed that a significant negative correlation existed between protein content and solubility (r= -0.66). A significant and moderate positive correlation was found between acidic and basic subunits with 11S protein (r= 0.72 and 0.47 respectively). The 11S and 7S proteins also showed positive and negative correlation with 11S/7S ratio (r= 0.70 and -0.85 respectively). On the other hand, acidic subunits were characterized by significant positive and negative relationship with 11S/7S ratio and some anti-nutrients protein respectively. Thereupon, these results suggested that the development of new genotypes of soybean with high level of acidic subunits of 11S protein can be notable in increasing seed storage protein quality in soybean breeding programs.

Keywords: Soybean, Seed storage proteins, β-conglycinin, Glycinin, Anti-nutrient proteins.
Abbreviation: 7S, β-conglycinin protein; 11S, Glycinin protein; KTI, Kunitz protein; BBI, Bowman-Birk protein; BSA, bovine serum albumin.

Introduction

The nutritional values of soybean play a prominent role for human and livestock nutrition. Soy protein is increasingly consumed by humans and it also makes a relatively inexpensive protein source for livestock. However, protein composition of soybean seed is not ideal because of its low levels of the sulfur containing essential amino acids, methionine and cysteine (Fukushima, 1991). Soybean storage proteins mainly consist of globulins, which are classified to 2S, 7S, 11S and 15S according to their sedimentation properties (Osborne and Campbell, 1898). β-Conglycinin (7S) and glycinin (11S) are two major proteins consisting of about 70% of the total seed protein content (Kitamura, 1995). Functional properties of soybean based storage protein are mainly reflected on their composition and structure (Barac *et al*, 2004). Due to abundance of 7S and 11S, these proteins were the main responsible factors for soybean protein quality. The 7S is a glycoprotein and composed of ά, α and β subunits. On the other hand, 11S a hexamer consists of acidic and basic polypeptide linked by disulfide bonds (Mori *et al*. 1981). The 11S proteins have three to four times more sulfur containing amino acids (particularly methionine) than does 7S protein (Beilinson *et al*. 2002). Furthermore, the β subunit of 7S protein is known to be void of methionine and cysteine (Krishnan, 2000). On the other hand, raw soybeans contain a number of allergenic proteins such as Gly m Bd 60 K (α subunit of 7S), Gly m Bd 28 K, Gly m Bd 30 K (Lecin) and protease inhibitors (Kunitz and Bowman–Birk proteins) that can possibly alter the body metabolism of consumers (Krishnan *et al*. 2009; Liener, 1994; Ogawa *et al*. 2000; Norton, 1991). Allergic symptoms to soybean include skin, gastrointestinal, and respiratory reactions and in some cases anaphylaxis (Sicherer and Sampson, 2006). Gly m Bd were recognized by IgE antibodies from soybean sensitive patients with atopic dermatitis (Ogawa, 2000; Krishnan *et al*. 2009). The anti- nutritional Lectin activity is related to its ability to recognize and specifically link to carbohydrates in the membranes of the epithelium cells of the digestive tract. Lectin can produce structural change in the intestinal epithelium and resist gut proteolysis (Pusztai *et al*, 1990). Protease inhibitors, served as storage proteins and as regulator of endogenous proteases, can certainly interfere with protein digestion and consequently exert a negative impact on the utilization of soybean-based protein products (Liener and Kakde 1980). Several protease inhibitors were identified in soybean storage proteins, but most of their activity was thought to be due to Kunitz (KTI) and Bowman–Birk (BBI) proteins, which represents the majority of the bioactive proteins that strongly inhibits trypsin and trypsin- chymotrypsin respectively (Norton, 1991). The inhibitors have been shown to induce pancreatic enzymes, hyper secretion and a fast stimulation of pancreas growth, which is histologically described as pancreatic hypertrophy and hyperplasia (Liener, 1995).

It has been suggested that the levels of 11S and 7S protein and their subunits (Fehr *et al*. 2003; Panthee *et al*. 2004; Taski-Ajdukovic *et al*. 2010) as well as anti-nutrient proteins (lin *et al*. 2008; Gu *et al*. 2010) vary among genotypes. Thus different cultivars may have dissimilar protein products. The aim of the present work focused on assessing 7S and 11S proteins, their subunits and anti-nutrient protein contents of seed storage soybean proteins of six cultivars which are currently cultivated in Iran, and after that assessing the relationship between these proteins and subunits. Awareness of the relationship between these characters among Iranian varieties could be useful for further and facilitate the ongoing efforts on improving the quality of protein in breeding programs of soybean proteins.

Materials and methods

Plant material

The six adopted Iranian soybean cultivars: Hill, Sahar, 032, 033, DPX and JK, provided by Mazandaran Agriculture Research Center in the north of Iran, were planted in the experimental field of Genetic and Agriculture Biotechnology Institute of Tabarestan (GABIT). After complete growth, plants were evaluated for seed protein content, protein solubility and the composition of seed storage protein.

Measurement of protein content and solubility

About 10 mg seed powder was added to 500 μl of protein extraction buffer (92 mM Tris base pH 8.1, 23 mM $CaCl2$) and then centrifuged at 14000 rpm for 20 min in $4^{o}C$, the supernatant was used for the SDS-PAGE analysis. Protein concentration was determined by Bradford method (Bradford, 1976) with bovine serum albumin (BSA) as standard. The protein content of seed was also determined by Kjeldahl method and the amount of total protein was estimated from percent nitrogen content using a conversion factor of 6.25.

SDS-PAGE analysis

Protein electrophoresis was performed by a vertical slab gel apparatus according to Laemmli, (1970). The staking gel consisted of 6%

and the separation gel constituted of 14% polyacrylamid respectively. Amount of 80 μg protein was loaded in the well for all the samples. The molecular weight of the polypeptides was calculated from the standard graph plotted Rf vs. Log. Mol. Wt. of marker proteins electrophoresed (SMO431 Fermentas Co.) along with the samples. To investigate the varietal effect, electrophoresis of the storage proteins in six cultivars was performed in triplicate. Videlicet three aliquots of the same sample were analyzed at the same time. The gels were run simultaneously in the same electrophoretic cell.

Quantifying the composition of seed storage protein profile

Scanned protein gels (Bio-Rad Calibrated Densitometer GS-800) were analyzed with Melanie 6 software. The quantity of each protein band was calculated with the basis of percent volume definition which is equal to

$$\frac{volume}{\sum_{s=1}^{n} volumes} \times 100.$$ The volume of a spot

is calculated as the volume above the spot outline, which is situated at 75% of the spot height (as measured from the peak of the spot).

Statistical analysis

The assays were carried out using completely randomized design (CRD) with three replications. Statistical analyses were done using SPSS software. The Significant differences between cultivars means were determined by the Duncan's multiple range tests (P < 0.05), after the analysis of variance test (ANOVA) for independent samples. Pearson's correlation coefficients were used to determine the degree and significance of association traits.

Results

Seed protein content and solubility

There were significant differences among cultivars for seed protein content and solubility (Table 1). The highest protein content belonged to 033 and the lowest was found for DPX cultivars (42.45 % and 34.90 % respectively). The DPX was also characterized by the highest pro-

tein solubility (76.86 mg g-1), whereas the lowest was related to 032 when compared to other cultivars (63.38 mg g-1).

Profiling of seed storage proteins

To identify variants of storage protein subunits in the six adopted Iranian soybean varieties Hill, Sahar, 032, 033, DPX and JK, protein extracts were analyzed by SDS–PAGE. Typical electrophoretic patterns obtained from total proteins are illustrated in Figure 1.

Table 1. Protein content and solubility in seeds of six soybean cultivars.

Genotypes	Protein content (%)	Protein solubility (mg g^{-1})
Hill	41.74 ± 0.07 [ab]	70.91 ± 0.52 [b]
Sahar	40.94 ± 1.01 [ab]	64.80 ± 0.17 [c]
033	42.45 ± 0.28 [a]	65.58 ± 0.24 [c]
DPX	34.90 ± 0.07 [c]	76.58 ± 0.46 [a]
JK	39.29 ± 0.09 [b]	65.58± 0.33 [c]
032	41.04 ± 0.26 [ab]	63.38 ± 0.06 [d]

Mean± Standard Deviation followed by the same letter within the same column are not significantly different at P≤0.05 probability.

The patterns among cultivars were similar, containing the most basic polypeptides. The low level of protein polymorphism could be attributed to conservative nature of the seed protein (Bonfitto et al. 1999). However, these patterns could be used as a general biochemical fingerprint for the soybean. The protein banding has the subunits of the major storage proteins, 11S and 7S proteins, including the bulk storage proteins. The 7S subunits separated on SDS–PAGE into three bands of 78, 75 and 47 kDa, corresponding to the ά, α and β subunits of this storage protein respectively. The subunits of 11S separated into five acidic (A) and basic (B) bands. The 11S bands from 34 to 35 kDa correspond to the acidic polypeptide chains A1-4. A5, the smaller acidic polypeptide of 11S is designated with 15 kDa.

The cluster polypeptide that separated on the SDS–PAGE gels from 22 to 23 kDa was the basic polypeptide, designated as B1-4. These results are consistent with Adachi et al., (2003) and Maruyama et al., (2001) reports. In addition

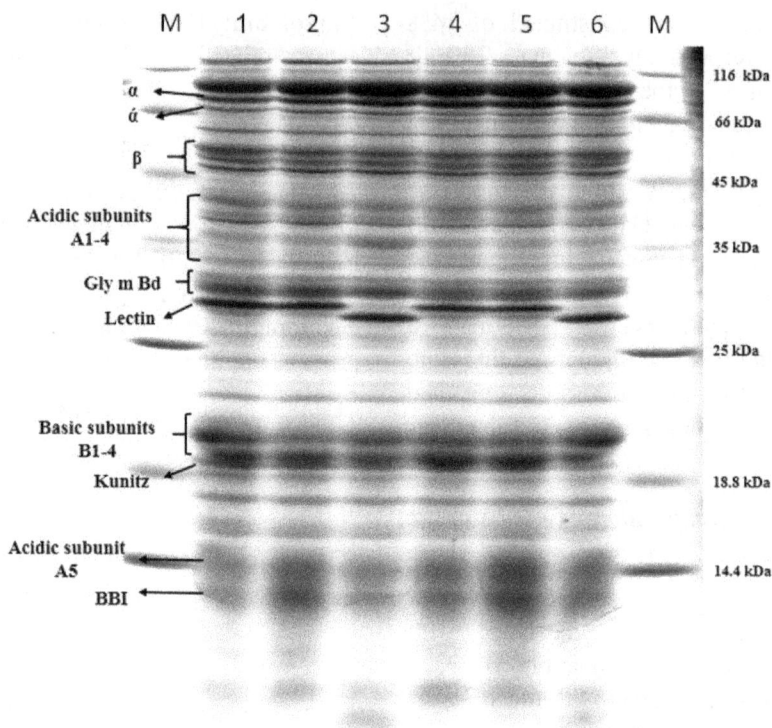

Figure 1- SDS-PAGE analysis of the protein extraction from six soybean cultivars. Lane M, protein molecular marker weight. Lane 1- JK, lane 2, 032. Lane 3, DPX. Lane 4, 033. Lane 5, Sahar. Lane 6, Hill.

to these components of 11S and 7S subunits, there were four other proteins that were separated as single bands in the SDS– PAGE gel, including the Gly m Bd 28 K, Lectin (Gly m Bd 30 K), Kunitz and Bowman-Birk which are known food allergens of soybeans with 30, 33, 21 and 8 kDa, respectively. The findings of this study are comparable to those reported previously by Yagasaki *et al*. (1997); Yaklich, (2001); Ogawa *et al*. (2000) and Schenk *et al*. (2003) for soybean isogenic lines with different 11S subunit composition.

Quantifying the composition of seed storage protein profile

Densitometry analysis of seed protein profiles was used to quantify the two major storage proteins and their subunits from these cultivars separated on SDS- PAGE (Table 2). The β-conglycinin (7S) content was derived by summation of the original scanned value of ά, α, and β subunits and the glycinin (11S) content were derived by summation of the acidic and basic components (Figure 1).

The values obtained from the densitometer scans were converted to percentage of the total protein in each lane. The obtained results indicated that the 033 and Sahar exhibited the highest concentration of ά (10.17 and 9.52 % respectively) and α subunit (4.45 and 4.34 % respectively) of 7S fraction. Whereas, low concentration of ά subunit were found for 032 and Jk cultivars (6.91 and 6.64 % respectively), and the lowest α subunit was found in Hill and JK (3.45 and 3.66 % respectively). High level of β subunit was found in 032 and the lowest was found in JK and DPX cultivars (10.04 and 10.00 % respectively). Among the cultivars, Hill cultivar flowed by Sahar and 033 were characterized by high concentration of 7S protein (25.82, 25.36 and 27.22 % respectively), while Jk considered by low level of this protein (20.35 %). All cultivars identified by high level of acidic subunits except for Hill (11.51 %). Whereas, 032 recognized by high concentration of basic subunits (16.79 %) and

Table 2- Comparison of the subunit composition of β-conglycinin (7S) and glycinin (11S) and individual composition of total extractable seed storage proteins by SDS-PAGE and Qualification by densitometry (expressed as percent relative volume of each spot(s)) of six soybean cultivars.

Genotypes		Hill	Sahar	033	DPX	JK	032
β-Conglycinin	ά	9.08 ± 0.33^{ab}	9.52 ± 0.92^{a}	10.17 ± 0.97^{a}	7.39 ± 0.10^{bc}	6.64 ± 0.02^{c}	6.91 ± 0.55^{c}
	α	3.45 ± 0.05^{b}	4.34 ± 0.01^{a}	4.45 ± 0.02^{a}	3.88 ± 0.20^{ab}	3.66 ± 0.02^{b}	3.82 ± 0.42^{ab}
	β	13.27 ± 0.01^{ab}	11.49 ± 0.07^{bc}	12.59 ± 1.21^{ab}	10.00 ± 0.52^{c}	10.04 ± 0.78^{c}	13.72 ± 0.40^{a}
	7S[1]	25.82 ± 0.38^{a}	25.36 ± 1.00^{a}	27.22 ± 2.15^{a}	21.28 ± 0.62^{bc}	20.35 ± 0.80^{c}	26.46 ± 1.38^{ab}
Glycinin	Total acidic subunits	11.51 ± 0.26^{b}	17.81 ± 1.05^{a}	16.70 ± 1.66^{a}	16.16 ± 0.95^{a}	17.53 ± 0.97^{a}	14.95 ± 0.38^{a}
	Total basic subunits	15.71 ± 0.80^{ab}	14.48 ± 1.13^{bc}	12.74 ± 0.68^{cd}	13.58 ± 0.11^{bcd}	11.60 ± 0.42^{d}	16.79 ± 0.51^{a}
	11S[2]	27.22 ± 1.07^{a}	32.29 ± 2.19^{a}	29.44 ± 2.35^{a}	29.75 ± 0.84^{a}	29.14 ± 1.40^{a}	31.74 ± 0.79^{a}
11S+7S		53.05 ± 1.46^{bc}	57.65 ± 1.19^{a}	56.66 ± 0.19^{ab}	51.03 ± 1.46^{c}	49.49 ± 2.23^{c}	56.21 ± 0.49^{ab}
11S/7S		1.05 ± 0.02^{b}	1.28 ± 0.13^{ab}	1.12 ± 0.17^{ab}	1.39 ± 0.01^{a}	1.43 ± 0.01^{a}	1.31 ± 0.11^{ab}
Gly m Bd 28 K		8.27 ± 0.06^{a}	5.30 ± 0.12^{b}	5.73 ± 0.07^{b}	8.03 ± 0.31^{a}	5.49 ± 0.76^{b}	6.14 ± 1.05^{b}
Lectin		8.24 ± 0.63^{a}	7.38 ± 0.43^{a}	6.94 ± 0.35^{a}	8.60 ± 1.28^{a}	6.33 ± 0.45^{a}	7.47 ± 0.39^{a}
KTI[3]		1.17 ± 0.04^{bc}	0.94 ± 0.17^{c}	1.14 ± 0.01^{bc}	1.96 ± 0.19^{ab}	2.28 ± 0.52^{a}	1.99 ± 0.35^{ab}
BBI[4]		3.36 ± 0.29^{c}	4.01 ± 0.20^{b}	4.00 ± 0.14^{b}	1.33 ± 0.05^{d}	8.83 ± 0.08^{a}	1.12 ± 0.22^{d}

Mean ± Standard Deviation followed by the same letter within the same row are not significantly different at P≤0.05 probability.
1- β-Conglycinin protein, 2- Glycinin protein, 3- Kunitz protein, 4- Bowman-Birk protein.

the lowest was found in JK (11.60 %). No significant difference was found for 11S among the under evaluated cultivars. Nonetheless the Sahar was considered by high concentration of 7S+11S proteins (57.65 %) and the lowest was found in DPX and JK (51.49 and 49.49 %). The DPX and JK cultivars characterized by the high level of 11S/7S ratio (1.39 and 1.43 % respectively), and the lowest was found in Hill cultivar (1.05 %). Among the cultivars, Hill and DPX were considered by significantly high concentration of Gly m Bd 28 K (8.27 and 8.03 % respectively). No significant difference was found between cultivars for Lectin. The JK cultivar was characterized by high concentration of KTI and BBI proteins (2.28 and 8.83 % respectively), whereas low concentration of KTI was found in Sahar (0.94 %) and in DPX and 032 cultivars for BBI (1.33 and 1.12 % respectively).

Correlation analysis

To investigate the relationship between composition of seed storage proteins and their subunits, correlation coefficient analyses were carried out (Table 3). Protein content showed negative correlation with protein solubility and 11S/7S ratio, but had positive correlation with β-subunit and 7S protein. Protein solubility showed no correlation between 7S and 11S proteins and their subunits and 11S/7S ratio, except to 11S+7S and Gly m Bd 28 K. The ά and β subunits of 7S protein had a significant positive correlation with 7S and significant negative correlation with 11S/7S ratio. Also, ά and α subunits of 7S protein showed negative correlation with KTI and Gly m Bd 28 K anti-nutrient proteins. The 7S protein showed positive and negative correlation with 11S+7S and 11S/7S respectively, whereas the 11S protein demonstrates positive correlation with 11S/7S ratio.

Only acidic subunits of 11S protein showed positive correlation with 11S and 11S/7S ratio. On the other hand, basic subunits of 11S protein showed negative correlation with the BBI protein. Moreover, 11S/7S ratio showed negative correlation with KTI protein.

Dendrogram studies

Dendrogram of six adopted Iranian soybean cultivars based on seed storage protein composition densitometry using Ward's method showed that cultivars were divided into three clusters (Figure 2). The first clusters consisted of Sahar, 033 and 032 cultivars. Second clusters contained the cultivar JK only, and tertiary cluster included the Hill and DPX cultivars. The lowest genetic distance was recorded between the two cultivars Sahar and 033 and indicated that these cultivars were closely related to each other.

On the other hand the highest genetic distance was recorded between the two cultivars Sahar and DPX indicating that these cultivars were genetically distant genotypes.

Discussion

More recently, soybean cultivars have been bred on the seed yield and oil content basis. This regime has resulted in a narrow genetic base that potentially could limit the concurrent improvement in seed yield, protein and oil contents (Kisha et al. 1998). Soybean seed storage proteins have a good balance of the essential amino acids required by humans and animals and are mainly used as a source of protein for animal husbandry. With the current increase in meat consumption, the request for protein in animal husbandry has increased. Furthermore, it is relatively inexpensive compared to other protein sources used for livestock. As a result

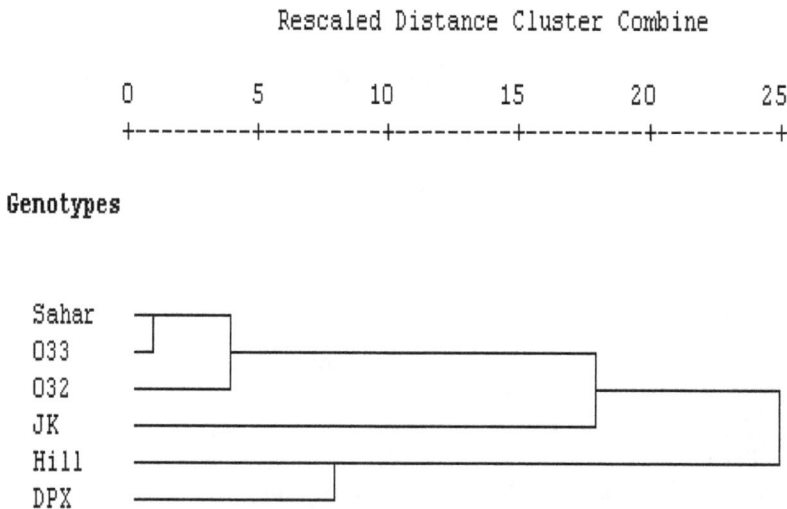

Figure 2. Dendrogram of six soybean cultivars based on seed storage proteins.

of the extensive use of soybean proteins in the animal industry, humans are also increasingly consuming soybeans and soy products. Earlier studies have shown that the high protein cultivars accumulate higher amounts of both 11S and 7S proteins (Krishnan et al. 2007; Yaklich, 2001). A broad range of variability exists among cultivars for all subunit components of 7S, and for both the acidic and basic polypep-

tide chains of 11S proteins (Table 2). These results are in accord with those of Taski-Ajdukovic et al. (2010); Pantee et al. (2004); Fehr et al. (2003) and Yaklich (2001), who also investigated the individual components of these proteins in various cultivars.

Today, increasing quality of seed storage proteins is the most important goal in soybean breeding programs. In comparison with meat,

Table 3. Correlation coefficients between β-conglycinin (7S), glycinin (11S) and their subunits and individual components of total extractable seed storage proteins.

Traits	Protein solubility	ά	α	β	7S¹	Acidic subunits	Basic subunits	11S²	11S+7S	11S/7S	Gly m Bd 28 K	Lectin	KTI³	BBI⁴
Protein content	-0.66**	0.46	-0.03	0.61**	0.60**	-0.30	0.17	-0.15	0.38	-0.49*	-0.16	-0.05	-0.26	0.15
Protein solubility	1	-0.09	-0.26	-0.37	-0.30	-0.27	-0.09	-0.31	-0.49*	0.02	0.72**	0.51*	0.11	-0.33
ά		1	0.50*	0.42	0.85**	-0.27	-0.17	-0.37	0.49*	-0.81**	-0.12	0.11	-0.65**	-0.13
α			1	0.09	0.47*	0.46	-0.23	0.25	0.59**	-0.18	-0.60**	-0.41	-0.54	-0.10
β				1	0.81**	-0.51*	0.47*	-0.12	0.58*	-0.67**	-0.05	-0.05	-0.47*	-0.39
7S					1	-0.37	0.14	-0.23	0.65**	-0.85**	-0.18	-0.03	-0.70**	-0.31
Acidic subunits						1	-0.26	0.72**	0.25	0.69**	-0.55*	-0.58*	-0.04	0.34
Basic subunits							1	0.47*	0.49*	0.09	0.23	0.13	0.21	-0.61**
11S								1	0.58*	0.70**	-0.33	-0.44	-0.19	-0.12
11S+7S									1	-0.16	-0.42	-0.37	-0.73**	-0.36
11S/7S										1	-0.07	-0.22	0.43	0.19
Gly m Bd 28 K											1	0.54*	0.28	-0.40
Lectin												1	0.22	-0.46
KTI													1	0.15

* and** Significant different at 5% and 1% probability level respectively.

1- β-Conglycinin protein, 2- Glycinin protein, 3- Kunitz protein, 4- Bowman-Birk protein.

soybean proteins are deficient in sulfur contain-ing essential amino acids such as methionine followed by cysteine and possibly threonine (Zarkadas *et al.*, 1999). Due to abundance of 7S and 11S, these proteins were the main fac-tors responsible for soybean protein quality (Kitamura, 1995). The ά and α subunits of 7S protein contain traces of methionine and cys-teine. In addition, the β subunit of this fraction is known to be void of methionine and cyste-ine (Krishnan, 2000). Therefore, selection for a low level of the β subunit and high levels of ά and α subunits of 7S protein could help increase the total sulfur containing amino acids. Never-theless, 11S protein is a better source of sulfur amino acids than 7S due to 3-4.5% more per unit of protein (Beilinson *et al.* 2002; Krishnan, 2000). By comparison the sulfur amino acids of 7S protein account for less than 1% of the amino acid residues (Burton *et al.* 1982). No significant difference was found for 11S among the cultivars under evaluation. A significant and moderate positive correlation was also found between acidic and basic subunits with 11S pro-tein. In this way, Taski-Ajdukovic *et al* (2010) estimated the accumulation of the main seed storage protein subunits, 11S and 7S proteins, among high protein soybean cultivars, to deter-mine how these genotypes preferentially accu-mulate specific polypeptides in different matu-rity groups. These authors indicated significant positive correlation between acidic and basic subunits with 11S protein. This indicates that both subunits increase in seeds simultaneously and suggested that selecting the genotypes with high concentration of both the acidic and basic subunits could increase the 11S protein. All cul-tivars identified by high level of acidic subunits except for Hill, besides 032 were recognized by high concentration of basic subunits.

In a nutritional view the ά and α subunits of 7S protein with an allergenic potential (Krishnan *et al.*, 2009) contain much less sulfur amino acids (Kitamura, 1995). Thus, increase in the 11S to 7S ratio should lead to improvement in protein quality (Poysa *et al.* 2006). This ratio of current Iranian soybean cultivars was from 1.05 to 1.43 (Table 2). Reported data revealed that proportion of 11S to 7S ratio in soybean cultivars was varied from 1.26 to 3.40 (Zilic *et*

al. 2010; Mujoo *et al.* 2003; Fehr *et al.* 2003). Due to differences in the gelation properties of soybean storage protein fractions, many re-searchers have attempted to correlate 7S and 11S proteins with tofu quality. Yagasaki *et al.* (1997) indicated that decreasing the 11S to 7S ratios due to the lack of specific subunits, had a negative effect on tofu quality and the food pro-cessing properties of soybeans. As reported by Mujoo *et al.* (2003) tofu textural quality can be positively determined using 11S, 11S/7S ratio and negatively by 7S fraction. Among the culti-vars, DPX and JK which were characterized by low concentration of 7S protein and high level of 11S/7S ratio could be a suitable cultivar to gain the higher tofu quality.

On the other hand, Fehr *et al.* (2003) suggest-ed that 11S can be increased at the expense of 7S protein. Ogawa, (1989) also reported a typical inverse relationship between 7S and 11S concentrations. However, Panthee *et al.* (2004) indicated significant positive correla-tion between 11S with 7S. While, in our study no correlation was found between 7S and 11S proteins. It has been suggested that the 11S, 7S and 11S/7S ratio were influenced by the en-vironment; however no significant differences were expressed among years or locations of these traits (Fehr *et al.* 2003). Therefore it can be the lack of significant differences among the locations which indicated that no site with-in the test area would be expected consistently different from 11S, 7S and 11S/7S ratio. The difference in the correlation between 7S and 11S in the different studies may be due to the studied cultivars and the number of environ-ments used to evaluate their performance. The results indicated that the importance of genet-ic improvement on protein quality in soybean breeding program.

In addition, 11S and 7S proteins of the cultivars have a significant positive and negative correla-tion with the 11S/7S ratio respectively. These results are in agreement with those reported by Fehr *et al.* (2003). However, Pantee *et al.* (2004) reported no relationship between 11S and 11S/7S, but demonstrated significant nega-tive correlation between 7S fraction and 11S/7S ratio. On the other hand, the acidic subunit of glycine protein showed significant positive correlation with 11S/7S ratio. However, Pantee *et al.* (2004) reported positive correlation between basic subunits and 11S/7S. These different results may be due to the genetic differences of cultivars.

More recently, soybean cultivars have been bred to increase seed yield and oil content, while protein meal is mainly used as a source of protein for animal husbandry. A major impediment to increasing soybean protein through selective breeding lies in the inverse relationship between protein content and yield (Helms and Orf, 1998). Nevertheless, it is important to find a balance between protein content and specific protein composition. Taski- Ajdukovic *et al.* (2010) studied forty genotypes of different majority groups of soybean and reported protein content is independent of protein subunits of storage proteins. Fehr *et al.* (2003) also observed no correlation between soybean protein subunits and protein content. These researchers suggested that it is possible to select soybean genotypes for desired protein composition without influencing protein content. However, Yaklich, (2001) and Krishnan *et al.* (2007) showed that high protein cultivars accumulated higher amounts of 11S and 7S proteins. In this study protein content showed positive correlation whit β-subunit and 7S protein but negative correlation with 11S/7S ratio. Several studies have shown that the accumulation of the β-subunit is promoted by excess application of nitrogen or by sulfur deficiency, while the application of sulfur fertilization increases the synthesis of 11S (Krishnan, 2000). Krishnan *et al.* (2005) have shown that nitrogen application to soybean plants favored the accumulation of β- subunit while decreasing the accumulation of BBI, a protein rich in cysteine. Based on these results, it seems that an inverse relationship exists between protein content and sulfur amino acids content.

Presence of allergenic proteins like α subunit of 7S (Gly m Bd 60 K), Lectin (Gly m Bd 30 K), Gly m Bd 28 K, KTI and BBI makes the deterioration of the soybean protein quality that can possibly alter the body metabolism of consumers (Krishnan *et al.* 2009; Liener, 1994;

Ogawa *et al.* 2000; Norton, 1991). An inverse relationship between acidic subunits and Gly m Bd 28 K and Lectin, and also between basic subunits and BBI suggested that genotypes of soybean with high concentration of acidic and basic subunits of 11S protein could decrease these allergenic factors in soybean protein.

In conclusion, results from our study show that, similar seed storage protein pattern exists between six cultivars of soybean which are currently cultivated in Iran. The low level of protein polymorphism could be attributed to conservative nature of the seed protein. However, concentration of 7S and 11S proteins and respective subunits was statistically different among the soybean cultivars. Considering the cultivars, Sahar and 033 with high concentration of ά, α and 7S showed the lowest genetic distance. Moreover, the JK cultivar with low concentration of 7S and basic subunits of 11S proteins and high level of KTI and BBI proteins allocated in one cluster alone. According to the results, JK, which has the lowest concentration of 7S as well as the best 11S/7S protein ratio, could be used as a parent to improve soybean protein quality. On the other hand, 032 with the high level of acidic and basic subunits of 11S fraction and low level of anti-nutrient proteins could also be a suitable cultivar to gain higher seed protein quality. Moreover, the results suggested that development of new genotypes of soybean with high level of acidic subunits of 11S protein which has significant positive correlation with 11S/7S ratio and inverse relationship to some anti-nutrient proteins can be notable in increasing seed storage protein quality in soybean breeding programs.

Acknowledgment

This research was conducted with the financial support of the Genetics and Agricultural Biotechnology Institute of Tabarestan (GABIT), Sari, Iran.

References

1. Adachi, M., Kanamori, J., Masuda, T., Yagasaki, K., Kitamura, K. and Mikami, B. 2003. Crystal structure of soybean 11S globulin: glycinin A3B4 homohexamer. Proc Natl Acad Sci USA, 100: 7395-7400. Barac, M.B., Sladana, P.S., Snezana,

T.J. and Mirjana, B.P. 2004 Soy protein modification. BIBLID, 35:3-16.

2. Beilinson, V., Chen, Z., Shoemaker, R.C., Fischer, R.L., Goldberg, R.B. and Nielsen, N.C. 2002. Genomic organization of glycinin genes in soybean. Theo and Appl Gen,104: 1132-1140.

3. Bonfitto, R., Galleschi, L., Macchia, M., Saviozzim, F. and Navari-Izzo, F. 1999. Identification of melon cultivars by gel and capillary electrophoresis. Seed Sci. Tech, 27: 779-83.

4. Bradford, M.M. 1976. A rapid and sensitive method for quantification of microgram quantities of protein of utilizing the principle dye binding. Analyz Biochem, 72: 680-685.

5. Burton, J.W., Purcell, A.E. and Walter, W.M. 1982. Methionine concentration in soybean protein from populations selected for increased percent protein. Crop Sci, 22: 430-432.

6. Fehr, W.R., Hoeck, J.A., Johnson, S.L., Murphy, P.A., Nott, J.D., Padilla, G.I. and Welke, G.A. 2003. Genotype and environment influence on protein components of soybean. Crop Sci. 43:511–514.

7. Fukushima, D. 1991. Recent progress of soybean protein foods: chemistry, technology and nutrition. Food Review International, 7: 323-351.

8. Gu, C., Pan, H., Sun, Z. and Qin, G. 2010. Effect of soybean variety on anti- Nutritional factors content and growth performance and nutrients metabolism in rat. Int J Mol Sci, 11: 1048-1056.

9. Helms, T.C. and Orf, J.H. 1998. Protein, oil, and yield of soybean lines selected for increased protein. Crop Sci, 38: 707-711.

10. Kisha, T.J., Diers, B.W., Hoyt, J.M. and Sneller, .C.H. 1998. Genetic diversity among soybean plant introductions and North American germplasm. Crop Sci. 38: 1669-1680.

11. Kitamura, K. 1995. Genetic improvement of nutrition and food processing quality in soybean. Japan Agric Res Qua, 29: 1-8.

12. Krishnan, H.B., Kim, W., Jang, S. and Kerley, M.S. 2009. All three subunits of soybean β-Conglycinin are potential food allergens. J Agric Food Chem, 57: 938- 943.

13. Krishnan, H.B. 2000. Biochemistry and molecular biology of soybean seed storage proteins. Journal of New Seeds, 2: 1-25.

14. Krishnan, H.B., Bennett, J.O., Kim, W.S., Krishnan, A.H. and Mawhinney, T.P. 2005. Nitrogen lowers the sulfur amino acid content of soybean [Glycine max (L.) Merr.] by regulating the accumulation of Bowman–Birk protease inhibitor. J Agri Food Chem, 53: 6347-6354.

15. Krishnan, H.B., Savithiry, S.N., Ahmed, A.M. and Randall, L.N. 2007. Identification of Glycinin and β-conglycinin subunits that contribute to the increased protein content of high-protein soybean lines. J Agri Food Chem, 55: 1839-1845.

16. Laemmli, U.K. 1970. Cleavage of structural proteins during the assembly of the head of bacteriophage T4. Nature, 277: 680-685.

17. Liener, I.E. and Kakde, M.L. 1980. Protease inhibitors in toxic constituents of plant food stuff. In: I. Liener (ed) 2rd edn, Academic Press, New York, pp 7–71.

18. Liener, I.E. 1994. Implications of antinutritional components in soyabean foods. Crit Rev Food Sci Nutr. 34: 31-67.

19. Liener, I.E. 1995. Possible adverse effects of soybean anticarcinogens. J Nutr. 125: 744-750.

20. Lin, P., Ye, X. and Ng, T. 2008. Purification of melibiose-binding lectins from two cultivars of Chinese black soybeans. Acta Biochim Biophys Sin, 40(12): 1029-1038. Maruyama, N., Adachi, M., Takahashi, K., Yagasaki, K., Kohno, M. and Takenaka, Y. 2001. Crystal structures of recombinant and native soybean b-conglycinin homotrimers. Europ J Biochem, 268: 3595-3604.

21. Mori, T., Utsumi, S., Inaba, H., Kitamura, K. and Harada, K. 1981. Differences in subunit composition of glycinin among soybean cultivars. J Agric Food Chem. 29: 20-23.

22. Mujoo, R., Trinh, D.T. and Perry, K.W. 2003. Characterization of storage proteins in different soybean varieties and their relationship to tofu yield and texture. Food Chem, 82: 265-273.

23. Norton, G. 1991. Proteinase inhibitors, In FJP D'Mello, CM Duffus, JH Duffus eds, Toxic substances in crop plants, The royal society of chemistry, Cambridge UK, pp 68-106

24. Ogawa, A., Samoto, M. and Takahashi, K. 2000. Soybean allergens and hypoallergenic soybean products. J Nutr Sci Vitamin, 46(6): 271-279.

25. Ogawa, T., Tayama, E., Kitamura, K., and Kaizama, N. 1989. Genetic improvement of seed storage proteins using three variant alleles of 7S globulin sub-unit in soybean (Glycine max L.). Japan J Breed, 39: 137-147.

26. Osborne, T.B. and Campbell, G.F. 1898. Proteins of the pea. J Am Chem Soc, 20: 348-362.

27. Panthee, D.R., Kwanyuen, P., Sams, C.E., West, D.R., Saxton, A.M. and Pantalone, V.R. 2004. Quantitative trait loci for β- conglycinin (7S) and glycinin (11S) fractions of soybean storage protein. JAOCS, 81: 1005-1012.

28. Poysa, V., Woodrow, L., and Yu, K. 2006. Effect of soy protein subunit composition on tofu quality. Food Res Inter, 39: 309- 317.

29. Pusztai, A., Ewen, S.W.B., Grant, G., Peumans, W.J., van Damme, E.J.M., Rubio, L. and Bardocz, S. 1990. Relationship between survival and binding of plant lectins during small intestinal passage and their effectiveness as growth factors. Digestion, 46 (suppl. 2): 308-316.

30. Schenk, G., Neidig, M., Zhou, J., Holman, T.R. and Solomon, E.I. 2003. Spectroscopic

characterization of soybean lipoxygenase-1 mutants: the role of second coordination sphere residues in the regulation of enzyme activity. Biochemistry, 42: 7294-7302.

31. Sicherer, S. H. and Sampson, H.A. 2006. Food allergy. J Allergy Clin Immunol, 117: 470-475.

32. Taski-Ajdukovic, K., Djordjevic, V., Vidic, M. and Vujakovic, M. 2010. Subunit composition of seed storage proteins in high-protein soybean genotypes. Pesq Agrop Brasil, 45: 721-729.

33. Yagasaki, K., Takagi, T., Sakai, M. and Kitamura, K. 1997. Biochemical characterization of soybean protein consisting of different subunits of glycinin. J Agri Food Chem, 45: 656-660.

34. Yaklich, R.W. 2001. β-Conglycinin and glycinin in high-protein soybean seeds. J Agri Food Chem, 49: 729-735.

35. Zarkadas, C.G., Voldeng, H.D., Yu, Z.R. and Choi, V. 1999. Assessment of the protein quality of nine northern adapted yellow and brown seed coated soybean cultivars by amino acid analysis. J Agric Food Chem, 47: 5009-5018.

36. Zilic, S.M., Barac, M.B., Pesic, M.B, Drinic, S.D.M., Ignjatovic-Micica, D.D. and Srebric, M.B. 2010. Characterization of proteins from kernel of different soybean varieties. J Sci Food Agric, 91: 60-67.

12

Fingerprinting and genetic diversity evaluation of rice cultivars using Inter Simple Sequence Repeat marker

Behzad Shahin Kaleybar[1], Sara Kabirnattaj[1], Ghorban Ali Nematzadeh[1], Seyyed Kamal Kazemitabar[2] & Seyede Mona Salim Bahrami[3]

1. Genetics and Agricultural Biotechnology Institute of Tabarestan (GABIT), Sari Agricultural Sciences and Natural Resources University, Sari, Iran.
2. Department of Agronomy and Plant Breeding, Sari Agricultural Sciences and Natural Resources University, P.O. Box 578, Sari, Iran.
3. Department of Biology, University of Mazandaran, Babolsar, Iran.
   *Corresponding Author, Email: shahin.bio65@gmail.com

Abstract

Rice as one of the most important agricultural crops has a putative potential for ensuring food security and addressing poverty in the world. In the present study, in order to provide basic information to improve rice through breeding programs, Inter Simple Sequence Repeat marker (ISSR) was used For DNA fingerprinting and finding genetic relationships among 32 different cultivars. In this study, 12 out of 17 used primers amplified 184 distinct and reproducible fragments with high value of polymorphism (88%). also, for fingerprinting the cultivars 29 loci were used that generated high polymorphic bands among the cultivars. Results indicated that similarity index varied between 39% and 88.4 %, furthermore, PIC value with an average of 23% ranged between 0.1 (for primer #3) to 0.34 (for primer #2). Clustering based on Jaccard coefficient similarity index and UPGMA algorithm divided the cultivars into 6 main sub-clusters in cut-off point of 64% similarity index. The two Italian rice cultivars 'Ribe' and 'Roma' were the closest cultivars in addition, 'Vialone nano' and 'Anbarbu' showed the highest dissimilarity. In total, high genetic divergence was observed among the cultivars also, poly (GA)-containing 3-anchored primers amplified the highest number of bands. According to similarity and cluster analysis, it could be inferred that crosses involving Anbarbu cultivar are the most promising ones to improve rice through breeding programs. In fact, results of this study would be promising as a genetic marker for the identification of rice cultivars and an important source of knowledge for subsequent rice researches.

Keywords: Dendrogram, Genetic Similarities, Genome analysis, PCR based Marker, Polymorphism.

Introduction

Rice (*Oryza sativa* L.) as the world's most important crop for more than 2 billion people belongs to Gramineae or grass family and subfamily of *Orazoidea* (1). The people get 60-70% of their daily energy requirements from this cereal grain as a source of complex carbohydrates, nutrients, vitamins and minerals (7, 16). Assessment of rice cultivars to analyze their genetic relationships not only provides basic information to improve productivity and traits related to agronomic parameters but it is also important to variety registration systems or uniformity and stability testing (10, 12).

DNA fingerprinting and assessment of genetic relationships have become an important tool for varietal identification in plant breeding programs (15, 22). There are several systems including: morphological, chemical, and biochemical markers for evaluation of similarity among genotypes. Nevertheless, these characterizing systems have some limitations, such as being time-consuming, limited to a few characteristics and influenced by factors like temperature, humidity, light or the age of plants which can modify classification based on these systems (24). In contrast, DNA-based marker systems are highly heritable and obtainable in high numbers. They also show enough polymorphism and by simultaneous elimination of environmental impacts, provide a reliable and powerful tool for assessing differences among closely related organisms with high accuracy (20). Several DNA-based marker systems are available for genetic fingerprinting of plants. RAPD (random amplified polymorphic DNA), AFLP (amplified fragment length polymorphism) and SSRs (simple sequence repeats or microsatellites) are generally PCR (polymerase chain reaction)-based DNA marker systems (18) but, ISSR as a DNA-based marker is a powerful, fast and low cost method that requires small amounts of DNA as the template. In addition, because of using longer primers with high annealing temperature, it provides reproducible and reliable results. This PCR-based DNA marker does not requires any prior sequence information for implementation and uses a single primer targeting microsatellite motifs generating abundant polymorphisms (13). ISSR has been used successfully to assess variation in a vast range of plants, such as blueberry (5), lingo berry (4), *Auricularia polytricha (25), Dioscorea opposite Thunb (26),*

Leonurus cardiaca (11) and *Oryza Sativa* (9) which was described as a powerful technique to assess genetic diversity and similarities between and within species.

The objective of this study was evaluation of ISSR marker ability to assess genetic diversity among rice cultivars in order to fingerprint and estimate genetic relationships among the cultivars. Actually, the results can generate basic information that could be useful for rice breeding programs, such as cultivar identification, parental choice for developing heterotic hybrids or germplasm preservation.

Materials and methods

Preparation of plant materials and genomic DNA isolation

The seeds of 32 rice cultivars were obtained from the Genetics and Agricultural Biotechnology Institute of Tabarestan (GABIT) (Table 1) and cultivated in hydroponic medium in controlled condition; the nutrients for rice hydroponic medium have been described previously by Yoshida *et al.*, (23). Genomic DNA was extracted from approximately 0.1 g leaf materials powdered by mortar and pestle using Dellaporta *et al,.* (6) method with some modifications, for quantification and qualification of the isolated DNA were used spectrophotometer and 0.7% agarose gel electrophoresis.

Table 1. The cultivar names of rice used in this study.

Number	Genotype	Number	Genotype	Number	Genotype	Number	Genotype
1	Hasan Sarai 1	9	Onda	17	Rashti Sard	25	Mosa Tarom
2	Ahlami Tarom	10	Ribe	18	Sadri	26	Mir Tarom
3	Hasan Sarai 2	11	Dular	19	Baldo	27	IR 58
4	CH2-1	12	Roma	20	Amol 1	28	Abji Boji
5	Lemonimo	13	Jasmin 85	21	Bajar	29	Anbar Bou
6	Hasani	14	Zir Band P	22	Vialone nano	30	Domsiah
7	Manjing	15	Amol 2	23	Ringo	31	Amol 3
8	Fujiminori	16	Salari	24	Cripto	32	Sepidroud

ISSR amplification

In the present study, seventeen ISSR primers (Alpha DNA, Canada) were analyzed for detecting polymorphism among the cultivars. Finally, the PCR reactions were performed in a volume of 20 µl reaction mixture containing 40 ng of template DNA, 0.6 mM of each dNTPs, 1x PCR buffer (10 mM Tris- HCl, 50 mM KCl, 0.08 Nonidet P40), 2 mM MgCl2, 1 U of Taq

DNA polymerase enzyme and 0.6 µM of each ISSR primers. To reduce background amplification, 2 % (v/v) formamide was added to the reaction mixture. Amplification reactions were carried out using a AB Applied Biosystems thermal cycler (Bio-Rad, USA) with an initial denaturation step of 5 min at 95˚C, followed by 35 cycles at 94° C for 40s, 55° C for 1 min, 72° C for 2 min and a final extension step at 72°

C for 7 min. Two independent PCRs were carried out for each primer, and the products were fractionated on separate gels. A reaction containing all PCR components except DNA (negative control) was used in each experiment to test DNA contamination of the reagents. PCR amplified fragments were separated on 2.5% agarose gels by electrophoresis at 90 V in 0.5x TBE (Tris/Boric Acid/EDTA) buffer. Gels were stained with ethidium bromide and imaged in Doc-Print VX5 (VIBER LOURMAT) gel documentation system.

Data analysis

ISSR reproducible fragments were scored as present (1) or absent (0). Since ISSR is a dominant marker, the presence of a band was interpreted as either a heterozygote or dominant homozygote and the absence of a band was interpreted as recessive homozygote. The data matrix structure was assembled by binomial (0/1) data and was used as input data for further analysis applying NTSYS version 2.02

software program. To test whether clusters in the dendrogram agrees with the information from similarity matrix, cophenetic matrices were created from the dendrogram and were compared with the similarity matrix. Similarity of ISSR data was computed using Jaccard's similarity index and cluster diagrams generated by UPGMA algorithm and the resulting clusters were expressed as dendrograms.

Results

In the present study, polymorphic information content (PIC) varied ranging from 0.1 (primer #3) to 0.34 with an average of 0.23. The primer ISSR #2 showed the highest value of polymorphism that referred to the efficiency of the primer to detect polymorphism within a population depending on the number of alleles and frequency distribution. Among 17 ISSR primers used, 12 primers that produced distinct and reproducible fragments were selected for fingerprinting and diversity analysis (Table 2).

Table 2. List of 12 ISSR primer sequences used for analysis of the rice cultivars with annealing temperature, size variation, number of bands, polymorphism and PIC value.

Primer	Sequence (5' to 3')	Annealing temperature	Fragment size length	Total no. of bands	Polymorphic bands	Polymorphism (%)	PIC
ISSR #2	(GA) 9C	57° C	220-2300	34	30	88	0.34
ISSR #3	(GA) 9T	55° C	420-1150	17	15	88	0.10
ISSR #5	(GT) 8C	51° C	400-1200	17	14	82	0.14
ISSR #8	(CT) 8G	51° C	600-1200	11	11	100	0.10
ISSR #9	(AG) 8C	52° C	500-1620	16	14	87	0.22
ISSR #13	(TC) 8C	51° C	700-1250	7	7	100	0.31
ISSR #15	(AC) 8G	51° C	500-1150	12	10	83	0.31
ISSR #16	(TG) 8A	49° C	680-1920	15	13	86	0.22
ISSR #17	(AC) 8C	51° C	500-1500	17	15	88	0.29
ISSR #18	(ATC) 6T	49° C	550-1150	11	10	90	0.25
ISSR #19	(ATC) 6C	50° C	800-1750	14	12	85	0.23
ISSR #21	(ATA) 6T	42° C	600-1500	13	11	84	0.30

Within the selected primers, there were nine di-nucleotide repeat primers and three tri-nucleotide repeat primers with 3' anchors and the other primers produced faint or no distinct bands. Totally, using the 12 selected ISSR primers, 184 fragments with the size ranging between 400-2300 bp were amplified that 162 loci of which were polymorphic and no band was detected in negative controls. Primers ISSR #8 and #13 showed the highest polymorphism value (100%) and the lowest one was produced by ISSR #5.

Fingerprinting analysis

DNA fingerprinting with the aim of cultivar or varietal identification, has become an important tool for genetic identification, germplasm management and registration systems

(21). In the present study, 29 polymorphic loci were amplified by more informative primers. The obtained banding patterns were reproducible for each primer and genotype that were used for finger printing the cultivars. The selected primers can be used for genetic distance estimation and identification of the cultivars that are difficult to characterize because of their similar morphological properties (Figure 1). Two-dimensional PCA plot analysis showed that the first, second and third components represented 15%, 13% and 12% of variation respectively, and the first 10 coordinates explained 78% of the total variation that represented appropriate sampling of the primers and scattering over different parts of the genome.

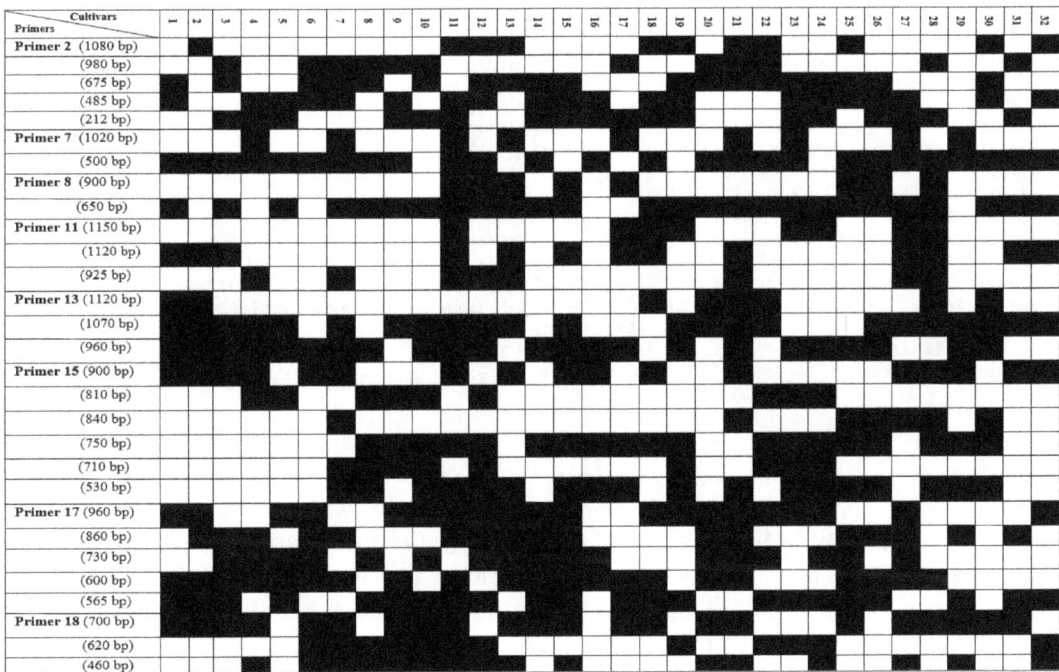

Figure 1. The fingerprinting profile of 32 rice varieties obtained by ISSR markers. The black blocks indicate presence of band and white blocks indicate absence of band.

Genetic relationships

Knowledge about genetic constitute differences of genotypes is highly important in breeding programs. The importance is because of the need to identify parents with high genetic distances to produce progenies with higher heterosis (2). In this study, generated similar-

ity matrix by ISSR markers exhibited a range of similarity indexes between 39%- 88.4% (Table 3). The two Italian rice cultivars 'Ribe' and 'Roma' were the closest genotypes with the highest similarity index (88.4 %) that were followed by two Iranian rice cultivars 'Hasan Sarai 1' and 'Hasan Sarai 2' with 87%

similarity index. The high level of similarity could be due to having common ancestors and or selecting similar traits during breeding programs in Italian or Iranian rice cultivars. The highest distance (61%) was between 'Vialone nano' and 'Anbarbu' followed by distance between 'Lemonino' and 'Anbarbu' cultivars with 58.7% dissimilarity index.

Cluster analysis

Evaluation similarity coefficients or dendrogram helps to select superior individuals to ob-

Table 3. Similarity index values among the rice cultivars using Jaccard coefficient method.

	1	2	3	4	5	6	7	8	9	10	11	12	13	14	15	16	17	18	19	20	21	22	23	24	25	26	27	28	29	30	31	32
1	1																															
2	0.86	1																														
3	0.88	0.84	1																													
4	0.74	0.68	0.79	1																												
5	0.81	0.74	0.75	0.7	1																											
6	0.76	0.72	0.78	0.77	0.69	1																										
7	0.71	0.65	0.73	0.77	0.63	0.77	1																									
8	0.64	0.59	0.67	0.62	0.67	0.65	0.68	1																								
9	0.76	0.7	0.73	0.68	0.76	0.72	0.74	0.79	1																							
10	0.73	0.65	0.71	0.7	0.76	0.67	0.7	0.81	0.85	1																						
11	0.68	0.65	0.68	0.68	0.63	0.69	0.76	0.62	0.71	0.69	1																					
12	0.7	0.65	0.68	0.68	0.75	0.67	0.71	0.78	0.8	0.88	0.73	1																				
13	0.68	0.64	0.7	0.74	0.59	0.71	0.78	0.59	0.67	0.66	0.77	0.66	1																			
14	0.81	0.75	0.84	0.73	0.74	0.77	0.7	0.73	0.79	0.76	0.67	0.74	0.69	1																		
15	0.76	0.68	0.78	0.75	0.67	0.7	0.74	0.7	0.72	0.76	0.73	0.73	0.8	0.79	1																	
16	0.68	0.66	0.72	0.72	0.68	0.64	0.71	0.58	0.62	0.62	0.64	0.62	0.72	0.67	0.76	1																
17	0.61	0.62	0.59	0.52	0.62	0.57	0.52	0.51	0.61	0.54	0.49	0.53	0.49	0.68	0.52	0.58	1															
18	0.71	0.64	0.66	0.55	0.59	0.64	0.58	0.56	0.65	0.58	0.6	0.57	0.57	0.72	0.61	0.52	0.61	1														
19	0.7	0.62	0.64	0.59	0.73	0.62	0.67	0.72	0.76	0.82	0.71	0.82	0.6	0.69	0.67	0.57	0.51	0.59	1													
20	0.76	0.7	0.76	0.75	0.69	0.82	0.77	0.64	0.77	0.7	0.7	0.7	0.78	0.83	0.75	0.64	0.58	0.65	0.65	1												
21	0.77	0.71	0.77	0.81	0.66	0.75	0.87	0.65	0.71	0.71	0.76	0.7	0.86	0.76	0.82	0.75	0.52	0.6	0.62	0.83	1											
22	0.59	0.54	0.59	0.57	0.64	0.56	0.55	0.68	0.72	0.74	0.53	0.71	0.5	0.65	0.59	0.51	0.52	0.62	0.69	0.59	0.55	1										
23	0.68	0.62	0.65	0.67	0.7	0.62	0.67	0.76	0.78	0.82	0.66	0.79	0.6	0.73	0.69	0.59	0.57	0.59	0.82	0.65	0.66	0.81	1									
24	0.66	0.59	0.64	0.66	0.71	0.63	0.65	0.79	0.78	0.87	0.65	0.84	0.6	0.74	0.74	0.6	0.52	0.56	0.85	0.65	0.67	0.74	0.86	1								
25	0.67	0.61	0.72	0.6	0.61	0.64	0.68	0.66	0.66	0.64	0.64	0.66	0.65	0.76	0.73	0.58	0.54	0.7	0.59	0.68	0.7	0.66	0.61	0.6	1							
26	0.76	0.72	0.76	0.64	0.77	0.68	0.72	0.68	0.72	0.7	0.7	0.74	0.67	0.8	0.77	0.69	0.65	0.65	0.69	0.73	0.73	0.59	0.69	0.65	0.78	1						
27	0.72	0.69	0.77	0.78	0.64	0.73	0.73	0.61	0.69	0.69	0.72	0.67	0.81	0.73	0.73	0.68	0.56	0.6	0.6	0.76	0.83	0.57	0.66	0.65	0.65	0.67	1					
28	0.69	0.63	0.67	0.56	0.56	0.59	0.64	0.53	0.64	0.59	0.61	0.6	0.6	0.66	0.64	0.49	0.49	0.58	0.56	0.66	0.67	0.52	0.56	0.54	0.66	0.7	0.63	1				
29	0.51	0.52	0.54	0.53	0.41	0.52	0.6	0.51	0.49	0.51	0.56	0.48	0.56	0.51	0.51	0.46	0.43	0.46	0.46	0.51	0.57	0.39	0.48	0.44	0.49	0.51	0.53	0.54	1			
30	0.72	0.64	0.68	0.61	0.61	0.59	0.67	0.62	0.64	0.66	0.6	0.66	0.63	0.69	0.67	0.56	0.49	0.54	0.64	0.67	0.68	0.53	0.63	0.6	0.67	0.74	0.62	0.67	0.65	1		
31	0.62	0.58	0.62	0.57	0.53	0.54	0.57	0.52	0.57	0.55	0.51	0.51	0.58	0.59	0.57	0.55	0.52	0.57	0.54	0.59	0.57	0.53	0.58	0.51	0.52	0.54	0.57	0.47	0.56	0.7	1	
32	0.64	0.6	0.66	0.59	0.59	0.56	0.61	0.56	0.59	0.54	0.53	0.53	0.64	0.63	0.61	0.66	0.56	0.52	0.51	0.61	0.64	0.47	0.57	0.53	0.54	0.65	0.67	0.51	0.58	0.71	0.75	1

tain hybrids with greater segregation and heterotic effects during recombination (3). In this study, clustering based on Jaccard's coefficient similarity index and UPGMA algorithm divided the cultivars into 6 main subclusters. The cultivars were grouped based on the cut-off point of 64% similarity equivalent to the mean genetic similarity of all the cultivars Figure 2. Cophenetic correlation created from comparing similarity and output matrix of the dendrogram was 0.91, that indicates the used similarity coefficient and cluster analysis method were reliable for grouping the rice cultivars. Majority of cultivars were placed in cluster I that consisted of five subgroups with a range of similarity between 73%- 0.79%. This range of similarity suggests the existence of relatively less divergence in group I, that may be because of the origin of some of the cultivars from closely related ancestors, For example, 'Mosa Tarom' with 'Mir Tarom' and 'Hasan Sari1' with 'Hasan Sari2' are actually identical varieties that were separated from each other during cultivation in different regions.

Discussion

This study is on the use of ISSR markers for creating genetic fingerprints and analyzing relationships among domestic and foreign rice cultivars in Iran, which allows breeders to identify cultivars by combined use of the generated bands. According to Ferreira et al., (8) heterosis and combining ability of parents depend directly on the genetic diversity between them, and the chance of finding promising combinations is more when more

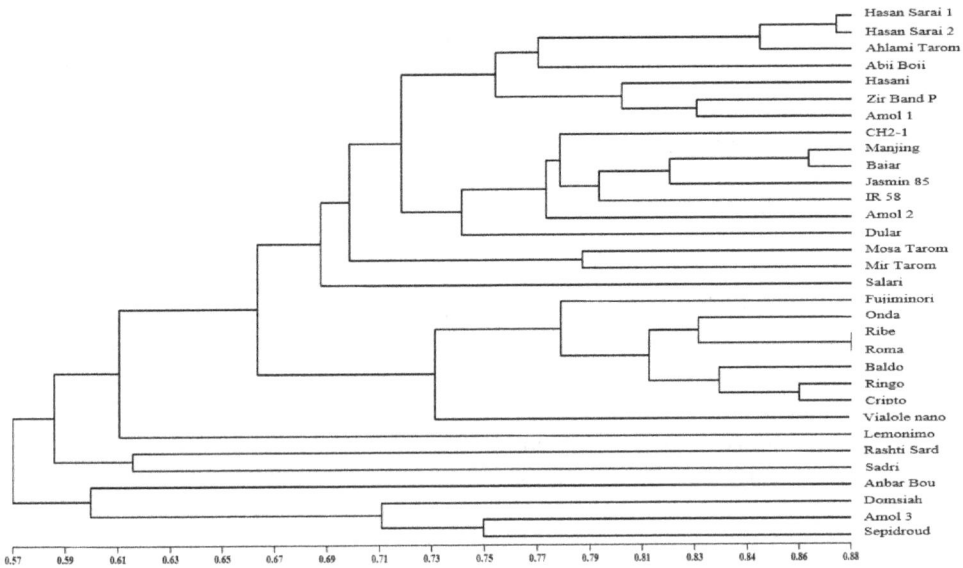

Figure 2. UPGMA dendrogram based on the Jaccard's similarity index calculated from an ISSR analysis of different rice cultivars.

divergent materials are used. The similarity matrix results indicated a considerable level of genetic variation and a wide range of genetic distance among the cultivars, which indicates the importance of studying relationships and genetic distance among the cultivars. In the established fingerprint (Figure 1) the position of bands was different among cultivars, and differed from bands that were unique for some cultivars in comparison with those which were present in approximately 50% of them. It is important to distinguish the identical cultivars that are called by different names in different regions of a country and different cultivars with identical names and or germplasm preservation. It has been approved by Prevost and Wilkinson (19) that generated profiles by ISSR markers are a quick, reliable and highly informative system for DNA fingerprinting. In the present study, the "Anbarbu" cultivar showed the highest genetic distance in comparison with other cultivars, that can be considered the most interesting and promising cultivar for producing progenies with higher heterosise. Based on the similarity coefficients and dendrogram results, it could be inferred that crossing between "Vialone nano" and "Anbarbu" cultivars was the most interesting and promising cross to produce progenies with

higher heterosise because of their greatest genetic distance (60.9 %) and valuable traits for breeding programs such as, long and highly aromatic grains for "Anbarbu" and very tasty grains for "Vialole nano" (Table 3, Figure 2). The result is in accordance with the studies conducted by Manonmani et al., (14) in Rice (Oryza sativa L) and Orlovskaya et al., (17) in spring triticale accessions. They reported the existence of positive correlation between genetic distances derived from PCR based markers and heterosis. Furthermore, the increase in genetic distance value between parental components leads to higher probability of obtaining heterotic hybrids. Using 12 ISSR primers, different numbers of loci were amplified with the range of 7-34 bands, which poly (GA)-containing 3- anchored primers produced the highest number of bands. The result suggests the existence of a high frequency of dinucleotide simple sequence repeats especially poly (GA) motifs in genome of rice, thus primer ISSR #2 was introduced as an efficient primer to detect polymorphism and fingerprinting within rice population because of its high value of polymorphism, PIC, frequency distribution and poly (GA)- containing 3- anchors. Overall, the results indicated that there is a high degree of diversity among the cultivars for genetic

relationship evaluation and the ISSR marker is an informative, quick and reproducible
approach that generates sufficient polymorphisms for large-scale DNA fingerprinting
purposes in rice. The results would be promising as the genetic markers for identification
of rice cultivars and gleaned critical data are
considered as a source of knowledge for future rice research, such as identification and
genotyping of the cultivars, germplasm improvement and parental selection for breeding purposes. For a more detailed review,
any given collection of rice can be sampled
in different regions and used molecular techniques for characterizing and fingerprinting
the varieties.

References

1. Babaei, A., Nematzadeh, G.A. and Hashemi, H. 2011. An evaluation of genetic differentiation in rice mutants using semirandom markers and morphological characteristics. Aust J Crop Sci, 5: 1715-1722.
2. Cruz, C.D. and Regazzi, A.J. 1997. Modelos biométricos aplicados ao melhoramento genético Vol. 2. Ed. UFV.
3. Cruz, C.D., Regazzi, A.J. and Carneiro, P.C.S. 2003. Biometric models applied to genetic improvement. Editora UFV, Viçosa, MG, Brazil.
4. Debnath, S.C. 2007. Inter simple sequence repeat (ISSR) to assess genetic diversity within a collection of wild lingonberry (Vaccinium vitis-idaea L.) clones. Can J Plant Sci, 87: 337-344.
5. Debnath, S.C. 2009. Development of ISSR markers for genetic diversity studies in Vaccinium angustifolium. Nord J Bot, 27: 141–148.
6. Dellaporta, S.L., Wood, J. and Hicks, J.B. 1983. A plant DNA minipreparation: version II. Plant Mol Biol Rep, 1: 19-21.
7. FAO (2008) Food Consumption: Pattern of main food items-share in total dietary energy consumption. FAO Statistic Division. Bulletin.
8. Ferreira, D.E., Oliveira, A.C., Santos, M.X. and Ramalho, M.A.P. 1995. Métodos de avaliação da divergência genética em milho e suas relações com os cruzamentos dialélicos. Pesq Agropec Bras, 30: 1189-1194.
9. Girma, G., Tesfaye, K., and Bekele, E. 2013. Inter Simple Sequence Repeat (ISSR) analysis of wild and cultivated rice species from Ethiopia. Afr J Biotechnol, 9: 5048-5059.
10. Ichii, M., Hong, D.L., Ohara, Y., Zhao C.M. and Taketa, S. 2003. Characterization of CMS and maintainer lines in indica rice (Oryza sativa L.) based on RAPD marker analysis. Euphytica, 129: 249-252.
11. Khadivi-Khub, A. and Soorni, A. 2014. Comprehensive genetic discrimination of Leonurus cardiaca populations by AFLP, ISSR, RAPD and IRAP molecular markers. Mol Biol Rep, 41: 4007-4016.
12. Kwon, Y.S., Lee, J.M., Yi, G.B., Yi, S.I., Kim, K.M., Soh, E.H., Bae, K.M., Park, E.K., Song, I.H. and Kim, B.D. 2005. Use of SSR markers to complement tests of distinctiveness, uniformity and stability (DUS) of pepper (Capsicum annuum L.) varieties. Mol Cell, 19: 428-435.
13. Liu, B. and Wendel, JF. 2001. Inter simple sequence repeat (ISSR) polymorphisms as a genetic marker system in cotton. Mol Ecol Notes, 1: 205-208
14. Manonmani, S., Senthlivel, S., Fazlullahkhan, A.K. and Maheswaran, M. 2004. RAPD and Isozyme markers for genetic diversity and their correlation with heterosis in Rice (Oryza sativa L). In Proceedings of the 4th International Crop Science Congress Brisbane, Australia. www. cropscience. org. au.
15. McGregor, C.E., Lambert, C.A., Greyling, M.M., Louw, J.H. and Warnich, L. 2000. A comparative assessment of DNA fingerprinting techniques (RAPD, ISSR, AFLP and SSR) in tetraploid potato (Solanum tuberosum L.) germplasm. Euphytica, 113: 135-144.
16. Matsumoto, T., Wu, J., Namiki, N., Kanamori, H., Fujisawa, M. and Sasaki, T. 2006. Completion of rice genome sequencing -a paradigm shift of rice biology. J Agric Res, 40: 99-105.
17. Orlovskaya, O.A., Koren, L.V. and Khotyleva, L.V. 2012. The impact of parental Genetic Divergence on the Heterosis of F1-Hybrids of Spring Triticales. Russ J Genet, 3: 405-411.
18. Pradeep Reddy, M., Sarla, N. and Siddiq, E.A. 2002. Inter simple sequence repeat (ISSR) polymorphism and its application in plant breeding. Euphytica, 128: 9–17
19. Prevost, A. and Wilkinson, MJ. 1999. A new system of comparing PCR primers applied to ISSR fingerprinting of potato cultivars. Theor Appl Genet, 98: 107-112.
20. Wang, Y., Xue, Y., Li, J. 2005. Towards molecular breeding and improvement of rice in China. Trends Plant Sci, 10: 610-614.
21. Wu, Y.G., Guo, Q.S., He, J.C., Lin, Y.F., Luo, L.J. and Liu, G.D. 2010. Genetic diversity analysis among and within populations of Pogostemon cablin from China with ISSR and SRAP markers. Biochem. Syst Ecol, 38: 63-72.
22. Wünsch, A. and Hormaza, J.I. 2002. Cultivar identification and genetic fingerprinting of temperate fruit tree species using DNA markers. Euphytica, 125: 59-67.
23. Yoshida, S., Forno, D.A. and Cock, J.H. 1971. Laboratory manual for physiological studies of

rice. Laboratory manual for physiological studies of rice.

24. Żebrowska, J.I. and Tyrka, M. 2003. The use of RAPD markers for strawberry identification and genetic diversity studies. Food Agr Environ, 1: 115-117.

25. Zhang, D., Shen, F., Liu, J. and Falandysz, J. 2014. Studies on germplasm resources of Auricularia polytricha by Inter-Simple Sequence Repeat (ISSR). Med Engin Bioin, 19: 1.

26. Zhou, Y., Zhou, C., Yao, H., Liu, Y. and Tu, R. 2008. Application of ISSR markers in detection of genetic variation among Chinese yam *(Dioscorea oppositaThunb)* cultivars. Life Sci J, 5: 6-12.

13

PCR-based markers for identification of some allelic variation at *Glu-1* and *Glu-3* loci in common wheat

E. Mehrazar[1], A. Izadi-Darbandi[1*], M. Mohammadi[2] & G. Najafian[2]

1. Department of Agronomy and Plant Breeding Sciences, College of Aburaihan, University of Tehran, P.O.Box, Tehran, Iran.
2. Seed and Plant Improvement Institute, Cereals Research Department, Karaj, Iran.
   *Corresponding Author. aizady@ut.ac.ir

Abstract

Marker assisted selection (MAS) is a tool for breeding, screening, and genetic characterization of germplasm. Allelic variation of both high and low molecular weight glutenin subunits (HMW/LMW-GS) is associated with the rheological properties of wheat flour. In this study, we investigated glutenin pattern using SDS-PAGE and their PCR based on DNA markers in 60 advanced wheat lines and cultivars with different origins. Specific DNA markers regarding to *Glu-1* loci, such as 1319 bp, 669 bp and 450 bp fragments were respectively validated for 2*, 17+18, 5+10 alleles. These alleles showed the highest allelic percentage in *Glu-1* loci in studied cultivars. However the Null, 7+8 and 5+10 alleles showed the highest allelic percentage in advanced lines. In this study, 23%, 40% and 37% of cultivars respectively, got good (10), moderate (8-9) and weak (4-7) quality scores. In advanced lines, 18%, 44% and 38% got good, moderate and weak quality scores respectively. Ten specific DNA PCR markers were also detected for genotyping *Glu-B3* alleles. The most frequent *Glu-B3* alleles in wheat cultivars were *i, a, b* and *d* with 24%, 21%, 20% and 12%, respectively. Specific PCR markers regarding to the reported *Glu-B3* alleles were produced as 621bp, 1095bp, 1570 bp and 662bp consequently. The most frequent *Glu-B3* alleles in advanced lines belonged to *a, i* and *d* alleles with 35%, 26% and 21% respectively. The results provided useful information for breeding program to improve breadmaking quality and develop new cultivars.
Key words: *Glu-1, Glu-B3*, Wheat.

Introduction

Wheat (*Triticum aestivum* L.) quality is mainly determined by the seed storage proteins in the grain's endosperm (Shewry and Halford 2002; Peymanpour *et al.* 2012; Majzoobi *et al.* 2012). These proteins composed of two major fractions, gliadin and glutenin that play a main role in rheological properties of bread wheat dough. Glutenins principally consist of two types of subunits: high-molecular-weight glutenin subunits (HMW-GS) and low-molecular-weight glutenin subunits (LMW-GS), which are cross-linked to form glutenin polymer by intermolecular disulphide bonds. The HMW-GS depict nearly 10% of the seed storage proteins. LMW-GS depict about one-third of seed stor-

age proteins of the total grain and also 60% of glutenins (Luo *et al.* 2001; Azizi *et al.* 2006). HMW-GS are encoded by the *Glu-A1*, *Glu-B1* and *Glu-D1locithat* are located on the distal of the long arm of wheat chromosomes 1A, 1B and 1D, respectively (Payne *et al.* 1987). While the LMW glutenin subunits (LMW-GS) encoded by the *Glu-A3*, *Glu-B3* and Glu- D3loci, are located on the distal of the short arm of these chromosomes (Wang *et al.* 2009). HMW-GS include less number of subunits and extensive studies have been done on them, while LMWGS include a larger number of polypeptides and their relationship to grain processing quality have not been studied to the same degree as for the HMW-GS yet (D'Ovidio and Masci

2004). Although the role of HMW-GS on bread making quality (Dough strength) was recognized obviously, LMW glutenin subunits also play a significant role on dough viscosity and formation of large polymers. It is found that some allelic forms of LMW-GS have greater effects on the quality than HMW-GS (Gupta *et al.* 1991; Luo *et al.* 2001).Studies have shown that the allelic variation of both HMW-GS and LMWGS are associated with the rheological properties of wheat flour (Payne *et al.* 1987). Most types of LMW-GS are in group B in which the *Glu-B3* locusis located between *Gli-B1* locus (at 2cM) and the centromere (Wang *et al.* 2009). Afterwards, six protein alleles were found for the *Glu-A3* locus, nine for the *Glu-B3* locus and five for the *Glu-D3* locus. Dough strength or Rmax (maximum dough resistance), is mostly controlled by *Glu-3*loci with the following ranking order: $i > b = a > e = f = g = h > c$ (Gupta *et al.* 1991). To identify different HMW-GS and LMW-GS subunits, sodium dodecyl sulphate-polyacrylamide gel electrophoresis (SDS-PAGE) or reversed phase high performance liquid chromatography (RP-HPLC) (Singh *et al.* 1991) were used in the past, while nowadays to overcome some conflicts on allelic detection of glutenin subunits, the development of specific PCR-based markers are used (Xu *et al.* 2008; Wang *et al.* 2009). Different studies showed the usefulness of marker assisted selection for identification of variant ω- gliadin genes (Chen *et al.* 2011), allelic variants encoded at the *Glu-D3* locus (Appelbee *et al.* 2009), allelic variation at the *Glu-A3* and *Glu-B3* loci (Zhang *et al.* 2004; Si *et al.* 2012), allelic variation at the *Glu-1* and *Glu-3* loci and the presence of 1B-1R translocation lines (Liu *et al.* 2005). Therefore an important goal in wheat quality improvement is the identification of specific HMW-GS and LMW-GS alleles (Gupta *et al.* 1991; Gale 2005). It has been recognized that three alleles comprise 5+10, 2* and 17+18 which are allelic variants, are connected with good quality characteristics in bread wheat cultivars and lines (Payne *et al.* 1987; Mohammadi *et al.* 2013). Studies by Uthayakumaran et al (2006) and Kuchel et al (2007) showed that the marker assisted selection (MAS) for HMW-GS and LMW-GSusing sequence tagged

sites (STS) DNA markers can speed up breeding programs. In this study we used STS-PCR method (Abdel-Mawgood 2008; Izadi- Darbandi and Yazdi-Samadi 2012; Goutam *et al.* 2013; Mohammadi *et al.* 2013) for screening of *Glu-1* and *Glu-B3* loci in wheat genotypes.

Materialsand Methods

Materials: Thirty-four advanced wheat lines and 26 hexaploid wheat cultivars originated from CIMMYT, Iran, Canada, Australia, USA, Turkey, France and Russia, which is kindly provided by the GenBank of Seed and Plant Improvement Institute of Iran were investigated in this study (Table 1).

SDS-PAGE analyses: For the glutenin extraction, the sequential extraction procedure described by Singh *et al.* (1991) was employed.A gel system consisting of two layers i.e., stacking and gradient acrylamide separating layers was used. A linear gradient acrylamide gel from 8.1% to 12.5% with 1% crosslinker concentration (bisacrylami - de/acrylamide ratio) allowed better visualization of HMW and LMW glutenin subunits (Izadi-Darbandi *et al.* 2010).). Payne nomenclature system (Payne *et al.*1988) was applied to detect HMWGS.

DNA extraction: Genomic DNA was isolated from fresh seedling leaves by modified CTAB procedure as reported by Murray and Thompson (1980).

PCR amplification: Polymerase chain reaction (PCR) was done on some wheat genotypes based on their protein banding patterns to validate their correspondence specific DNA markers. Then nine genotypes containing our interesting alleles at protein level was selected. Ten primer sets (Table 2) were used to amplify different *Glu-B3* alleles based on the detected SNPs (Wang et al 2009). The complete information for primers and their correspondence allele-specific PCR product are shown in Table 2. Each PCR reaction was performed in $25\mu l$ final volume, consisting of 1U Taq DNA polymerase, $2.5\mu l$ PCR buffer, 1.5 mM MgCl2 and 2.5mM of each dNTP, 0.4pmol of each primer and 50ng genomic DNA. PCR amplifications of *1Ax2**, *1Bx17*, *1Dx5* and*Glu-B3*loci within germplasm were tested by using primers and protocols reported in previous researches

Table 1. Advance lines evaluated using SDS-PAGE and allele-specific PCR marker with a genome score based on Payne (1988).

No. advance lines	Advance Line Pedigree	Origin	Analysis allele-specific PCR marker				Analysis SDS-PAGE Protein			
			Glu-A1b	HMW-GS Glu-B1i	Glu-D1d	LMW-GS Glu-B3	Glu-A1	HMW-GS Glu-B1	Glu-D1	quality score
1	SOOTY-9/RASCON-37	CIMMYT	2*	17+18	*	C	2*	17+18	*	6
2	AUK/GUIL//Green	CIMMYT	Non-2*	17+18	*	d	null	17+18	*	4
3	RASCON_37/BEJAH-7	CIMMYT	2*	Non-17+18	*	d	2*	6+8	*	2
4	ZAGBA_2/BICHENA	CIMMYT	Non-2*	17+18	*	d	null	17+18	*	4
5	CNDO/PRIMADUR//HAI-OU_17/3/SNITAN	CIMMYT	2*	17+18	*	d	2*	17+18	*	6
6	SULA/AAZ_5//CHEN/ALTAR84/3/AJAIA_12/F3LO	CIMMYT	Non-2*	Non-17+18	*	d	null	7+8	*	4
7	Seri82//Shuha"S"/4/Rbs/Anza/3/Kvz/Hys//Ymg/Tob	CIMMYT	Non-2*	17+18	5+10	i	1	17+18	5+10	10
8	Alvd//Aldan"s"/las58/4/kal/Bb/cj"s"/3/Hork"s"	Iran	Non-2*	Non-17+18	Non-5+10	f	1	7+8	2+12	6
9	1-66-22/5/1-66-31/4/Anza/3/Pi/Nar/Hys/6/M-75-7	Iran	Non-2*	Non-17+18	5+10	a	1	7+8	5+10	10
10	1-66-22/5/1-66-31/4/Anza/3/Pi/Nar/Hys/6/M-75-7	Iran	Non-2*	17+18	Non-5+10	e	1	17+18	2+12	8
11	Hereward/Siren/5/Gov/Az//Mus/3/DoDo/4/Bow	Iran	Non-2*	Non-17+18	Non-5+10	e	null	7+8	2+12	6
12	KAUZ//ALTAR84/AOS/3/PASTOR	CIMMYT	Non-2*	Non-17+18	5+10	a	1	7+9	5+10	9
13	PRL/2*PASTOR	CIMMYT	Non-2*	Non-17+18	5+10	d	1	7+9	5+10	9
14	KAUZ/PASTOR	CIMMYT	Non-2*	Non-17+18	5+10	a	1	7+9	5+10	9
15	Alvd//Nanjing8343/Kauz	Iran	Non-2*	Non-17+18	5+10	a	1	7+8	5+10	10
16	Alvad//Aldan/las/3/Rsh	Iran	2*	Non-17+18	5+10	i	2*	7+9	5+10	9
17	Alvd//Nanjing8343/Kauz	Iran	2*	17+18	5+10	d	2*	17+18	5+10	10
18	PBW343*2/CHAPLO	CIMMYT	Non-2*	Non-17+18	5+10	a	1	7	5+10	8
19	PBW343*2/KUKUNA	CIMMYT	Non-2*	17+18	5+10	a	1	17+18	5+10	10
20	CHAPLO	CIMMYT	2*	Non-17+18	Non-5+10	a	2*	20	2+12	6
21	Alvd//Aldan"s"/las58/4/kal/Bb/cj"s"/3/Hork"s"	Iran	Non-2*	17+18	5+10	f	null	17+18	5+10	8
22	GF-gy54/Attila	Iran	Non-2*	Non-17+18	Non-5+10	a	null	7+8	2+12	6
23	Sakha8/Darab#2//1-66-22	Iran	2*	17+18	5+10	i	2*	17+18	5+10	10
24	GV/D630//Ald"s"/3/Azd/4/1-75-104	Iran	Non-2*	Non-17+18	Non-5+10	a	1	7+9	2+12	7
25	Hmd//1-66-22//Inia	Iran	Non-2*	Non-17+18	Non-5+10	i	null	7+8	2+12	6
26	Hmd//1-66-22//Inia	Iran	Non-2*	Non-17+18	Non-5+10	i	null	7+8	2+12	6
27	1-66-22/3/Alvd//Aldan/Los	Iran	Non-2*	Non-17+18	5+10	a	null	7+8	5+10	8
28	Desprez80/Rsh//1-66-22/Inia	Iran	Non-2*	Non-17+18	5+10	i	null	7+8	5+10	8
29	v82.187/1-66-22/5/Kvz/cgn/4/Hys/Drc*z/7c/3/2*Rsh	Iran	2*	Non-17+18	5+10	i	2*	7+9	5+10	9
30	snb"s"//Emu"s"/Tjb84-1543/3/kauz/stm	Iran	2*	Non-17+18	Non-5+10	b	2*	7+8	2+12	8
31	Alborz/5/K62909/4/Cno//k58/Tob/3/wa/5/ehen	Iran	2*	Non-17+18	Non-5+10	i	null	7+8	2+12	8
32	Kau2*2/Opata//kauz/3/sakha8/4/kauz/srkhtm	Iran	Non-2*	17+18	Non-5+10	i	null	17+18	2+12	6
33	GV/D630//Ald"s"/3/Azd/4/Flt	Iran	Non-2*	17+18	5+10	a	null	17+18	5+10	8
34	1-66-22/passarinho/3/Vee/Nac//1-66-22	Iran	Non-2*	Non-17+18	Non-5+10	a	null	7+8	2+12	6

Table 1. continued.

Cultivar	Pedigree	Country								
Ac-Barrie	Neepawa/Columbus//BW90	Canada	2*	Non-17+18	5+10	h	2*	7+8	5+10	10
Inia	SUMAI-2[1747]; FUNO/TAIWANMAI[2959];	CIMMYT	Non-2*	Non-17+18	5+10	d	1	13+16	5+10	10
Ac-Crystal	HY377/L8474-D1	Canada	Non-2*	Non-17+18	5+10	i	1	7+8	5+10	10
argentine	LP-585-67/KLEIBER[667][851][1258][2846];	USA	Non-2*	Non-17+18	Non-5+10	e	null	7	2+12	4
aqoa	(S)CRIMEAN[37][39][1111];	Turkey	Non-2*	Non-17+18	Non-5+10	b	1	7+8	2+12	8
bezostaya	Import cultivar32338	Russia	2*	Non-17+18	5+10	d	2*	7+9	5+10	9
Cadet	(S)BEZOSTAYA-4[37][80][104][10][11];	Canada	2*	Non-17+18	5+10	b	2*	7+8	5+10	10
Aroona	MERIT/THATCHER[37][616][39][1111];	Australia	Non-2*	Non-17+18	Non-5+10	b	1	7+9	2+12	7
Tobari-66	WW-15/RAVEN[113][626][851];	CIMMYT	Non-2*	Non-17+18	Non-5+10	i	null	7+8	2+12	6
panjamoo	Tzpp/Sn64A	CIMMYT	2*	Non-17+18	Non-5+10	f	2*	7+8	2+12	8
soisson	FKN/N10B	France	2*	Non-17+18	5+10	b	2*	7+8	5+10	10
Glenlea	IENA(JENA)/(HYBRIDE-NATUREL).HN-35[1346][1413][1665][1764][1790][2845];	Australia	2*	Non-17+18	Non-5+10	g	2*	7	2+12	6
Katepwa	UM-530//(MEX)CB-100[39][1323][2331];	Canada	2*	Non-17+18	5+10	h	2*	7+9	5+10	9
Norstar	NEEPAWA*6/RL-2938/3/NEEPAWA*6//CI-8154/2*FROCOR;	Canada	Non-2*	Non-17+18	5+10	b	1	7+8	5+10	10
Ac-Vista	HY44/ 7915QX76B2)LOSPROUT//HY358*3/BW553;	Canada	Non-2*	Non-17+18	Non-5+10	i	1	7+8	2+12	8
Celtic	ANGUS/LEN[1318];	Canada	2*	Non-17+18	5+10	b	2*	7+9	5+10	9
Ac-Reed	ANGUS/LEN[1318];	Canada	2*	Non-17+18	Non-5+10	b	2*	20	2+12	6
Ac-Foremost	HY-320*5/BW-553//HY-320*6/7424BW-5-B-4[1323];	Canada	Non-2*	Non-17+18	Non-5+10	i	1	7+8	2+12	8
Ac-Taber	HY-320*3/BW-553[1125][1315][1323][113];	Canada	2*	Non-17+18	5+10	i	2*	7+9	5+10	9
Hope	YAROSLAV-EMMER/MARQUIS[47][1102][1111]; VERNAL-EMMER(TR.DN)/MARQUIS[39];	USA	2*	Non-17+18	Non-5+10	b	2*	6+8	2+12	6
Marquis	HARD-RED-CALCUTTA/RED-FIFE[10][201][1446];	Canada	Non-2*	Non-17+18	5+10	b	1	7+9	5+10	9
Anza	LERMA-ROJO-64//NORIN-10/BREVOR/3/3*ANDES-ENANO;	USA	Non-2*	Non-17+18	Non-5+10	b	1	7+8	2+12	7
Selkirk	MCMURACHY/EXCHANGE//3*REDMAN,CAN[39]	Canada	Non-2*	Non-17+18	5+10	g	1	6+8	5+10	8
Laura	BW-15/BW-517[229][592][1318][1323][113];	Canada	Non-2*	Non-17+18	5+10	h	1	7+8	5+10	10
Pasqua	BW-63*2/COLUMBUS[1323][1429];	Canada	2*	Non-17+18	5+10	g	2*	7+9	5+10	9
Ac-Eatonia	LEADER/LANCER[1323][1411];	Canada	2*	Non-17+18	5+10	h	2*	7+8	5+10	10

*: Don't have D genome and advance lines are durum.

(Ma *et al.* 2003, Vjell, 1998 and Wang *et al.* 2009). PCR reactions for *Glu-1* alleles were performed using an initial denaturing step at 94°C for 2min, followed by 35 cycles of denaturation at 94°C for 45s, annealing at [51-58]°C for 45s, an extension at 72°C for 90s. The PCR conditions for *Glu-B3* allele-specific markers consisted of an initial denaturing step at 94°C for 5min, followed by 35 cycles of denaturation at 94°C for 45s, annealing at [55-62]°C for 45s, an extension at 72°C for 90s. All PCR reactions were terminated by a final extension at 72°C for 10min. The purity of studied advanced lines was checked for 1AL-1RS and 1BL- 1RSwheat-rye translocation lines that will present 1RS alleles byusing one pair of specific primers as: Forward:5'- TGACAACCCCCTTTCCCTC-

GT-3'and Reverse:5'- TCATCGACGCTAAG-GA - GGACCC-3' (Saal and Wricke 1999).

A touch down PCR condition with an initial denaturation at 94C° for 3min, followed by 40 cycles, 15 cycles of which were performed for 45s at 94C°, 40s at 65C°, 40s at 72C° and were reduced by 1C° per cycle in annealing temperature, at the end, the PCR was continued by 25 cycles of 94C° for 45s, 40s at 50C°, 40s at 72C° and then final extension of 72C° for 5min. PCR products (10μl) were separa-ted by electrophoresis on 1.5% Agarose gels using TAE buffer and then stained with Ethidium bromide (0.5 $\mu g/ml$ final concentration) before being visualized under UV light.

Table 2. Allele-specific molecular markers and PCR conditions used in this study.

Name Marker	Sequence (5'-3')	Allele	Product Size (bp)	Annealing Temperature (°C)
gluB3a	F: CACAAGCATCAAAACCAAGA R: TGGCACACTAGTGGTGGTC	a	1095	55
gluB3b	F: ATCAGGTGTAAAAGTGATAG R: TGCTACATCGACATATCCA	b	1570	56
gluB3c	F: CAAATGTTGCAGCAGAGA R: CATATCCATCGACTAAACAAA	c	472	56
gluB3d	F: CACCATGAAGACCTTCCTCA R: GTTGTTGCAGTAGAACTGGA	d	662	58
gluB3fg	F: TATAGCTAGTGCAACCTACCAT R: CAACTACTCTGCCACAACG	fg	812	62
gluB3g	F: CCAAGAAATACTAGTTAACACTAGTC R: GTTGGGGTTGGGAAACA	g	853	60
gluB3h	F: CCACCACAACAAACATTAA R: GTGGTGGTTCTATACAACGA	h	1022	60
gluB3i	F: TATAGCTAGTGCAACCTACCAT R: TGGTTGTTGCGGTATAATTT	i	621	58
gluB3bef	F: GCATCAACAACAAATAGTACTAGAA R: GGCGGGTCACACATGACA	bef	750	60
gluB3e	F: GACCTTCCTCATCTTCGCA R: GCAAGACTTTGTGGCATT	e	669	58
Ax2*	F: ATGACTAAGCGGTTGGTTCTT R: ACCTTGCTCCCCTTGTCTTT	2*	1319	53
Bx17	F: CGCAACAGCCAGGACAATT R: AGAGTTCTATCACTGCCTGGT	17+18	669	58
Dx5	F: GCCTAGCAACCTTCACAATC R: GAAACCTGCTGCGGACAAG	5+10	450	51

Results

HMW-GS identification by SDS-PAGE: In order to confirm the STS-PCR markers, protein

allelic patterns of advanced lines, containing all of our target alleles were detected through SDS-PAGE system for *1AX2**, *1BX17* and

1DX5 alleles. The presence of each specific PCR band was confirmed by the expression of desired alleles at protein level (Figure 1).

PCR analysis: Specific primers (Table 2) for

2*, 17+18, and 5+10 alleles in wheat advanced lines produced 1319 bp, 669 bp and 450 bp fragments, respectively (Figure 2a, Figure 2b, Figure 2c).

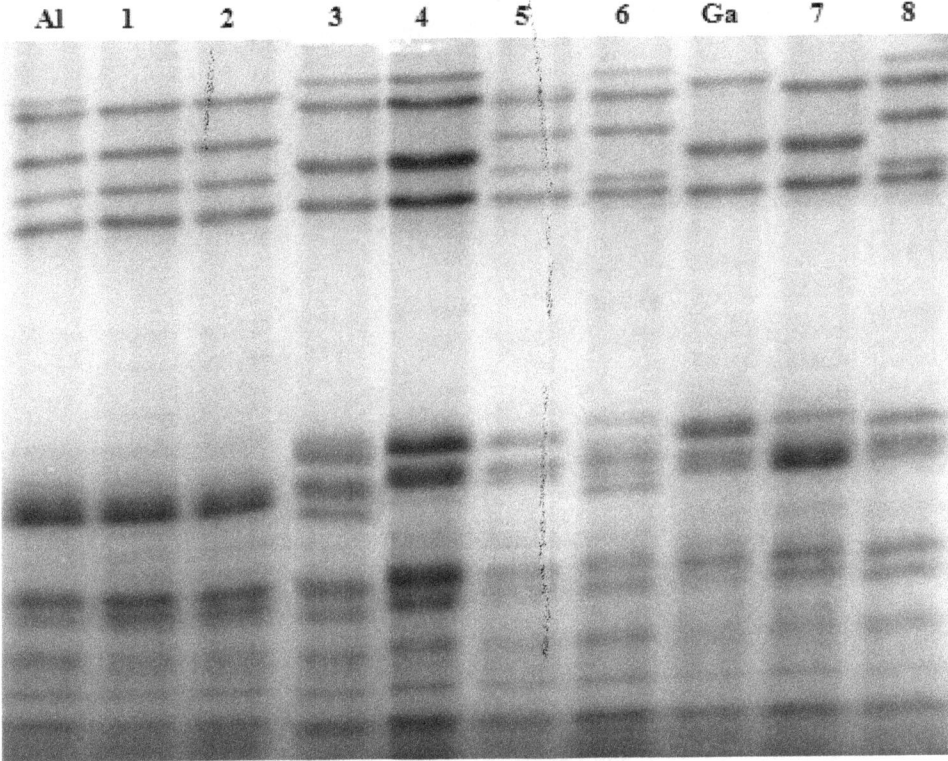

Figure 1. SDS-PAGE patterns of glutenin subunits of wheat advanced lines.1 no. 34, 2 no. 25, 3 no. 19, 4 no. 7, 5 no. 28, 6 no. 13, 7 no. 33, 8 no. 14 in Table 1. Gabo (Gb) and Alvand (Al) were used for the identification of their banding patterns.

Among 34 wheat advanced lines, only a 1,095-bp PCR fragment was amplified in three advanced lines (No.9, 19, and 33) using specific primer set for *Glu-B3a* allele (Figure 3a). One PCR fragment as 1,549-bp was detected for the *Glu-B3b* allele in advanced lines (Figure 3b). For the *Glu-B3c* allele in one line (No 1), a unique 472 bp PCR product was generated (Figure 3c). Primer set *Glu- B3d* was used to identify the d allele in seven lines, producing a 662-bp band (Figure 3d). For advanced lines containing the e allele (No.10 and 11 of 34), a specific 669-bp PCR product was generated using the primer set *Glu-B3e* (Figure 3e). Primer set *Glu-B3i* Specifically amplified a 621 bp PCR fragment in 9 lines (Figure 3h). Since it

was difficult to design a specific primer set for *Glu-B3f*, primer set *Glu-B3fg* was used to amplify *Glu-B3f* and *Glu-B3g* in two advanced lines (Figure 3f). In combination with *Glu-B3f*, this primer set can be used to identify *f* allele. Additionally, primer set *Glu-B3b*ef was used to amplify *Glu-B3b*, *e* and *f* in 5 advanced lines producing a 750-bp band (Figure 3g). This set can be used to verify the former primer sets. The *Glu- B3g*primer set was used to detect the g allele in lines with a 853-bp PCR fragment. Primer set *Glu-B3h*, which generated a 1022bp band in lines, was used to discriminate the h allele from others. The results of Table 1 indicates that none of these 34 advanced lines showed the *Glu-B3g* and h alleles that were present in commercial cultivars. The absent of specific PCR products attributed to the 1RS of Rye representing their purity of advanced lines without any translocation.

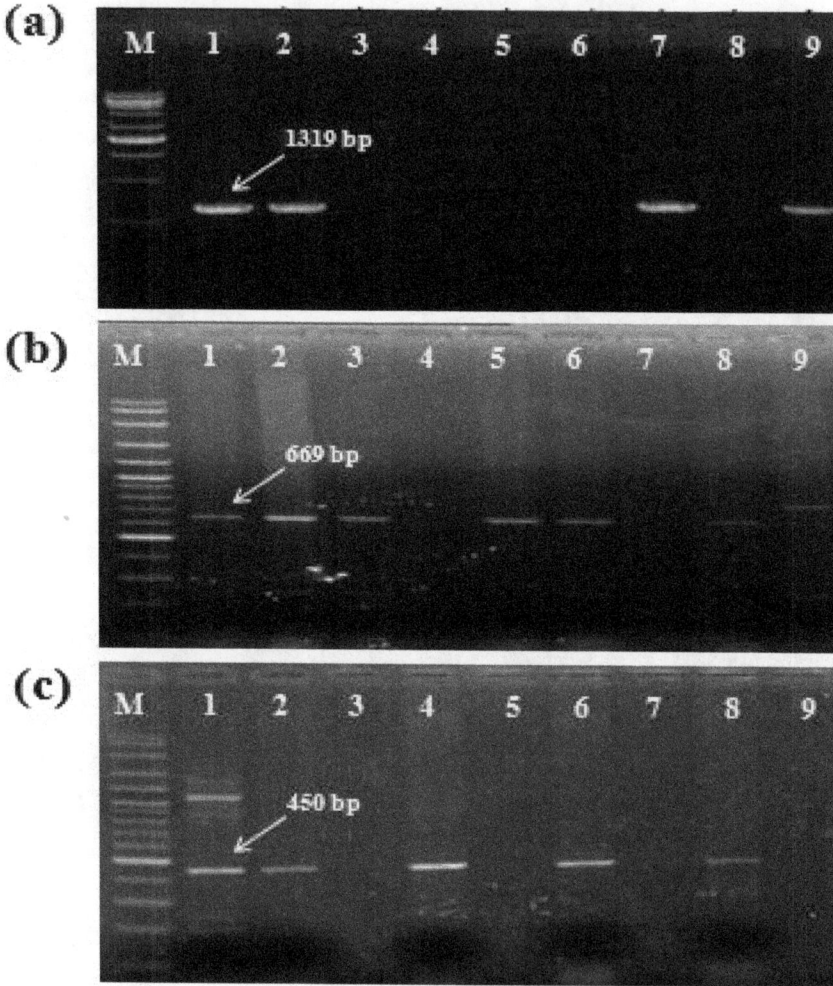

Figure 2. Electrophoresis of PCR products of three gene-specific primer sets on agarose gels. Wheat are used as PCR templates are as listed in Table 1 as lanes 1-9: numbes;.23, 17, 10, 9, 21, 19, 30, 33 and 1. (a) Primer Ax2* for the 2*allele. M, 1kb DNA ladder (b) primer Bx17 for the 17+18 allele M, 100 bp DNA ladder (c) primer Dx5 for the 5+10 allele M, 100 bp DNA ladder.

Discussion

We dissected the allelic variation of *Glu- 1* and *Glu-B3* glutenin loci in 34 wheat advanced lines and 26 cultivars. In these study allelic variations at loci were identified using SDS-PAGE and were nominated by using Gabo and Alvand standard genotypes based on their known banding patterns (Izadi-Darbandi *et al.* 2010). The advantages of PCR-based assay compared with SDS-PAGE for selecting HMW-GS and LMW-GS alleles have been reported previously by Abdel-Mawgood (2008) and Wang *et al.* (2009) respectively. In this study the exact and fine looking of HMW-GS composition at *Glu-A1*b (2*), *Glu-B1i* (17+18) and *Glu-D1d* (5+10) loci and 9 alleles of *Glu-B3* locus from LMW-GS were validated by using a set of specific applied primers. DNA markers that had been used in this study were able to identify common HMW-GS alleles with a high quality ranking and *Glu-B3* alleles in wheat breeding programs Wheat dough propertiesis related to its rheology as well as its maximum resistance. Research has shown that HMW-GS components increase the dough strength (Gupta *et al.* 1991). While LMW-GS components, in comparison with HMWGS components, have more important role in maximizing the dough elasticity (Gupta *et al.* 1991). It has been proven that by increasing the amount of protein, those LMW-GS with more subunits are screened as a result of their higher effects on dough and this has been accepted as a general strategy (Payne *et al.* 1987). Among the LMW-GS, *Glu-B3* locus has the highest allelic diversity and major amount related to the *Glu-D3* and *Glu-A3* loci. Thus, allelic variation in loci associated with this group of chromosomes can be helpful in identi - fying varieties and their phylogenetic relationship (Long *et al.* 2005). Although stud-

ies of LMW-GS proteins have several difficulties, such as gliadin contamination and co-migration on gel, however, Singh *et al.* (1991) proposed the sequential extraction methods and the use of gradient gels for this purpose. The overall implementation of this method is, however, time consuming and hard working. Moreover, due to the large number of bands, the analysis is very difficult and sometimes wrong, though STS-PCR markers have solved such problems and are very informative. The results showed that only 38% of the studied genotypes had 2* alleles and the remaining showed; 1 or null alleles (Figure 4a). 20% of wheat genotypes had the 17+18 allele and 80% of the remaining showed other related alleles at *Glu-B1* locus. In the case of the 5+10 allele, only 52% of the studied advanced lines showed this allele

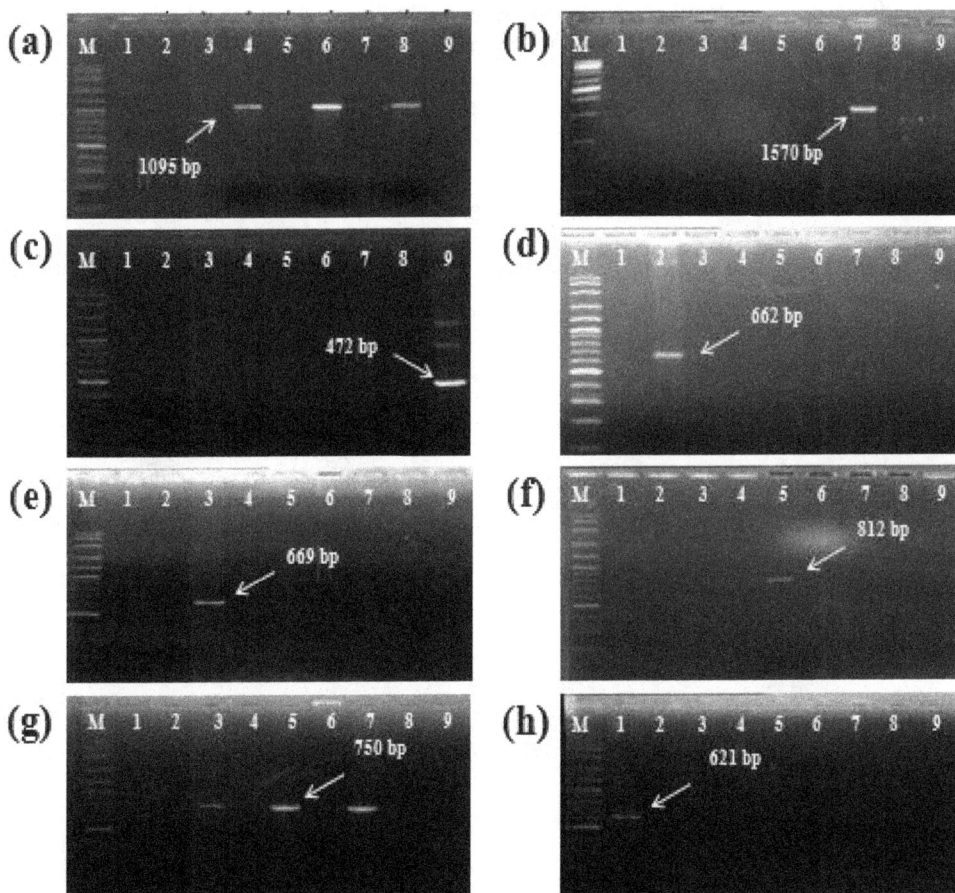

Figure 3. Electrophoresis of ten allele-specific PCR markers for genotyping Glu-B3 alleles in nine wheat advance lines on agarose gels. (a) glu-B3a, (b) glu-B3b, (c) glu-B3c, (d) glu-B3d, (e) glu-B3e, (f) glu-B3fg, (g) glu-B3bef and (h) glu-B3i. Wheat are used as PCR templates were as lanes 1-9 as listed in Table 1: numbers:.23, 17, 10, 9, 21, 19, 30, 33 and1M, 1kb and 100 bp DNA ladder.

(Figure 4a). The results of PCR with the specific primers for *Glu-B3* locus, confirmed the existence of these nine alleles, *i* > *a* > *b* >*d* > *h* > *g* = *f* = *e* > consequently with 24%, 21%, 20%, 12%, 6%, 5%, 5%, 5% and 2% of allelic composition. Allelic frequency at *Glu-B3* loci, showed that there are a bimodal or trimodal distribution in our studied cultivars. Results showed that the alleles of i and a are more frequent in Iranian genotypes and *b, g* and halleles in this loci were got higher frequency in Canadian genotypes. howeverthe *d* allele was the most frequent in CIMMYT genotypes.

The most frequent *Glu-B3* alleles in advanced lines of breeding programs of Iran belonged to a, i and d alleles with 35%, 26% and 21%, respectively. This type of distribution and a high percentage of alleles *a, i, b* and *d* in the studied advanced lines and cultivars are consistent with other results (Izadi- Darbandi *et al.* 2010) (Figure 4a). Gupta *et al.* (1991) were ranked the quality of *Glu-B3* alleles as following order: *i* > *b* = *a* > *e* = *f* = *g* = *h* > *c*. and therefore existing

of the *a* and *i* as the frequent alleles with high ranking on breadmaking quality in studied advanced lines is expected to show their good potential in breeding programs. The existence of *d* allele in our advanced lines is also expected to show their potential for being used as noodle consumes (Gale *et al.* 2005). Wang *et al.* (2009) reported that alleles *Glu-B3d* and *Glu-A3*d are better application for Chinese noodles. Wheat cultivars and advanced lines showed the highest frequency (38%) for 2* or b allele at *Glu-A1*locus. At the *Glu-B1*locus, the highest frequency was detected for the 7+8 allele with 43%, 7+9 allele with 22% and 17+18 allele with 20%. The same frequencies were reported at *Glu-A1* locus in Australian wheat and existing of the null allele (c) represents poor quality in studied genotypes (Gupta *et al.* 1991). Allele 1By was Null in the line No. 18 and cultivars of Argentina, Glenlea containing 1Bx (7) allele. Furthermore line No. 20 and AcReed cultivar expressed 1Bx (20) alone without their 1By linked subunits (Figure 4b). Studies showed

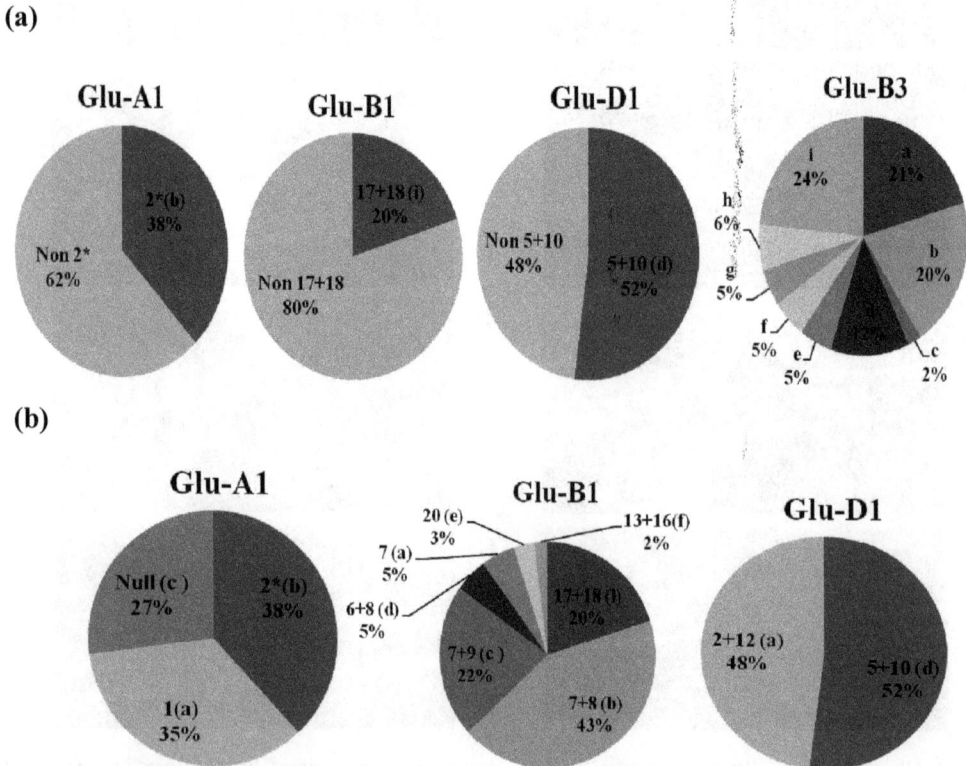

(a)

(b)

Figure 4. (a) Allelic frequency distributions for *Glu-1* and *Glu-B3* loci are shown in the DNA level (b) Allelic frequency distributions for *Glu-A1, Glu-B1*and *Glu-D1* loci are shown by SDS-PAGE.

that in common wheat, 1Bx, 1Dx and 1Dy alleles are always expressed, while 1Ax and 1By are not always expressed. 1Ay subunit mostly is off in hexaploid wheats whereas it is usually expressed in tetraploid and diploid (Jiang *et al.* 2009; Izadi-Darbandi *et al.* 2010).

Payne *et al.* (1988) showed that the presence of subunits 7+8 compared with subunits 7+9 which is coded at *Glu-B1* locus is associated with higher dough strength. At Glu-D1 locus, the highest frequent allele was 5+10 (d) allele with 52% frequency which is among the most valuable alleles in the bakery. The frequency of this allele was lower than 2+12 in Iranian bread wheat cultivars Figure 4b). In fact, the majority of Iranian wheat cultivars showed null allele at locus *Glu-A1*, thus it can be mentioned that they have a lower rating quality (Payne *et al.* 1987). As it can be seen, only 23% of the studied advanced lines got good quality score about 40% of lines were well worth to the bakery because their genome scores were around 8-9 and 37% of lines are considered weak with genome scores 4-7 (Table 1). In various studies, the positive effect of subunits 7+8 and 5+10 on the baking quality has been reported. The frequencies of these two subunits in the studied lines were higher than other alleles which indicate their high bread making quality. Alleles 2+12 and 5+10 at Glu-D1 locus, respectively accompany with the weakest and strongest impact on the brad making quality. Screening for some useful alleles such as 5+10 is highly recommended in Iranian wheat breeding programs. However, screening at early generation of breeding program can increase the frequency of suitable alleles lead to improving bread making quality. Nine *Glu-B3* and three *Glu-1* allele-specific markers with high effects on bakery usage were validated in 60 wheat advanced lines and cultivars (Table 1). The results of allelic variation at *Glu-B3* and *Glu-1* loci obtained by PCR based markers was quite verified with those of detected by SDS-PAGE. The overall results of this study showed that the protein mobility alleles determined by SDS-PAGE were consistent with the screening results obtained using the allele specific markers in 60 wheat advanced lines and cultivars. SDSPAGE is one method for identification of allelic components in qual-

ity scoring of wheat cultivars, but the mobility of subunits in this system does not exactly correspond with the size and sometimes makes the interpretation of banding pattern difficult. However, marker assisted selection can help avoid misinterpretation of results from SD-SPAGE. Wheat quality identification can be done in early stage of growth development without having to wait for seed set and analysis of their glutenin composition. In this study, STS-PCR markers verified for the sequences of three and nine different alleles is located on the *Glu-1* and *Glu-B3* loci respectively. We can use these markers as an alternative method in wheat quality breeding program for detecting poor or good qualities in early stage of growth. Marker assisted selection by DNA markers which have verified in this study can be used for both quality classification and accelerating breeding program for bread-making quality.

Reference

1. Abdel-Mawgood, A.L. 2008. Molecular markers for predicting end-products quality of wheat (*Triticum aestivum* L.). *Afr. J. Biotechnol.*, 7: 2324-2327.

2. Appelbee, M.J., Mekuria, G.T., Nagasandra, V., Bonneau, J.P., Eagles, H.A., East - wood, R.F. and Mather, D.E. 2009. Novel allelic variants encoded at the Glu- D3 locus in bread wheat. *J. Cereal Sci.*, 49: 254-261.

3. Azizi, M.H., Sayeddin, S.M. and Payghambardoost, S.H. 2006. Effect of flour extraction rate on flour composition, dough rheological characteristics and quality of flat bread. *J. Agric. Sci. Technol., 8:* 323-330.

4. Chen, F., Yang, L., Zhao, F., Min, H. and Xia, G. 2011. Molecular cloning and variation of ω-gliadin genes from a somatic hybrid introgression line II-12 and parents (*Triticum aestivum* cv. Jinan 177 and *Agropyron elongatum*). *J. Genet.*, 90: 137-142.

5. D'Ovidio, R. and Masci, S. 2004. The lowmolecular-weight glutenin subunits of wheat gluten. *J Cereal Sci.*, 39:321-339.

6. Gale, K.R. 2005. Diagnostic DNA markers for quality traits in wheat. *J Cereal Sci,* 41:181-192.

7. Goutam, U., Kukreja, S., Tiwari, R., Chaudhury, A., Gupta, R.K., Dholakia, B.B. and Yadav, R. 2013. Biotechnological approaches for grain quality improvement in wheat: Present status and future possibilities. *AJCSAust. J. Crop Sci.*, 7. 469-483.

8. Gupta, R.B., Bekes, F. and Wrigley, C.W. 1991. Prediction of physical dough properties from

glutenin subunit composition in bread wheats. *Cereal Chem.*, 68: 328-333.

9. Izadi-Darbandi, A. and Yazdi-Samadi, B. 2012. Marker assisted selection of highmolecular weight glutenin alleles related to bread-making quality in Iranian common wheat (*Triticum aestivum* L.). J. Genet., 91: 193-198.

10. Izadi-Darbandi, A., Yazdi-Samadi, B., Shahnejat-Boushehri, A.A. and Mohammadi, M. 2010. Allelic variations in *Glu- 1* and *Glu-3*loci of historical and modern Iranian bread wheat (*Triticum aestivum* L.) cultivars. *J. Genet.*, 89: 193-199.

11. Jiang, Q.T., Wei, Y.M., Wang, F., Wang, J.R., Yan, Z.H. and Zheng, Y.L. 2009. Characterization and comparative analysis of HMW glutenin *1Ay*alleles with differential expressions. *BMC Plant Biol.* 6: 9-16.

12. Kuchel, H., Fox, R., Reinheimer, J., Mosi - onek, L., Willey, N., Bariana, H. and Jefferies, S. 2007. The successful application of a markerassisted wheat breeding strategy. *Mol. Breed.* 20: 295- 308.

13. Liu, L., He, Z., Yan, J., Zhang, Y., Xia, X. and Pena, R. 2005. Allelic variation at the Glu-1 and *Glu-3*loci, presence of the 1B. 1R translocation, and their effects on mixographic properties in Chinese bread wheats. *Euphytica.* 142: 197-204.

14. Long, H., Wei, Y. M., Yan, Z.H., Baum, B., Nevo, E. and Zheng, Y.L. 2005. Classification of wheat low-molecular-weight glutenin subunit genes and its chromosome assignment by developing LMWGS groupspecific primers.*Theor. Appl. Genet.*, 111: 1251-1259.

15. Luo, C., Giyn, W. B., Branlard, G. and McNeil, L. 2001. Comparison of low and high molecular-weight wheat glutenin allele effects on your Quality.*Theor. Appl. Genet.*, 102: 1088-1098.

16. Ma, W., Zhang, W. and Gale, K.R. 2003. Multiplex-PCR typing of high molecular weight glutenin alleles in wheat. *Euphytica*, 134: 51-60.

17. Majzoobi, M., Farhoodi, S., Farahnaky, A. and Taghipour, M.J. 2012. Properties of dough and flat bread containing wheat germ.*J. Agric. Sci. Technol.*,14: 1053- 1065.

18. Mohammadi, M., Mehrazar, E., Izadi- Darbandi, A. and Najafian, G. 2013. Marker assisted selection for complementation of *Glu-A1b*, *Glu-B1i*, and *Glu-D1d*alleles in early segregating generations of wheat. *Wheat Inf. Serv.* 116:3- 8. www.shigen.nig.ac.jp/ewis.

19. Murray, M. and Thompson, W.F. 1980. Rapid isolation of high molecular weight plant DNA. *Nucleic Acids Res.*, 8: 4321- 4325.

20. Payne, P.I., Holt, L.M., Krattiger, A.F. and Carrillo, J.M. 1988. Relationship between seed quality characteristics and HMW glutenin subunit composition determined using wheats grown in Spain. *J. Cereal Sci.* 7: 229-235.

21. Payne, P.I., Nightingale, M.A., Krattiger, A.F. and

22. Holt, L.M. 1987. The relationship between HMW glutenin subunit composition and the bread-making quality of british-grown wheat varieties.*J. Sci. Food. Agric.*,40: 51-65.

22. Peymanpour, Gh., Rezaei, K., Sork-hilalehloo, B., Pirayeshfar, B. and Najafian, G. 2012. Changes in rheology and sensory properties of wheat bread with the addition of oat flour.*J. Agr. Sci. Tech.*, 14: 339-348.

23. Saal, B. and Wricke, G. 1999. Development of simple sequence repeat markers in |rye (*Secale Cereale*L). *Genome*, 42: 964-972.

24. Shewry, P.R. and Halford, N.G. 2002. Cereal seed storage proteins: structures, properties and role in grain utilization. *J. Exp. Bot.*, 53: 947-958.

25. Si, H., Gao, Y., Liu, F., Li, Z. and Ma, C. 2012. Distribution of low-molecularweight glutenin subunit *Glu-B3* alleles in mini core collections of Chinese wheat germplasms. *Aus. J. Crop. Sci.* 6: 1390- 1394.

26. Singh, N.K., Shepherd, K.W. and Cornish, G.B. 1991. A simplified SDS-PAGE proce-dure for separating LMW subunits of glutenin.*J. Cereal Sci.*, 14: 203-208.

27. Uthayakumaran, S., Listiohadi, Y., Baratta, M., Batey, I.L. and Wrigley, C.W. 2006. Rapid identification and quantitation of high-molecular-weight glutenin subunits. *J. Cereal Sci.*, 44: 34-39.

28. Vjell, P. 1998. Vyuzitl Genetickych Markeru Pro Tvorbu Dihaploidn Psenice Obecne (*Triticum aestivum* L.). [Disertatacn prace.] CZU, *praha*, 344s.

29. Wang, L.H., Zhao, X.L., He, Z.H., Ma, W., Appels, R., Pena, R.J. and Xia, X.C. 2009. Characterization of low-molecularweight glutenin subunit *Glu-B3*genes and development of STS markers in common wheat (*Triticum aestivum* L.). *Theor Appl Genet*, 118: 525539.

30. Xu, Q., Xu, J., Liu, C.L., Chang, C., Wang, C.P., You, M.S., Li, B.Y. and Liu, G.T. 2008. PCR-based markers for identification of HMW-GS at *Glu-B1x*loci in common wheat.*J. Cereal Sci.*,47: 394-398.

31. Zhang, W., Gianibelli, M.C., Rampling, L.R. and Gale, K.R. 2004. Characterization and marker development for low molecular weight glutenin genes from *Glu-A3* alleles of bread wheat (*Triticum aestivum.* L). *Theor Appl Genet*, 108: 1409-1419.

14

Assessment of genetic diversity and identification of SSR markers associated with grain iron content in Iranian prevalent wheat genotypes

N. Moradi[1*], H. Badakhshan[2], H. Mohammadzadeh[3], M.R Zakeri[3], Gh. Mirzaghaderi2

1. Ph.D student of Agricultural Biotechnology, Genetics and Agricultural Biotechnology Institute of Tabarestan. (GABIT), Sari, Iran.
2. Department of Biotechnology, Faculty of Agriculture, University of Kurdistan, Sanadaj, Iran.
3. Master Degree in Agricultural Biotechnology.
   [*]Corresponding Author, Email: moradinamdar@gmail.com

Abstract

MarkerIron is one of the most important nutrients in the human diet. According to the high consumption of staple foods such as wheat, the deficiency of iron in these crops would lead to nutritional disorders and related complications. To identify microsatellite markers associated with wheat grain iron content,38Iranian prevalent wheat genotypes were assessed using 30 pairs of genomic and EST microsatellite markers. Based on field experiments, significant difference was observed among studied genotypes for grain iron content which ranged from 34-53 mg/Kg. in the molecular experiment, the range of alleles per SSR locus was 2-9 with a mean of 4.5 and the mean of polymorphism information content (PIC) was 0.55. The stepwise regression analysis has been used for estimating the relationship between microsatellite markers and grain iron content. The results indicated that Xwmc617 (4A, 4B, 4D), Xgwm160 (4A) and Xbarc146 (6D,6B,6A) were significantly correlated with wheat grain iron content. The results of this research can be used in further studies and marker assisted breeding of wheat to increase grain iron content.

Keywords: micronutrient, microsatellite marker, stepwise regression, wheat.

Introduction

Micronutrient deficiency and its consequences such as related malnutrition (hidden hunger) is one of the most important issues in the human health, especially in the poor and developing countries. Nearly half of the world's population is suffering from micronutrient deficiency. Unfortunately the main efforts in cereal breeding activities have been focused on increasing the yield without considering their nutritional quality (Bouis, 2003; Tiwari *et al.*, 2009; Buis and Welch, 2010). Among the micronutrients, iron is important due to its key role in fundamental physiological processes, so its deficiency could cause nutritional disorders such as anemia, mental retardation, immune weakness, and even death (Raboy, 2009;

Tiwari *et al.*, 2009). Wheat is one of the most important cereal crops and plays a major role in the world's energy and protein supply. Although limited, increasing iron content in wheat, where have no negative impact on yield performance, play a significant role in reducing malnutrition caused by the lack of this element (Raboy, 2009). Assessment of the genetic diversity and search for the relationship between the existing genotypes and desired traits and consequently identifying the traitpromised genotypes are the first most important steps in improving plants for various purposes, especially in the field of micronutrient content (Chacmak *et al.*, 2004). Among different methods used for evaluating genetic diversity and relationship between the traits and existing diversity, molecular markers have particular importance. Among the molecular markers, microsatellite markers due to many advantages such as codominant effect, vast genome coverage, ease of detection, polymorphism and discrimination power are used in different studies such as, association between markers and traits as well as identification of genes and QTLs (Pirseyedi *et al.*, 2006; Ahmadi and Fotokian, 2011; Tiwari *et al.*, 2009; Peleg *et al.*, 2009).

Identifying the correlation between markers and wheat traits, such as resistance to biotic and abiotic stresses, yield and nutritional quality, is one of the important fields of study in plant breeding (Qi *et al.*, 2010). In several studies using SSR markers on different types of wheat genotypes ranging from wild diploid wheat to tetra and hexaploid genotypes including emmer, durum and bread wheat, several QTLs controlling concentration of micronutrients especially Fe, Zn and protein have been identified on wheat genome (Tiwari *et al.*, 2009; Peleg *et al.*, 2009).

The identification of markers associated with agronomic and morphological traits is another trend for research in molecular marker studies. In a study on evaluation of genetic diversity and identification of markers related to yield and plant height, 23 SSR loci have indicated a significant correlation with aforesaid traits (Wu *et al.*, 2012). In Iran, however several studies have been performed on different plant attributes such as biotic and abiotic stresses using SSR markers but no study have been reported on wheat grain iron content yet.

The objective of the present study was to identify microsatellite markers that illustrate a significant correlation with wheat grain iron content using Iranian prevalent wheat genotypes and consequent identification of the chromosome regions controlling this trait according to the specified chromosomal location of SSR markers.

Material and methods

Plant materials and growth condition

A number of 38 genotypes of prevalent bread wheat cultivars prepared from the cereal part of "Karaj seed and plant improvement institute" Karaj, Iran were used. Field cultivation was performed in the Kurdistan University's research farm during 2010-2011 in a completely randomized block design with three replications of each cultivar and three rows per replication. When completely ripened, middle spikes from the middle row of each plot were randomly picked and kept separately in paper bags.

Grain iron content estimation

Grain iron content was measured using atomic absorption spectrometry Model (SpectrAA220) VARIAN Inc. which includes, the preparation of the meal, sample digestion for preparation of extract, preparation of standards and finally, sample readings. For the preparation of wheat meal, 15 to 20 grams of purified and isolated wheat was milled for 30 to 40 seconds. Extracts and iron standards were prepared according to Singh *et al.*, (1999) with slight alterations.

DNA extraction and PCR amplification

Genomic DNA was extracted from fresh leaves, based on Saghai-Maroof *et al.* (1984) with slight modifications. The quality and quantity of extracted DNA samples were determined using 0.8% agarose gel electrophoresis and spectrophotometry.

To assess the genetic diversity of genotypes, 30 polymorphic genomic microsatellite primer pairs of Xgwm, Xbarc and Xwmc SSR marker types were selected from different chromosomes based on previous studies and their genome coverage.

A ten-microliter volume, polymerase chain reaction was used according to CIMMYT protocol (Warburton, 2005). Thermal cycling consisted of an initial temperature of 94°C for five minutes, 35 denature cycles at 94°C each for 30 seconds, 30 seconds minute at 50 to 67°C for primer annealing, 40 seconds at a temperature of 72°C and the final extension at 72° C for 7 minutes.

Gel electrophoresis and scoring PCR products

PCR products were separated using 6% denature polyacrylamide gel electrophoresis. Polyacrylamide gels were stained with silver nitrate according to ambionet1. Band patterns amplified by SSR markers, were scored according to the marker band positions compared to molecular weight marker, where

1 presented the presence, and 0 the lack of a band.

Data Analysis

Data analysis of grain iron content and analysis of variance for the trait data were performed using SPSS and XLSTAT software. The analysis of data obtained from microsatellite bands was performed with NTSYSpc-2.02, XLSTAT and Excel software. Diversity indices including the number of alleles, major allele frequency, and PIC were calculated for each marker using Excel software. Cluster analysis of relationships between genotypes based on SSR data was performed with Dice similarity coefficient and ward method using NTSYSpc-2.02 and XLSTAT software.

To identify markers associated with the grain iron content and ultimately determine the chromosomal regions associated with the trait, stepwise regression analysis was carried out taking into account the grain Fe content data as the dependent and marker data as independent variables using XLSTAT software.

Results and Discussions

Grain Fe content

After reading the prepared iron standards, extracts prepared from grain samples were read using atomic absorption spectrophotometer, the data were entered in excel software for analyses. The range of iron content in the grains of genotypes was 34- 53mg/kg with a mean and standard deviation of 43.57±5.12.

Table 1. The results of ANOVA on genotypes grain iron content using complete randomized block design.

Source of variation	Degree of freedom	The mean of squares of grain iron content (mg/kg)	F
Block	2	491.2	15.2**
Genotype	39	103.4	3.2**
Experimentalerror	78	32.3	

**significant at < 0.01 CV $= 9.26$ $R^2 = 0.714$

After testing the validity of the assumptions of the statistical model of analysis of variance based on the completely randomized block design, data were analyzed. The results of ANOVA showed a significant block effect at $P<0.01$, confirming the correctness of the experimental design based on the field conditions, it also revealed a significant difference between genotypes for grain iron content at $P<0.01$ (Table 1).

Because of the difference in environmental conditions and also the type of population used in different studies, wheat grain Fe content varies in terms of both range and mean. Monasterio and Graham (2000) estimated the wheat grain iron content in the range of 25 to 73 mg/kg based on a study on 324 selected wheat genotypes in the field conditions. In a study on

the Emmer wheat (*Triticum dicoccoides*), in both greenhouse and field conditions, Cakmak et al.(2004) have estimated the amount of grain iron content for aforesaid conditions in the range of 15 to 94 and 21to 91mg/kg, respectively. In a study on spring and winter wheat cultivars, Morgounov et al. (2007) have reported the range of the grain iron concentration from 65 to 25mg/kg. In another study on 150 hexaploid and 25 tetraploid genotypes performed in field condition in the Europe, Zhao et al. (2009) have reported wheat grain iron content in the range of 28.9-51 mg/kg. The range of Fe concentration of bread wheat in the present study was similar to those reported in earlier researches (Oury et al., 2006; Morgounov et al., 2007; Zhao et al., 2009; Badakhshan et al. 2013). As mentioned above, ANOVA showed

[1]http://www.cimmyt.org/ambionet

highly significant differences between wheat cultivars for grain Fe concentrations. Studies with rice and wheat, and preliminary studies with wild relatives and landraces of wheat have demonstrated a considerable variation in grain Fe concentration (Badakhshan *et al.*, 2013; Genc *et al.*, 2005; Gomez-Becerra *et al.*, 2010a, b).

Analysis of microsatellite data

After running and discriminating the PCR products on denature polyacrylamide gels, band patterns were scored according to the marker band positions based on 1, presence, and 0, lack of a band (Figure 1). In order to verify the results of microsatellite and diversity level in the studied population, microsatellite diversity indicators including allele number, polymorphism information content (PIC) and discriminative power (Tables 2 and 3) were calculated and cluster analysis was carried out using Dice dissimilarity coefficient and Ward method as a result of consistency with the pedigree and the origins of genotypes (Figure 2).

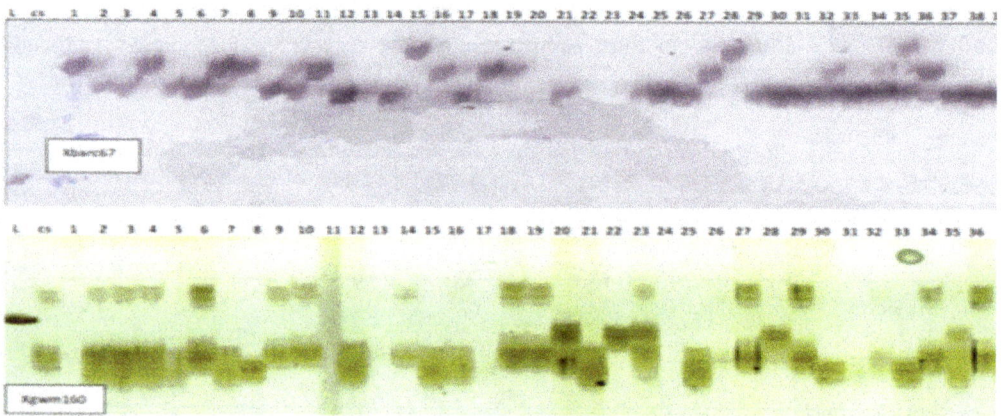

Figure 1. The profile of acrylamide gel band pattern of two SSR markers Xbarc67 (above) and Xgwm160 (below).

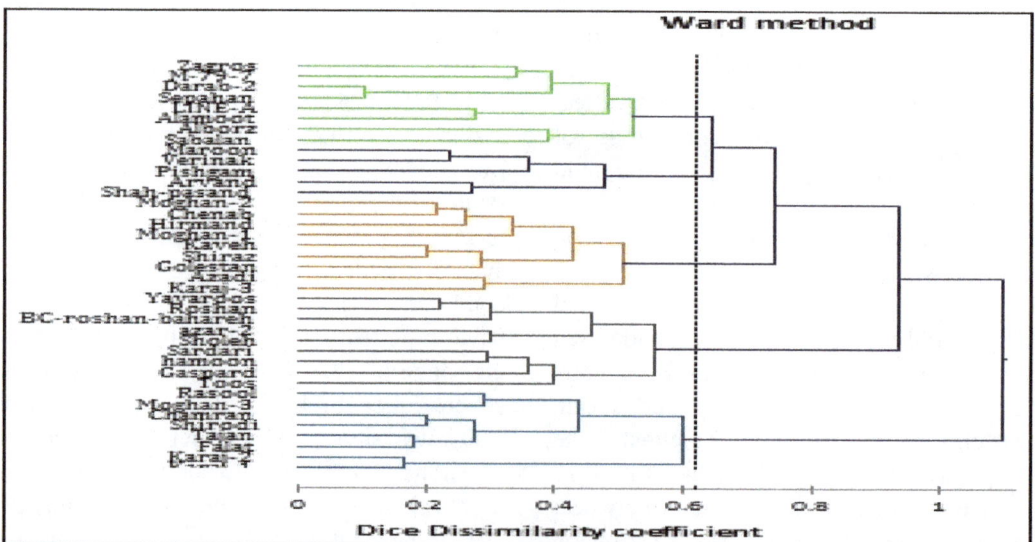

Figure 2. Dendrogram constructed for 38 wheat genotypes using SSR marker.

A number of 2-9 with an average of 4.5 alleles per locus resulted from microsatellite amplification among studied genotypes. The mean of major allele frequency and polymorphic information content (PIC) were 0.56 and 0.55, respectively. Although the main purpose of selecting markers used in the present study was to find genomic regions controlling grain iron content, these markers detected an appropriate level of genetic diversity in the population compared to other studies that aim to study genetic diversity specifically So that, genetic diversity index of three markers Xwmc617, Xbarc146 and Xgwm282 was up to 0.8, (Table 2). Therefore, we can offer these markers to study genetic diversity and kinship in near relatives and to separate such genotypes with high confidence using a fewer number of markers. Several different studies on

genetic diversity were searched and compared together, of which some agreed or closed to the present study. Mohammadi et al. (2008) have assessed the genetic diversity of Iranian wheat cultivars using microsatellites and reported the averages of the number of alleles per locus and gene diversity 8.53 and 0.74, respectively. Lee et al. (2006) with a study on 48 subspecies of T. turgidum in 16 genomic microsatellite loci have identified 96 with an average of 6.1 alleles per locus. In another done on winter wheat, a range of 2-14 with an average of 4.15 alleles per SSR locus and an average of PIC 0.56 have been reported (Wu et al. 2012).

The results of cluster analysis have clustered genotypes into five groups. The relationships among genotypes would also confirm the validity of the obtained results from SSR data.

Correlation analysis between SSR data and grain Fe content

Based on the results of stepwise regression analysis between microsatellite data (independent variable) and the grain Fe content (dependent variable), Xwmc617 (4A- 4B-4D) and Xgwm160 (4A) at the level of $P<0.01$ and Xbarc146 (6D-6B-6A) at the level of $P<0.05$ were significantly correlated to the trait. It have been reported in several studies that diploid and tetraploid species which carry A and B genomes, to be promising containers of grain Fe content and the other microelements correlated to Fe content such as Zn and protein. In a study on recombinant lines resulted from the cross between wild emmer and durum wheat, Peleg et al. (2009) have reported 11, 6 and 10 QTLs associated with grain iron content, zinc content and grain protein concentration respectively. The study also reported that two of the QTLs are located on chromosomes 7B and 4B. The marker Xwmc617 used in the present study carrying one of the amplification sites on the short arm of chromosome 4B, showed the most correlation with the wheat grain Fe content from among three aforementioned ones. In another study on A genome of diploid wheat, Tiwari et al. (2009) identified two QTLs on chromosomes 2A and 7A associated with grain iron content and another on 7A associated with grain zinc content. In another study,

Table 2. Diversity indicators calculated for genomic SSR mark.

Genomic SSRs	Number of alleles	PIC	D
Xbarc29	4	0.47	0.69
Xbarc67	4	0.51	0.69
Xbarc83	4	0.75	0.94
Xbarc146	7	0.81	0.97
Xbarc98	4	0.64	0.78
Xbarc124b	6	0.77	0.88
Xbarc48	3	0.19	0.32
Xgwm3	4	0.59	0.86
Xgwm6	6	0.67	0.93
Xgwm18	3	0.33	0.51
Xgwm149	3	0.63	0.90
Xgwm282	7	0.80	0.96
Xgwm397	3	0.38	0.59
Xgwm400	3	0.61	0.58
Xgwm473	2	0.18	0.31
Xgwm11	2	0.21	0.35
Xgwm46	8	0.77	0.97
Xgwm95	5	0.66	0.92
Xgwm160	4	0.72	0.93
Xgwm219	4	0.69	0.87
Xgwm312	4	0.29	0.52
Xgwm332	6	0.77	0.94
Xgwm368	5	0.31	0.53
Xwmc182	3	0.31	0.51
Xwmc289	4	0.55	0.90
Xwmc617	9	0.84	0.98
Mean	4.5	0.55	0.74
S.D	1.8	0.21	0.23

chromosomes 7A, 6B, 2A and 7B have showed a close relationship with grain protein content (Xu *et al.*, 2008). With a study on hexaploid wheat Genc *et al.* (2009) reported one QTL related to grain iron content on chromosome 3D and four others on chromosomes 3D, 4B, 6B and 7A, related to grain Zn content, note that the marker Xbarc146 in the mentioned study was also nearby the grain Zn content QTL. Using double haploid lines of two hexaploid wheat cultivars, Shi *et al.* (2008) identified seven QTLs on chromosomes 3A, 4A, 2D and 4D related to grain zinc content and concentration.

Table 3. Diversity indicators calculated for EST SSR markers and total mean and standard deviation calculated for genomic +EST SSRs.

EST SSRs	Number of alleles	PIC	D
edm16	2	0.32	0.51
edm28	4	0.5	0.62
edm96	4	0.57	0.9
edm54	3	0.22	0.38
edm80	5	0.73	0.94
Mean	3.6	0.47	0.67
S.D	1.14	0.2	0.24
Total mean (genomic+EST)	4.35	0.54	0.73
Total S.D (genomic+EST)	1.72	0.21	0.22

According to the results of different studies in this case, it has been demonstrated that di and tetraploid wheat species which possess genomes A and B, contain a significantly higher amount of grain Fe and Zn content and also the percent of protein than hexaploid wheat species (Tiwari 2009; Peleg 2009; Genc 2009; Badkhshan *et al.*, 2013). Thus, we can conclude that, chromosomes of genomes A and B have an important role in controlling micronutrients content, such as iron, in wheat grain and the significant correlation between these nutrients in wheat grain, allow the simultaneous improvement of these micronutrients in the wheat grain using traditional breeding techniques with the help of techniques such as marker-assisted selection and also using modern biotechnology techniques.

References

1. Ahmadi J, Fotokian MH (2011) Identification and mapping of quantitative trait loci associated with salinity tolerance in rice (Oryza Sativa) using SSR markers. Iranian Journal of Biotechnology 9:21-30.

2. Badakhshan H, Moradi N, Mohammadzadeh H, Zakeri MR, Mirzaghaderi G (2013)Genetic Variability Analysis of Grains Fe, Zn and Beta-carotene Concentration of Prevalent Wheat Varieties in Iran. International Journal of Agriculture and Crop Sciences 6(2):57- 62.

3. Bouis HE (2003) Micronutrient fortification of plants through plantbreeding: can it improve nutrition in man at low cost? Proceedings of the Nutrition Society 62:403–411.

4. Bouis HE, Welch RM (2010) Biofortification- a sustainable agricultural strategy for reducing micronutrient malnutrition in the global south. Crop Science 50:20-32.

5. Cakmak I, Turun A, Millet E, Feldman M, Fahima T, Korol A, Nevo E, Braun H, Ozkan H (2004) *Triticum dicoccoides*: An important genetic resource for Increasing Zinc and Iron concentration in modern cultivated wheat. Soil Science and Plant Nutrition 50:1047-1054.

6. Genc Y, Humphries JM, Lyons GH, Graham RD (2005) Exploiting genotypic variation in plant nutrient accumulation to alleviate micronutrient deficiency in populations. Journal of Trace Elements in Medicine and Biology 18:319-324.

7. Genc Y, Verbyla AP, Torun AA, Cakmak I, WillsmoreK, Wallwork H, McDonald GK (2009) Quantitative trait loci analysis of zinc efficiency and grain zinc concentration in wheat using whole genome average interval mapping. Plant Soil 314:49–66.

8. Gomez-Becerra HF, Erdem H, Yazici A, Tutus Y, Torun B, Ozturk L, Cakmak I (2010)a. Grain concentrations of protein and mineral nutrients in a large collection of spelt wheat grown under different environments. J Cereal Sci 52: 342-349.

9. Gomez-Becerra HF, Yazici A, Ozturk L, Budak H, Peleg Z, Morgounov A, Fahima T, Saranga Y, Cakmak I(2010)b. Genetic variation and environmental stability of grain mineral nutrient concentrations in *Triticum dicoccoides* under five environments. Euphytica 171:39-52.

10. Monasterio I, Graham RD (2000) Breeding for trace minerals in wheat. Food and Nutrition Bulletin 21:392–396.

11. Morgounov A, Gomez-Becerra H, Abugalieva A, Dzhunusova M, Yessimbekova M, Muminjanov H, Zelenskiy Y, Ozturk L, Cakmak I (2007) Iron and zinc grain density in common wheat grown in Central Asia. Euphytica 155:193-203.

12. Oury FX, Leenhardt F, Remesy C, Chanliaud E, Duperrier B, Balfourier F, Charmet G

(2006) Genetic variability and stability of grain magnesium, zinc and iron concentrations in bread wheat. European Journal of Agronomy 25:177-185.

13. Peleg Z, Cakmak I, Ozturk L, Yazici A, Jun Y, Budak H, Korol AB, Fahima T, Saranga Y (2009) Quantitative trait loci conferring grain mineral nutrient concentrations in durum wheat 3 wild emmer wheat RIL population. Theoretical and Applied Genetics 119: 353-369.

14. Pirseyedi M, Mardi M, Naghavi M, Irandoost HP, Sadeghzadeh D, Mohammadi SA (2006) Evaluation of genetic diversity and identification of Informative markers for morphological characters in"Sardari"derivative wheat lines. Pakistan Journal of Biological Sciences 9:2411-2418.

15. Qi X, Cui F, Li Y, Ding A, Li J, Chen G, Wang H (2010) Molecular tagging wheat powdery mildew resistance gene pm21 by EST-SSR and STS markers. Molecular Plant Breeding 1:1-5. Raboy V (2009) Progress in breeding low phytate crops. Journal of Nutrition 132:503S-505S.

16. Saghai-Maroof MA, SolimanK, Jorgensen RA, Allard RW (1984) Ribosomal DNA spacer-length polymorphisms in barley: Mendelian inheritance, chromosome location and population dynamics. Proceedings of National Academy of Sciences of the United States of America 81:8014–8018.

17. Shi R, LiH, Tong Y, Jing R, Zhang F, Zou C (2008) Identification of quantitative trait locus of zinc and phosphorus density in wheat(*Triticum aestivum* L.) grain. Plant and Soil 306:95-104.

18. Singh D, Chhankar PK, Pandey RN (1999) Soil plant water analysis: a methods manual. New Delhi. IARI (Indian Agricultrul Research Institute):120-121.

19. Tiwari VK, Rawat N, Chhuneja P, Neelam K, Aggarwal R, Randhawa GS (2009) Mapping of quantitative trait loci for grain iron and zinc concentration in diploid A genome wheat. Journal of Heredity 100:771–776.

20. Warburton, M (2005) Laboratory Protocols: CIMMYT Applied Molecular Genetics Laboratory. 3rd Edn. Me´xico, D.F., Mexico: CIMMYT Int. http://ideas.repec.org/p/ags/cimmma/561 84.html.

21. Wu YG, Wu CL, Qin BP, Wang ZL, Huang W, Yang M, Yin YP (2012) Diversity of SSR marker in 175 wheat varieties from Huang-Huai winter wheat region and its association with plant height and yield related traits. Acta Agronomica Sinica 38: 1018-1028.

22. Xu SS, Klindworth DL, Hareland GA, Elias LM, Faris JD, Chao S (2008) Chromosome location and characterization of genes for grain protein content in *Triticum dicoccoides*. 11th Int. Wheat Genet. Symp. vol. 2. Sydney University Press, Sydney, Australia. Pp 572-574.

23. Zhao FJ, Su YH, Dunham SJ, Rakszegi M, Bedo Z ,McGrath SP, Shewry PR (2009) Variation in mineral micronutrient concentrations in grain of wheat lines of diverse origin. Journal of Cereal Science 49: 290-295.

15

Association analysis for traits associated with powdery mildew tolerance in barley [*Hordeum vulgare* L.] using AFLP markers

Zeinab Mohammadi[1], Atefeh Sabouri[1*] & Sedigheh Mousanejad[2]

1. Department of Agronomy & Plant Breeding, University of Guilan, Rasht, Iran.
2. Department of Plant Protection, University of Guilan, Rasht, Iran.
   *Corresponding Author, Email: a.sabouri@guilan.ac.ir

Abstract

Association analysis is a useful method for evaluation of significant association between molecular marker and phenotype of trait. This study was performed to evaluate association between traits related with powdery mildew resistance and molecular markers. This investigation was performed using 77 barley genotypes and AFLP markers. In phenotypic evaluation, reaction of seedlings to powdery mildew was evaluated and the infection type and intensity were assessed based on 0-9 scale as the most important traits associated with resistance. Also in this study, the genetic diversity of genotypes was evaluated using seven combination primers *Eco*RI/*Mse*I. The average percentages of polymorphism and polymorphic information content were 92.37% and 0.43, respectively. General evaluation of the statistics of genetic diversity showed that among seven primer combinations, three combinations of E90-M160, E100-M160, and E100-M150 were higher value than others and had a more obvious effect in the detection and separation of barley genotypes. Association analysis was performed using four statistical models of GLM and MLM applying TASSEL software. In the complete MLM model, 33 markers showed significant association in the 5 percent probability level with traits and the highest coefficient of determination was related to marker E80-M150-3 that explained 14% of variations of infection intensity. E80-M510-3 and E80-M160-22 markers were showed significant association (pr<0.05) with both characteristic the severity and type of infection that can represent the effective role of this genomic region in resistance to powdery mildew. If the results are confirmed, it can be a suitable candidate for conversion to SCAR specific marker.

Keywords: *AFLP marker, Association analysis, Barley, Fungal disease, Powdery mildew.*

Introduction

Barley has an important role as a source of human food and is the most important nutrient following wheat, rice and maize.

Powdery mildew is one of the most important fungal diseases of barley which is caused by *Blumeria graminis* f. sp. *hordei* fungus. Powdery mildew reduces photosynthetic activity, increases respiration and transpiration, and reduces yield and the quality of harvested grain (20, 25).

In the past, improvement in quantitative traits of plants was performed with phenotypic evaluation but the information of the controlling loci of these traits was not obtained. Today improvement of quantitative traits is performed mainly to identify their controlling loci in the genome. The rapid advances in molecular techniques (particularly in DNA markers) have initiated a new era in genetic studies. DNA markers have increasingly been used in genetic analysis due to their advantages namely, they are highly polymorphic, randomly distributed in genome, least influenced by environmental factors, neutral to selection, stable across different developmental stages and show different individuals in the DNA molecule (18).

Association analysis has been proposed as a method for locating the quantitative traits in recent years. A linkage disequilibrium-based method is the one which evaluating the relationship between phenotypic and genotypic data (11). In association analysis, linkage disequilibrium in natural populations and germplasm collections is used. The linkage disequilibrium is the non- random association between two markers, two genes or marker-gene (19). Association analysis is based on the use of molecular markers. One of the most important markers of DNA is the AFLP marker that was introduced by Vos *et al.* (27). In addition to human genetics, association analysis is used in animal and plant populations.

Association analyses between agronomic traits and SSAP markers were performed in 108 durum wheat genotypes by Rashidimonfared *et al.* (21). In this study, 10 primer combinations of SSAP produced 74 polymorphic bands. The relationship between six agronomic traits and 74 polymorphic markers was measured using stepwise regression. 32 SSAP markers showed significant association with at least one of the six agronomic traits.

In another study, in order to identify genomic regions associated with root traits, association analysis of 100 winter barley (including 50 six-row barley and 50 two-row barley) was performed using 3964 SNPs markers. In phenotypic evaluation, the traits of root dry weight, root volume, average diameter and average secondary roots were assessed. 15 QTLs were detected for traits using MLM model. SNP-2981 marker with maximum value of the log probability was associated with the number of secondary roots. Results of this experiment and the mapped agronomic traits in previous studies showed that most QTLs of root traits were related to traits such as yield, kernels per spike, heading date, lodging and plant height (3). Roy *et al.* (22) performed association analyses of 14 agronomic traits in 55 wheat genotypes using 20 microsatellite primer pairs. In this study, 519 polymorphic markers were generated. 131 SSR, 43 SAMPL and 166 AFLP markers showed significant association with at least one of the 14 agronomic traits.

In another study, association analyses of 48 rice genotypes were performed for the related traits with drought stress using SSR markers. 82 markers showed high correlation with the traits of root length, root dry weight and root diameter. The markers RM170, RM572, RM318, RM3843, RM29, RM540, RM585 and RM36 were related to both root traits and yield under stress conditions (7).

Achleitner *et al.* (2) performed association analyses of 114 oats cultivars using eight AFLP primer combinations to identify markers associated with yield and yield components. Finally, 23 markers were introduced as markers with high potential and associated with complex traits for future breeding programs.

The present study was done to identify related markers with resistance controlling genes to powdery mildew that can be suitable to control diseases and reduce the utilization of fungicides. Utilization of resistant cultivars is the best way in order to adapt to the environment and protect human health.

Materials and methods

Plants materials

The plant materials consisting of 77 genotypes of barley were prepared from the Agricultural and Natural Resources Research Center of Golestan Province. The name and pedigree of barley genotypes are shown in Table 1.

This study was conducted with randomized complete block design in pots in greenhouse conditions in the Faculty of Agricultural Sciences, University of Guilan in 2013. Initially, samples contaminated with powdery mildew, were prepared. These samples were collected in June 2012 from the Varamin region and were stored in dry conditions in the Plant Protection Research Institute of Iran. Due to the presence of sexual stage of cleistothecium on the leaves of contaminated samples, which were able to survive under unfavorable environmental conditions for several years, the pathogen was activated and inoculated on the susceptible cultivar Afzal. The seedlings were placed at 20°C in moist conditions. After sporulation of the pathogen on the leaves surface, spores were collected for contamination. Five seeds of each genotype were planted in plastic pots, 14 cm apart. Seedlings were inoculated with spores of the fungus in two-leaf stage. It should be mentioned that seeds of six genotypes did not germinate in some replications therefore increasing precision was eliminated. Thus in phenotypic evaluation, 71 genotypes were evaluated. Eliminated genotypes are: NB5, Jonoob, EBYT-W-89-17, EBYT-W-89-18, EBYT-W-89-19 and EBYT-W-89-6. After 12 days, the traits of the infection type and infection intensity were assessed based on 0-9 (23) scale. In order to normalize the distribution of

experimental errors, data transformation was performed for the, infection type and infection intensity traits using equations (1) and (2), respectively.

$$\sqrt{x+0.5} \tag{1}$$

$$\text{Arc } \sin\sqrt{x+0.5} \tag{2}$$

Table 1. Name and pedigree of studied barley genotypes in this research.

Number	Name or Pedigree	Number	Name or Pedigree	Number	Name or Pedigree
1	Youssef	27	EB-88-3	53	EBYT-W-89-17
2	Izeh	28	EB-88-4	54	EBYT-W-89-18
3	NB17	29	EB-88-5	55	EBYT-W-89-19
4	NB5	30	EB-88-7	56	EBYT-W-89-4
5	L4shori	31	EB-88-10	57	EBYT-W-89-5
6	Nimroz	32	EB-88-14	58	EBYT-W-89-7
7	Kavir	33	EB-88-16	59	EB-88-20
8	Prodogtive	34	EB-88-19	60	EBYT-W-89-8
9	Bahman	35	Bomi	61	39Motadel
10	36Motadel	36	Rihane	62	EB-86-17
11	31Motadel	37	Arass	63	EB-87-7
12	28Garm	38	Goharjow	64	EB-88-13
13	24Garm	39	Karoon	65	Dasht
14	21Garm	40	EB-88-2	66	Makouee
15	EC-84-10	41	Jonoob	67	Nosrat
16	45Motadel	42	Shirin	68	EC-83-17
17	EC-82-11	43	Torsh	69	EBYT-W-79-10
18	EC-81-13	44	Fajre30	70	MB-83-14
19	MB-82-12	45	W-82-5	71	W-79-10
20	EB-86-14	46	EBYT-W-89-2	72	EBYT-W-89-3
21	EB-86-6	47	EBYT-W-89-9	73	EBYT-W-89-6
22	EB-86-4	48	EBYT-W-89-10	74	EB-88-11
23	EB-86-3	49	EBYT-W-89-11	75	EB-88-6
24	EB-85-5	50	EBYT-W-89-13	76	EB-88-8
25	EB-87-20	51	EBYT-W-89-15	77	EB-88-9
26	EB-88-1	52	EBYT-W-89-16		

AFLP analysis

DNA extraction from the fresh leaves of samples was performed using the CTAB method described by Saghai- Maroof *et al.* (24). The quantity and quality of extracted DNA were determined using agarose gel of 0.8 percent. The AFLP method was performed according to Vos *et al.* (27) method. Six μl of extracted DNA was digested with restriction enzymes of *Eco*RI and *Mse*I for 3 hours at 37 °C. The DNA fragments

were ligated to *Eco*RI and *Mse*I adopters for a period of 2 hours at 37 °C and 1 hour at 20 °C. The samples of the previous stage were used for pre-amplification with the *Eco*RI and *Mse*I primers with one selective nucleotide. In this stage, the thermal cycles were 30 times with the program of 94 °C for 30 seconds, 60 °C for 30 °C seconds and 72 °C for 60 seconds. The products of pre-amplification stage were diluted 5:1 and then selective amplifications were done with 10 primers combinations with three selective nucleotides (Table 2) in touch down thermal cycle including three stages of different temperatures. The PCR products were separated using polyacrylamid gel electrophoresis of 6% and were stained with silver nitrate. AFLP bands were scored as zero or one for absence or presence.

Table 2. Primer combinations in AFLP analysis.

EcoRI Primer		MseI Primer	
Name	Sequence	Name	Sequence
E060	GACTGCGTACCAATTCAAG	M140	GATGAGTCCTGAGTAAAAC
E070	GACTGCGTACCAATTCAAT	M150	GATGAGTCCTGAGTAAAGA
E080	GACTGCGTACCAATTCACG	M160	GATGAGTCCTGAGTAAAGT
E090	GACTGCGTACCAATTCACT		
E100	GACTGCGTACCAATTCAGT		
E110	GACTGCGTACCAATTCATC		

Statistical analyses

The polymorphic information content (PIC) was calculated using equation (3) by Excel software:

$$PIC_i = 2 f_i (1-fv) \qquad (3)$$

Equation above, PIC_i is the PIC of marker i, f_i is the frequency of presence of ith marker and $1-f_i$ is the frequency of absence of ith marker.

The marker index (MI) and the effective multiplex ratio (EMR) were calculated using equations (4) and (5) by Excel software:

$$MI = PIC \times EMR \qquad (4)$$
$$EMR = n_p \times f30 \qquad (5)$$

n_p and β are as:

n_p = The total number of polymorphic bands
β = Fraction of number of polymorphic bands to the total number of bands

Marker index and effective multiplex ratio were calculated by Excel software. other statistics of genetic diversity including the Nei's coefficient of variation and Shannon index were calculated using PopGene 32 (28) and PAST (14) software respectively.

The structure analysis and separation of population into subpopulations with different genetic structure were performed using STRUCTURE software. As previous information of population structure was not available, the number of subpopulations (K) was calculated with the simulation that was performed with 100000 Burn-in period and 100000 MCMC repetitions. The number of K was considered ranging from 2 to 10 and Evanno et al. (9) method was used to calculate the number of subpopulations.

In this way, columns of K and LnP (D) were used for calculations and the mean L (K) and standard deviation (STD) of repetitions were calculated for each K. Then the subtraction of mean repetitions L'(K) was calculated for the adjacent groups from difference between upper group and lower group and the subtraction of L'(K) for adjacent groups was calculated as L"(K) values. Finally, the values of ΔK were calculated using equation (6). Also a bilateral chart of K and ΔK was plotted.

$$\Delta K = L''(K) / Stdev \qquad (6)$$

The climax of the curve in bilateral chart of K and ΔK indicates the optimal number of K. Association analysis was performed between AFLP markers and phenotypic traits using four statistical models (Table 3) using TASSEL software.

Results and Discussion

Statistics of genetic diversity

A total of 245 bands were generated out of seven primer combinations of EcoRI and MseI, of which 227 bands were polymorphic and the average polymorphism obtained was 32.42 per polymorphic band. The average percentage of polymorphism in this study was 92.37 %. The high percentage of polymorphism obtained in this study showed that these markers can be used as powerful tools in the detection and separation of barley genotypes. Figure 1 shows AFLP banding patterns obtained from the amplification of primer combination of E100-M160 in barely as a typical example.

Table 3. The four used statistical models for doing of association analysis of AFLP markers and phenotypic traits.

Model	Used data set
1: GLM[a]	Phenotype + AFLP
2: GLM	Phenotype + AFLP + Q[b]
3: MLM[c]	Phenotype + AFLP + K[d]
4: MLM	Phenotype + AFLP + K + Q

a: General linear model. b: Population structure data. c: Mixed linear mode. d: Kinship data obtained from general similarity of individuals in genetic background arising the kinship.

Polymorphism information content for each primer combination separately is shown in Table 4. In this study, PIC index was variable between 0.39 and 0.48 with the average of 0.43. The highest rate of polymorphic information content was obtained in E100-M160, E100-M150 and E110-M150 primer combinations that were 0.48, 0.47 and 0.45 respectively. The polymorphic information content shows the resolution of a marker by the number of

Table 4. Genetic diversity statistics for seven primer combinations of AFLP.

Primer combination	Poly. Bands	Total. Bands	Poly. Percentag (%)	PIC	Marker index	Nei gene diversity	Shannon index
E80-M150	28	30	93.33	0.40	10.41	0.29	3.16
E90-M150	38	40	95.00	0.41	14.80	0.28	3.06
E100-M150	32	35	91.42	0.47	13.68	0.38	3.49
E110-M150	27	32	84.37	0.45	10.20	0.29	3.17
E80-M160	29	33	87.87	0.39	9.83	0.23	2.84
E90-M160	37	38	97.36	0.43	15.43	0.33	3.28
E100-M160	36	37	97.29	0.48	16.76	0.41	3.65
Total	227	245	646.46	3.05	91.11	2.25	22.65
mean	32.42	35	92.37	0.43	13.01	0.32	3.23

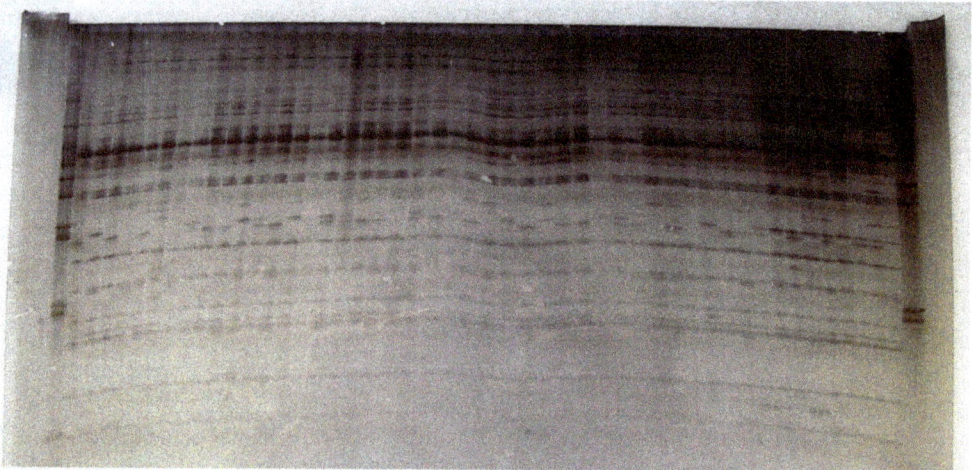

Figure 1. AFLP banding patterns obtained from the amplification of primer combination of E100-M160 in barely.

polymorphic alleles and their relative frequency in the population. Therefore, high values of PIC obtained for mentioned markers show their high effectiveness in distinguishing genotypes in the present study. The diversity index of Nei was variable between 0.23 and 0.41. Furthermore, three combinations E100-M160, E100-M150 and E90-M160 had the highest values, respectively. The Shannon index was variable between 2.84 to 3.65 and the combination of

E100-M160 had the highest value. The marker index was variable between 14.80 and 16.76. Also, the combination of E100-M160 had the highest value. This index considers the total number of bands and calculates potential of each primer to produce more bands on gel. General evaluation of the statistics of genetic diversity including the Nei's coefficient of variation, Shannon index and marker index showed that among seven primer combinations, three combinations of E90-M160, E100-M160, E100-M150 were the highest and in fact had a more powerful effect on distinguishing genotypes.

Structure analysis

The analysis of population genetic structure was performed using STRUCTURE software. Table 5 shows statistics to determine optimum K and Figure 2 shows a bilateral chart to determine the optimal number of K. Based on Figure 2, the climax of the curve is equal to 2. Therefore, the population structure is separable into two subpopulations with different genetic structures. Figure 3 shows inferred population structure by the STRUCTURE software The assignment of individuals into sub-populations was performed using Spataro et al. (26)

Table 5. Calculated statistics for optimum K using STRUCTURE software.

K	L(K)^a	Stdev^b	L'(K)^c	L''(K)^d	Δ K^e
1	8327.79	6583.403	-18095.2	18391.43	2.793605
2	-9767.41	54.69084	296.23	-8054.24	147.2685
3	-9471.18	157.0086	-7758.01	-6152.94	39.18854
4	-17229.2	13557.57	-13911	21729.27	1.602741
5	-31140.1	29473.96	7818.32	-26914.1	0.913147
6	-23321.8	12344.15	-19095.8	14921.43	1.208786
7	-42417.6	30632.26	-4174.32	10429.37	0.34047
8	-46591.9	55448.22	6255.05	-2523.93	0.045519
9	-40336.8	59592.69	3731.12	32874.6	0.551655
10	-36605.7	29774.71	36605.72	-36605.7	1.229423

a: The mean of LnP(D) of repetitions for each K. b: The standard deviation (STD) of repetitions. c: $L(K)_n - L(K)_{n-1}$, d: $L'(K)_n - L'(K)_{n-1}$, e: $|L'(K)| / stdev$

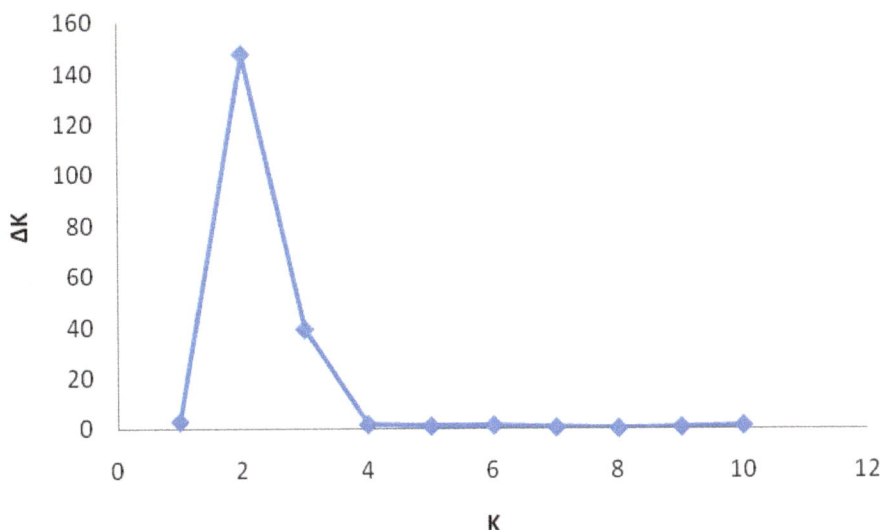

Figure 2. Bilateral chart to determine the optimal number of K. X axis: The number of sub-populations. Y axis: ΔK =.(L''(K))/Stdev. The method of calculation of ΔK are given in Table 5 and materials and method section.

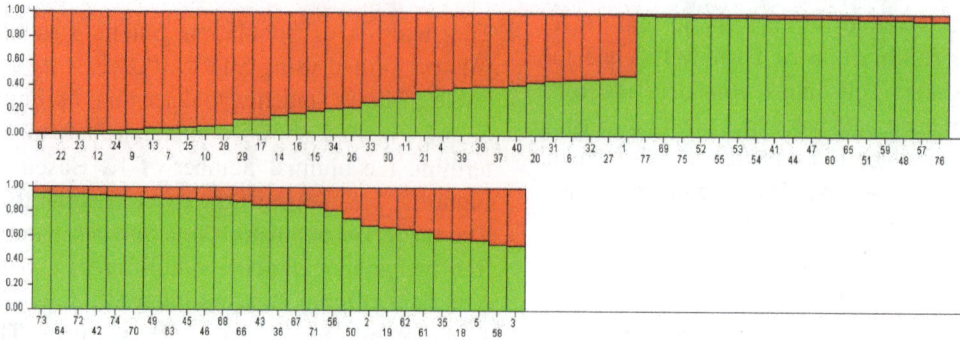

Figure 3. Inferred population structure out of STRUCTURE software obtained on the 227 AFLP markers data set, partitioned into K coloured segments at K=2. X axis: The genotypes numbers. The genotypes names with their numbers exist in the Table 1. Y axis: Proportion of membership of each genotype in each of the two clusters (sub-populations).

method. Also membership percentage for every individual in each group was calculated.

Table 6 shows proportion of membership of each genotype in each of the two clusters. According to this method, the assignment of genotypes into groups is possible when that membership percentage of a genotype is 0.7 or more than 0.7 and if its membership percentage is less than 0.69, it is considered as a mixed genotype. Thus, 21 genotypes were assigned to group 1 and 35 genotypes were assigned to group 2. Furthermore, 21 genotypes were identified as mixed ones. Genotypes belonging to each sub- population had the most similarity in terms of allele frequencies and genetic structures and were different from the other groups. The spurious associations will be identified between marker and QTL. As association analysis was performed between molecular markers and traits regardless of population genetic structure, determining the genetic structure of populations and germplasm collections which is very important (10, 13).

Casas *et al.* (5) performed association analysis for 225 barley accessions (including the 175 SBCC accessions) using SSR markers. Considering the population structure in this study, the number of significant associations was reduced.

Association analysis

Association analysis was carried out using TASSEL software. The results are shown in Table 7. According to this table, in the first model, (GLM: G+P) 35 were identified for the

disease severity and 3 markers were identified for infection type traits. In the second model, (GLM: G+P+Q), 36 markers showed significant association with the disease severity trait and one marker showed significant association with the infection type trait. In the model (MLM: G+P+K), 32 markers were identified for disease severity and 1 marker was identified for infection type traits. In the fourth model (MLM: G+P+Q+K), 31 markers were identified for disease severity and 2 markers were identified for infection type traits. In general, in the four mentioned models, 36 markers were identified for the disease severity trait and 3 markers for the infection type trait. In these four models, 13 markers were identified for the disease severity trait together that included: E80-M150- 1, E80-M150-2, E80-M150-3, E80-M150-4, E80-M150-5, E80-M150-6, E80-M150-7, E80-M150-8, E80-M150- 9, E80-M150-10, E80-M150-11, E80- M150-12, E80-M150-13, E80-M150- 14, E80-M150-15, E80-M150-16, E80- M150-17, E80-M150-18, E80-M150- 19,E80-M150-20,E80-M150-21, E80- M150-22, E80-M150-23, E80-M150-24, E80-M150-25, E80-M150-26, E80-M150-27, E80-M150-28, E110-M150- 25, E110-M150-27, E80-M160-22 and also one marker was identified for the infection type trait that was E100- M150-3.

E100-M160-34 marker in the three models (GLM: G+P), (GLM: G+P+Q), (MLM: G+P+K) furthermore, E110- M150-13, E80-M160-15 and E80-M160-24 markers in the two models

Table 6. Genotypes's membership based on extracted results of STRUCTURE software.

Name or pedigree	Percentage of group's membership 1	Percentage of group's membership 2	Name or pedigree	Percentage of group's membership1	Percentage of group's membership 2	Name or pedigree	Percentage of group's membership 1	Percentage of group's membership 2
Youssef	0.508	0.492	EB-88-3	0.535	0.465	EBYT-W-89-17	0.28	0.972
Izeh	0.31	0.69	EB-88-4	0.924	0.076	EBYT-W-89-18	0.028	0.972
NB17	0.467	0.533	EB-88-5	0.877	0.123	EBYT-W-89-19	0.025	0.975
NB5	0.633	0.367	EB-88-7	0.699	0.301	EBYT-W-89-4	0.189	0.811
L4shori	0.422	0.578	EB-88-10	0.556	0.444	EBYT-W-89-5	0.051	0.949
Nimroz	0.546	0.454	EB-88-14	0.54	0.46	EBYT-W-89-7	0.458	0.542
Kavir	0.944	0.056	EB-88-16	0.729	0.271	EB-88-20	0.038	0.962
Prodogtive	0.99	0.011	EB-88-19	0.785	0.215	EBYT-W-89-8	0.032	0.968
Bahman	0.961	0.039	Bomi	0.413	0.587	39Motadel	0.36	0.64
36Motadel	0.935	0.065	Rihane	0.147	0.853	EB-86-17	0.337	0.663
31Motadel	0.697	0.303	Arass	0.601	0.399	EB-87-7	0.093	0.907
28Garm	0.978	0.022	Goharjow	0.604	0.396	EB-88-13	0.058	0.942
24Garm	0.95	0.05	Karoon	0.611	0.389	Dasht	0.033	0.967
21Garm	0.841	0.159	EB-88-2	0.586	0.414	Makouee	0.115	0.885
EC-84-10	0.805	0.195	Jonoob	0.029	0.971	Nosrat	0.149	0.851
45Motadel	0.824	0.176	Shirin	0.067	0.933	EC-83-17	0.105	0.895
EC-82-11	0.876	0.124	Torsh	0.145	0.855	EBYT-W-79-10	0.015	0.985
EC-81-13	0.42	0.58	Fajre30	0.029	0.971	MB-83-14	0.083	0.917
MB-82-12	0.328	0.672	W-82-5	0.095	0.905	W-79-10	0.159	0.841
EB-86-14	0.568	0.432	EBYT-W-89-2	0.103	0.897	EBYT-W-89-3	0.058	0.942
EB-86-6	0.637	0.363	EBYT-W-89-9	0.03	0.97	EBYT-W-89-6	0.057	0.943
EB-86-4	0.984	0.016	EBYT-W-89-10	0.04	0.96	EB-88-11	0.074	0.926
EB-86-3	0.984	0.016	EBYT-W-89-11	0.087	0.913	EB-88-6	0.017	0.983
EB-85-5	0.966	0.034	EBYT-W-89-13	0.251	0.749	EB-88-8	0.054	0.946
EB-87-20	0.939	0.061	EBYT-W-89-15	0.036	0.964	EB-88-9	0.013	0.987
EB-88-1	0.778	0.222	EBYT-W-89-16	0.025	0.975			

Table 7. The results of association analysis between evaluated traits and AFLP markers using four statistical models.

Trait	GLM: G+P			GLM: G+P+Q			MLM: G+P+K			MLM: G+P+Q+K		
	Marker	R^2	P	Marker	R^2	P	Marker	R^2	P	Marker	R^2	P
II	E80MI501	0.130	0.0094	E80MI501	0.120	0.0138	E80MI501	0.100	0.0380	E80MI501	0.093	0.0161
	E80MI502	0.154	0.0037	E80MI502	0.146	0.0051	E80MI502	0.125	0.0178	E80MI502	0.118	0.0161
	E80MI503	0.170	0.0019	E80MI503	0.169	0.0020	E80MI503	0.141	0.0109	E80MI503	0.143	0.0161
	E80MI504	0.136	0.0075	E80MI504	0.129	0.0099	E80MI504	0.106	0.0317	E80MI504	0.102	0.0161
	E80MI505	0.136	0.0074	E80MI505	0.128	0.0102	E80MI505	0.107	0.0313	E80MI505	0.100	0.0161
	E80MI506	0.133	0.0083	E80MI506	0.124	0.0120	E80MI506	0.103	0.0345	E80MI506	0.099	0.0161
	E80MI507	0.131	0.0091	E80MI507	0.121	0.0131	E80MI507	0.101	0.0374	E80MI507	0.094	0.0161
	E80MI508	0.163	0.0026	E80MI508	0.150	0.0043	E80MI508	0.135	0.0132	E80MI508	0.126	0.0161
	E80MI509	0.130	0.0095	E80MI509	0.120	0.0135	E80MI509	0.100	0.0384	E80MI509	0.094	0.0161
	E80MI5010	0.139	0.0066	E80MI5010	0.127	0.0105	E80MI5010	0.110	0.0285	E80MI5010	0.099	0.0161
	E80MI5011	0.148	0.0047	E80MI5011	0.135	0.0078	E80MI5011	0.119	0.0213	E80MI5011	0.106	0.0161
	E80MI5012	0.130	0.0095	E80MI5012	0.120	0.0138	E80MI5012	0.100	0.0384	E80MI5012	0.093	0.0161
	E80MI5013	0.130	0.0095	E80MI5013	0.121	0.0133	E80MI5013	0.100	0.0384	E80MI5013	0.094	0.0161
	E80MI5014	0.132	0.0086	E80MI5014	0.120	0.0135	E80MI5014	0.103	0.0352	E80MI5014	0.094	0.0161
	E80MI5015	0.139	0.0068	E80MI5015	0.126	0.0108	E80MI5015	0.109	0.0289	E80MI5015	0.100	0.0161
	E80MI5016	0.142	0.0060	E80MI5016	0.128	0.0103	E80MI5016	0.112	0.0261	E80MI5016	0.102	0.0161
	E80MI5017	0.156	0.0034	E80MI5017	0.146	0.0051	E80MI5017	0.127	0.0167	E80MI5017	0.117	0.0161
	E80MI5018	0.134	0.0080	E80MI5018	0.124	0.0117	E80MI5018	0.105	0.0333	E80MI5018	0.097	0.0161
	E80MI5019	0.133	0.0085	E80MI5019	0.122	0.0130	E80MI5019	0.103	0.0349	E80MI5019	0.095	0.0161
	E80MI5020	0.136	0.0076	E80MI5020	0.127	0.0103	E80MI5020	0.106	0.0318	E80MI5020	0.100	0.0161
	E80MI5021	0.141	0.0061	E80MI5021	0.130	0.0094	E80MI5021	0.112	0.0264	E80MI5021	0.101	0.0161
	E80MI5022	0.130	0.0093	E80MI5022	0.120	0.0138	E80MI5022	0.100	0.0378	E80MI5022	0.093	0.0161
	E80MI5023	0.136	0.0074	E80MI5023	0.125	0.015	E80MI5023	0.107	0.0310	E80MI5023	0.099	0.0161
	E80MI5024	0.130	0.0095	E80MI5024	0.120	0.0138	E80MI5024	0.100	0.0383	E80MI5024	0.093	0.0161
	E80MI5025	0.130	0.0094	E80MI5025	0.120	0.0138	E80MI5025	0.100	0.0381	E80MI5025	0.093	0.0161
	E80MI5026	0.142	0.0059	E80MI5026	0.130	0.0093	E80MI5026	0.113	0.0258	E80MI5026	0.102	0.0161
	E80MI5027	0.133	0.0085	E80MI5027	0.121	0.0130	E80MI5027	0.103	0.0348	E80MI5027	0.095	0.0161
	E80MI5028	0.130	0.0095	E80MI5028	0.120	0.0138	E80MI5028	0.100	0.0384	E80MI5028	0.093	0.0161
	E110MI5013	0.064	0.0353	E110MI5013	0.068	0.0293	E110MI5025	0.069	0.0328	E110MI5025	0.061	0.0161
	E110MI5025	0.074	0.0228	E110MI5025	0.060	0.0415	E110MI5027	0.075	0.0265	E110MI5027	0.073	0.0161
	E110MI5027	0.098	0.0084	E110MI5027	0.094	0.0097	E80MI6022	0.074	0.0274	E80MI6022	0.085	0.0161
	E80MI6015	0.060	0.0403	E80MI6015	0.062	0.0378	E100MI6034	0.095	0.0444			
	E80MI6022	0.083	0.0155	E80MI6022	0.070	0.0270						
	E80MI6024	0.059	0.0424	E80MI6024	0.061	0.0388						
	E100MI6034	0.103	0.0264	E80MI6026	0.060	0.0402						
				E100MI6034	0.090	0.0430						
IT	E80MI503	0.101	0.0281	E80MI503	0.117	0.0152	E80MI503	0.104	0.0325	E80MI503	0.114	0.0161
	E80MI6022	0.057	0.0458							E80MI6022	0.064	0.0161
	E100MI6034	0.087	0.0476									

II: Infection intensity. IT: Infection type. R^2: Coefficient of determination. P: Significance probability level.

(GLM: G+P) and (GLM: G+P+Q) showed significant association in 5 percent probability level for the disease severity trait. Also, the E80-M160-22 marker in models (GLM: G+P) and (MLM: G+P+Q+K) showed significant association in 5 percent probability level with the infection type trait.

The new method of MLM considers the information of population structure (matrix Q) and kinship data (matrix K) in association analysis. Initially, the matrixes Q and K for doing MLM method should be prepared and then be used in association analysis to control the false associations between makers and traits. The MLM model with matrixes of Q and K will lead to better results in comparison with other ways in which the Q matrix or K matrix are used alone. However, the Q matrix can be replaced with P matrix (principal component analysis) which in this way, the MLM method is executed similarly and can be suggested as a potential for replacement (1).

In the fourth model, including four data sets and the MLM model, E80-M150-3 and E80-M160-22 markers showed significant association in 5 percent probability level with both traits. Based on the results, the absence and presence of the band respectively are representative of the resistance and susceptibility of cultivars. The highest coefficient of determination related to E80-M150-3 marker was 14% that explained variations of infection intensity trait.

Dadras et al. (6) performed association analysis of yields and seven important agronomic traits including leaf area index, plant height, leaf number, fresh leaf yield, dried leaf yield, length and the width of leaf in tobacco using AFLP markers. They used four statistical models of association analysis. According to their results, the combination E110-M160-23 was simultaneously significant for the leaf area index and fresh leaf yield traits. Also nine primer combinations were identified with the MLM model and four data sets including phenotypic, genotypic data, Q-matrix and K-matrix were significant for several traits. They proposed that if the effectiveness of these regions in genetic control of these traits is confirmed, they can be suitable candidates for conversion into the SCAR specific marker.

Association analysis of 115 genotypes of barley using 10 traits and 10 microsatellite markers was performed by Ebrahimi et al. (8), and a total of 70 polymorphic alleles were identified. The maximum number of markers was found for a number of nodes and the minimum number of markers was found for leaf number and radical length. The highest and lowest coefficients of determination were related to the grain width and germination traits, respectively. According to the results, markers HVM20, Gmsoo3, Bmaco36 and HVHVA1 were in controlling regions of agronomic traits more than other markers that explained more variations of studied traits. Some of the markers were associated with several traits simultaneously, that is considering the existence of a significant correlation among morphological traits can be due to genetic linkage or pleiotropic effects. In order to understand this subject, preparation of segregation generations and linkage maps is essential. In another study, association analysis of 35 barley genotypes was performed using microsatellite markers for traits related to freezing tolerance by Gangkhanlou et al. (12). A total of 62 alleles showed significant association with changes of 12 traits among 13 traits evaluated. The traits of crown moisture and relative water loss obtained maximum (10 allels) and minimum (one allele) number of alleles. Zhang et al. (29) performed association analysis of 26 agronomical traits with 204 SSR markers and 94 maize inbred lines. Using structure analysis, five sub- populations were obtained. Furthermore, using MLM model, 39 loci showed significant association in the five percent probability level with 17 agronomic traits in two years. Three loci with plant height, four with days to flowering, five with the number of kernel rows, and three with hundred kernels weight showed significant association simultaneously. They expressed that these results can be useful in genetic improvement and molecular breeding of maize. In another study, association analysis of 103 wheat germplasm was performed using 76 SSR and 40 EST-SSR markers. In the evaluation of phenotypes 6 traits were assessed in three places during three years. In this study, six sub-populations were obtained by population structure analysis based on 49 SSR and 40

EST- SSR markers. A total of 10 SSR markers on chromosome 4A showed significant association with six agronomic traits using the MLM model and by taking the Q and K matrixes (17). In another study, association analysis was performed on 40 durum wheat genotypes and SSR markers. In this study, six agronomic traits were evaluated. According to the results, 14 markers showed significant association with evaluated traits. Markers wmc54, wms118 and wmc165 on chromosomes 3B, 5B and 3A showed significant associations with several traits respectively (4). In another study, association analysis on a set of 160 *Brassica rapa* was performed using AFLP markers. In phenotypic evaluation, the amount of phytate and phosphate in seeds and leaves and some morphological traits were studied. Based on structure analysis, four subpopulations were obtained. In association analysis, 170 and 27 markers showed significant association with evaluated traits regardless of population structure and considering population structure (30).

As mentioned before, the fourth model including four data sets and MLM model, E80-M150-3 and E80-M160-22 markers showed significant associations in 5 percent probability level with both traits that can represent the effective role of this genomic region in powdery mildew resistance that may be due to genetic linkage or pleiotropic effects. If this experiment is performed in a few years and in several locations with different genotypes and these genomic regions are identified again, SCAR specific marker from these genomic regions can be provided. Julio *et al.* (15, 16) in their studies identified QTLs related to agronomic traits, leaf quality, chemical composition and characteristics of the smoke and QTLs associated with resistance to several diseases using QTL mapping in tobacco. They identified AFLP markers related to important QTLs. Some of these markers were converted to SCAR specific marker. They were also confirmed in recombinant inbred lines and doubled haploid populations.

Conclusion

According to the results, primer combinations of E90-M160, E100- M160 and E100-M150 achieved the highest amounts of the Statistics

of genetic diversity. Thus, they can be used as suitable and powerful combinations in breeding programs of barley. Based on the Structure analysis, the genotypes were separated into two groups with different genetic structures and one group was identified as mixed genotypes. In the MLM model, with the consideration of population structure and kinship data, 33 markers showed significant association in the 5 percent probability level with traits. E80-M510-3 and E80-M160-22 markers were identified in the 5 percent probability level linked to both severity and the type of infection traits that can represent the effective role of this genomic region in resistance to powdery mildew. Linkage of two markers and loci controlling of traits studied can be confirmed by further research which can be a suitable candidate for conversion into SCAR specific marker.

References

1. Abdurakhmonov, I.Y. and Abdukaimov, A. 2008. Application of association mapping to understanding the genetic diversity of plant germplasm resources. Int J Plant Genom, 1-18.
2. Achleitner, A., Tinker, N., Zechner, E. and Buerstmayr, H. 2008. Genetic diversity among at varieties of worldwide origin and associations of AFLP markers with quantitative traits. Theor Appl Genet, 117 (7): 1041-1053.
3. Ataei, R., Mohammadi, V., Talei, A.R. and Naghavi, M.R. 2013. Association mapping for root characteristics in barley (*rordeum vulgare*). Iran J Field Crop Sci, 44 (2): 347-357.
4. Bousba, R., Baum, M., Jighy, A., Djekoune, A., Lababidi, S., Benbelkacem, A., Iabhilili, M., Gaboun, F. and Ykhlef, N. 2013. Association analysis of genotypic and phenotypic traits using SSR marker in Durum wheat. OIIRJ, 3: 60-79.
5. Casas, A.M., Kopahnke, D., Habekub, A., Schwizer, G., Gracia, M.P., Lasa, J.M., Ciudad, F.J., Codesal, P., Moralejo, M.A., Molina-Cano, J.L., Igartua, E. and Ordon, F. 2006. Marker-trait association for disease resistance in the Spanish barley core collection. Eucarpia, Lleida, 81: 141-145.
6. Dadras, A.R., Sabouri, H., Mohammadinejad, G.h., Sabouri, A. and Shoai-Deylami, M. 2014 Association analysis, genetic diversity and structure analysis of tobacco based on AFLP markers. Mol Biol Rep, 41(5): 3317-3329.
7. Deshmukh, V.V. 2012. Genome-wide association mapping of drought resistance traits in rice (*Oryza sativa* L.). MSc. Biotechnology thesis. Tamil Nadu Agricultural University. India.
8. Ebrahimi, A., Naghavi, M.R., Sabokdast, M. and

Moradi Sarabshali, A. 2011. Association analysis with microsatellite markers for agronomic traits in Iranian barley landraces. Mod Genet J, 6(1): 35-43.

9. Evanno, G., Reganut, E. and Goudet, J. 2005. Detecting the number of clusters of individuals using the software STRUCTURE: a simulation study. Mol Ecol, 14: 2611-2620.

10. Falush, D., Stephens, M. and Pritchard, J.K. 2003. Inference of population structure using multilocus genotype data: linked loci and correlated allele frequencies. Genet, 164: 1567-1587.

11. Flint Garcia, S.A., Thornsberry, J.M. and Buckler, E.S. 2003. Structure of linkage disequilibrium in plants. Annu Rev Plant Biol, 54: 357-374.

12. Gangkhanlou, A., Mohammadi, S.A., Moghaddam, M., Ghasemi Golazani, K., Shakiba, M.R. and Yosefi, A. 2012. Genetic diversity in barley as revealed by microsatellite markers and association of these markers by traits related to freezing tolerance. Seed Plant Improv J, 28(1): 101-114.

13. Gupta, P.K., Rustgi, S. and Kulwal, P.L. 2005. Linkage disequilibrium and association studies in higher plants: Present status and future prospects. Plant Mol Biol, 57: 461-485.

14. Hammer, Ø., Harper, D.A.T. and Ryan, P.D. 2001. PAST: Paleontological statistics software package for education and data analysis. Palaeontologia Electronica, 4(1): 1-9.

15. Julio, E., Denoyes-Rothan, B., Verrier, J.L. and Dorlhac de Borne, F. 2006a. Detection of QTLs linked to leaf and smoke properties in Nicotiana tabacum based on a study of 114 recombinant inbred lines. Mol Breed, 18(1): 69-91.

16. Julio, E., Verrier, J.L. and Dorlhac de Borne, F. 2006b. Development of SCAR markers linked to three disease resistances based on AFLP within Nicotiana tabacum L. Theor Appl Genet, 112: 335-346.

17. Liu, L., Wang, L., Yao, J., Zheng, Y. and Zhao, C. 2010. Association mapping of six agronomic traits on choromosome 4A of wheat (mriticum aestivum L.). Mol Plant Breed, 1(5).

18. Mohammadi, S.A. 2006. Molecular analysis of genetic diversity of perspectives. Ninth Iran Crop Sci Congr, 27-29 Aug, Tehran.

19. Mohammadi, S.A. 2008. New methods of genetic structure analysis of quantitative traits in plant. Tenth Iranian Crop Sciences Congress. Seed and plant improvement research institute, 18-20 Aug, Karaj.

20. Pesaraklu, S., Soltanloo, H., Ramezanpour, S.S., Nasrollah Nejad Ghomi, A.A., Kalate Arabi, M. and Kia, Sh. 2013. Genetic analysis of powdery mildew resistance (Erysiphe graminis f. sp. hordei) in some barley lines. J Plant Prod, 20(3): 49-69.

21. Rashidimonfared, S., Moradi, M., Hosseinzadeh, A.H. and Naghavi, M.R. 2008. Association analysis between main agronomic traits and SSAP markers in durum wheat samples. Mod Genet J, 3(2): 29-35.

22. Roy, J.K., Bandopadhyay, R., Rustgi, S., Balyan, H.S. and Gupta, P.K. 2006. Association analysis of agronomically important traits using SSR, SAMPL and AFLP markers in bread wheat. Curr Sci, 90(5): 683-689.

23. Saari, E.E. and Prescott, J.M. 1975. A scale for appraising the foliar intensity of wheat diseases. Plant Dis Rep, 59(5): 377-380.

24. Saghai-Maroof, M.A., Biyashev, R.M., Yang, G.P., Zhang, Q. and Allard, R.W. 1994. Extraordinarily polymorphic microsatillate DNA in barely species diversity, choromosomal location, and population dynamics, Proceed of the Nat Acad of Sci. USA, 91: 5466-5570.

25. Silvar, C., Dhif, H., Igartua, E., Kopahnke, D., Gracia, M.P., Lasa, J.M., Ordon, F. and Casas, A.M. 2010. Identification of quantitative trait loci for resistance to powdery mildew in a Spanish barley landrace. Mol Breed, (25): 581-592.

26. Spataro, G., Tiranti, B., Arcaleni, P., Bellucci, E., Attene, G., Papa, R., Spagnoletti-Zeuli, P. and Negri, V. 2011. Genetic diversity and structure of a worldwide collection of Phaseolus coccineus L. Theor Appl Genet; 122: 1281-1291.

27. Vos, P., Hogers, R., Bleeker, M., Reijans, M., Lee, T.V.D., Hornes, M., Frijters, A., Pot, J., Peleman, J., Kuiper, M. and Zabeau, M. 1995. AFLP: a new technique for DNA fingerprinting. Nucleic Acids Res, 23 (21): 4407-4414.

28. Yeh, F.C., Yang, R.C., Boyle, T.J.B., Ye, Z.H. and Mao, J.X. 1997. POPGENE, The User-friendly Shareware for Population Genetic Analysis. Edmonton, Molecular Biology and Biotechnology Centre, University of Alberta, Canada.

29. Zhang, Q., Wu, C., Ren, F., Li, Y. and Zhang, C. 2012. Association analysis of important agronomical traits of maize inbread lines with SSRs. Aust J Crop Sci, 6(6): 1131-1138.

30. Zhao, J., Paulo, M.J., Jamar, D., Lou, P., Eeuwijk, F.V., Bonnema, G., Vreugdenhil, D. and Koornneef, M. 2007. Association mapping of leaf traits, flowering time, and phytate content in Brassica rapa. Genome, 50: 963-973.

16

Assessment of genetic diversity in Iranian wheat (*Triticum aestivum* L.) cultivars and lines using microsatellite markers

R. Mollaheydari Bafghi[1], A. Baghizadeh[2*] , Gh. Mohammadi-Nejad[3], B. Nakhoda[4]
Zeinab Mohammadi[1], Atefeh Sabouri[1*] & Sedigheh Mousanejad[2]

1. Department of Plant Breeding, Graduate University of Advanced Technology.
2. Department of Biotechnology, Institute of Sciences and High Technology and Environmental Sciences, Graduate University of Advanced Technology, Kerman-Iran.
3. Department of Agronomy and plant Breeding, College of Agriculture and Center of Excellence for Abiotic Stress in Cereal Crop., Shahid Bahonar University of Kerman-Iran.
4. Department of System Biology , Agricultural Biotechnology Research Institute of Iran , P. O. Box, 31535-1897, Karaj –Iran.
*Corresponding Author, Email: amin_4156@yahoo.com

Abstract

In this study, genetic diversity of 20 wheat genotypes was evaluated using 126 simple sequence repeats (SSR) alleles, covering all three wheat genomes. A total of 1557 allelic variants were detected for 126 SSR loci. The number of alleles per locus ranged from 4 to 19 and the allelic polymorphism information content (PIC) varied from 0.66 (*Xgwm*429) to 0.94 (*Xgwm*212 and *Xgwm*515). The highest polymorphism was observed in *Xgwm212* and *Xgwm515* primers with 19 alleles, while the lowest polymorphism belonged to *Xgwm429* with 4 alleles. The highest number of alleles per locus was detected in the genome A with 594, compared to 552 and 411 for B and D genomes, respectively. Dendrogram was constructed using Dice similarity coefficient and UPGMA algorithm by NTSYSpc2.0 software and genotypes were grouped in to six clusters. The knowledge about the genetic relationships of genotypes provides useful information to address breeding programs and germplasm resource management. This study also confirms the usefulness of SSR markers to study wheat genetic diversity.
Key words: Genetic diversity, Microsatellite markers, Polymorphism, Wheat (*Triticum aestivum* L.)

Introduction

Wheat (*Triticum aestivum* L.) is the most important and one of the oldest cultivated crops in the world, and understanding its genetics and genome organization using molecular markers is of great value for genetic and plant breeding purposes.

Molecular markers are a powerful tool to study the genetic structure of plant populations. In recent years, several molecular assays have been applied to assess genetic diversity among wheat cultivars (Chen *et al.*, 1994). These molecular methods are different in principle, application, type, the amount of detected polymorphism, task and time requirements. Various studies have used SSR markers to investigate genetic diversity in cultivated hexaploid wheat genotypes of *Triticum Aestivum* L. (Senturk Akfirat and Ahu Altinkut Uncuoglu 2013). Microsatellites (Tautz and Renz 1984; Tautz 1989)or simple sequence repeat (SSRs)-based molecular markers are now the marker of choice in most areas of plant genetics. Microsatellites are repeating sequences of 2–6 base pairs of DNA and are among the most stable markers of genetic variation and divergence among wheat genotypes because they are multiallelic, chromosome-specific and evenly distributed along chromosomes (Tautz, 1989). The advantages of SSRs are well documented (Powell *et al.* 1996) and these include: high information content, co-dominant inheritance, reproducibility and locus specificity. The improvement of wheat traits is mainly due to efficient use of wheat germplasm genetic diversity. Determination of genetic diversity is useful for plant breeding and hence production of more efficient plant species under different conditions. Accordingly, 20 of the most common wheat genotypes from different parts of Iran were selected and consequently analyzed for their genetic diversity by microsatellite markers. The aim of this research was to estimate the allelic variation

and evaluate the genetic diversity at the expressed sequences among Iranian extremes wheat genotypes and to provide information for wheat breeding and improvement in germplasm management of wheat.

Materials and Methods

Plant Material:

A total of 20 wheat genotypes including salinity tolerant, semi-salinity tolerant and non-tolerant genotypes were used (Table 1) as the source for evaluating genetic diversity and genomics coverage by microsatellite markers. All of them were hexaploid (*Triticum aestivum* L., AABBDD, 2n = 6x = 42), and known as materials of advanced lines and cultivars in Iran.

Table 1. Evaluated wheat genotypes.

1-Roshan	2- Arta	3- Moghan-3	4- S-78-11	5-N-83-3
6-MV-17	7-KRL-4	8- Arg	9-Shotordandan	10-Boolani
11- Shoele	12- Sorghtoghm	13-SNH-9	14- Sistan	15-107-PR-87
16-139-PR-87	17-140-PR-87	18-Kharchia	19- Mahooti	20- Gaspard

DNA Isolation:

Total genomic DNA was extracted from leaf tissue for each line and cultivar. Young leaves from four-week old plants were cut as tissue samples for DNA extraction. Genomic DNA was extracted by mini prep_ isolation method (Dellaporta 1983) with minor modifications. 0.2g of young leaves were frozen in liquid N2, mixed with 400 l of extraction buffer (50 mM Trisbase pH 8, 300 mM NaCl, 25 mM EDTA pH 8 and 1% SDS) and incubated at 65°C for 30 min. 200 l sodium acetate 5 mM was added to each tube and placed about 10 min on ice. 500 l chloroform/ isoamyl alcohol (24: 1) was added and mixed well. The mix was centrifuged at 12000 g for 15 min. The supernatant was precipitated with an equal volume of ice- cold isopropanol and centrifuged at

5000 g for 15 min. In this stage DNA was recovered by centrifuging. The pellet was hooked out by sterile pipettes, washed in 70% ethanol and air dried and suspended in 300 l of 1x TBE buffer. Both DNA quantity and quality were estimated using UV spectrophotometer (Carry 50) by measuring absorbencies at A260 and A280 nm and 1% agarose gel electrophoresis and comparing band intensity with DNA ladder of known concentrations. DNA samples were diluted to 50ng/ l for SSR reactions. (Dellaporta *et al.*, 1983).

Microsatellite Markers Analysis:

To test the genetic diversity of wheat genotypes, 126 SSR markers dispersed throughout the genome were used in this study. Genomic SSR primer information was obtained from two sources. The first primer set was obtained from Röder et al., (1998) from a conventional genomic library and designated as *GWM*, and the second one was obtained from Grain genes database (http:// graingenes.org). Microsatellite amplifications were carried out as reported by Röder et al. (Röder et al.,1998). Polymerase chain reaction (PCR) and fragment analysis were performed according to (Devos et al., 1995) and (Röder et al., 1998). PCR reactions were performed in a volume of 25 μL in Perkin-Elmer (Norwalk, CT) thermo cyclers. The reaction mixture contained 3 μL of each primer, 1.5 μL of each deoxy nucleotide, 1.5 μL MgCl2, 1 unit Taq polymerase, and 50–100 ng of template DNA. After 3 min at 94°C, 45 cycles were performed with 1 min at 94°C,. 1 min at either 50, 55, or 60°C (depending on the individual microsatellite), 2 min at 72°C, and a final extension step of 10 min at 72°C. Amplification products were separated on denaturing 8% polyacrylamide gel electrophoresis. Gel running times were adapted to fragment size, i.e. extended running times were used for the separation of larger fragments. The amplified fragments were detected using the silver staining methods and 100 bp size marker as described by (Bassam et al., 1993). The base material for the present study consisted of 126 microsatellite for all genomes.

Data analysis

The amplified bands were scored manually as 0 (absent) or 1 (present). Matrix similarity of genotypes was calculated using NTSYS-pc.2.1 (Rohlf Fj., 1998)) with Sanh-clustering using the UPGMA (Unweighted Paired Group Method Using Arithmetic Averages) method. We used the Dice genetic similarity coefficient (Dice Lr., 1945; Nei M, Li Wh ., 1979). The results are presented graphically in dendrogram. The term polymorphism information content (PIC) was originally introduced into human genetics by Botstein et al (1980). It refers to the value of a marker for detecting polymorphism within a population, depending on the number of detectable and the distribution of their frequency. The

polymorphic information content (PIC) was employed for each locus to assess the informative of each marker. The PIC for each marker was calculated according to formula of Nei (1973):

$$H_e = 1 - \sum_{i=1}^{n} p_i^2$$

where n is the total number of alleles detected for a locus of a marker and P_{ij} the frequency of the *j* th allele in the set of 20 investigated genotypes. The following parameters were estimated: the percentage of polymorphic loci and gene diversity, and other calculations were performed using the AlphaEaseFC4.0 software.

Results and Discussion

Microsatellite Polymorphism:

Twenty wheat cultivars of diverse origins were evaluated using 126 microsatellite markers. These microsatellites were selected on the basis of their known genetic locations to give a uniform coverage for all three wheat genomes (A, B and D) and a total of 1557 polymorphic alleles were detected at 126 loci (Table 2). A wide range of allelic variants was observed for each locus (Table 2). The number of alleles per locus ranged from 4 to 19, with the average number of 12.35 alleles per locus (Table 2). The largest number of alleles per locus occurred in the A genome which is accounted to be 594, compared to 552 for genome B and 411 for genome D (table 3). Microsatellite PIC values ranged from 0.66 to 0.94 (Table 2). Approximately 88.8% of microsatellite markers that used all chromosomes had a PIC value greater than 0.70, which indicates a high level of polymorphism for the majority of markers. The highest polymorphism was observed in *Xgwm212* and *Xgwm515* primers with 19 alleles at chromosome location 5D and 2A, respectively. The high percentage of polymorphism detected by microsatellites markers has been reported in Portuguese bread wheat cultivars (98.5%) (Carvalho et al., 2010), in Chinese barley accessions (98.13%) (Hou et al., 2005), and in Mediterranean faba bean cultivars (98.9%) (Terzopoulos and Bebeli, 2008).

Table 2: Wheat microsatellite marker name, chromosomal location, no. of alleles, and gene diversity for the microsatellite markers used in this study.

	Primer	Ch.	Allele No	Major Allele. Frquency	PIC	Gene Diversity
1	Xgwm135	1A	9	0.25	0.8272	0.845
2	Xgwm357	1A	11	0.15	0.8798	0.89
3	Xgwm550	1B	13	0.3	0.8551	0.865
4	Xgwm11	1B	12	0.2	0.8745	0.885
5	Xgwm18	1B	13	0.15	0.903	0.91
6	Xgwm498	1B	10	0.55	0.6609	0.675
7	Xgwm140	1B	10	0.55	0.6609	0.675
8	Xgwm153	1B	12	0.2	0.8804	0.89
9	Xgwm642	1D	13	0.25	0.876	0.885
10	Xgwm232	1D	13	0.2	0.8806	0.89
11	Xgwm636	2A	13	0.2	0.8921	0.9
12	Xgwm47.1	2A	8	0.5	0.6661	0.695
13	Xgwm339	2A	15	0.2	0.8982	0.905
14	Xgwm312	2A	17	0.1	0.9312	0.935
15	Xcfa2043a	2A	10	0.55	0.6609	0.675
16	Xbarc353.2	2A	11	0.2	0.8685	0.88
17	Xwmc261	2A	18	0.1	0.9367	0.94
18	Xcfa2058	2A	12	0.2	0.8691	0.88
19	Xgpw2206	2A	11	0.25	0.8579	0.87
20	Xwmc109d	2A	16	0.15	0.9202	0.925
21	Xgwm47.2	2A	13	0.2	0.8921	0.9
22	Xgwm294b	2A	10	0.25	0.8456	0.86
23	Xgwm515	2A	19	0.1	0.9422	0.945
24	Xwmc296	2A	14	0.3	0.8614	0.87
25	Xgwm95	2A	14	0.2	0.8979	0.905
26	Xgwm328	2A	15	0.2	0.9038	0.91
27	Xwmc170	2A	16	0.1	0.9256	0.93
28	Xgwm10	2A	18	0.1	0.9367	0.94
29	Xgwm512	2A	7	0.45	0.7069	0.735

Table 2: continued

	Primer	Ch.	Allele No	Major Allele. Frquency	PIC	Gene Diversity
30	*Xgwm372*	2A	14	0.3	0.8614	0.87
31	*Xcfa2121b*	2A	14	0.2	0.8923	0.9
32	*Xgwm249*	2A	8	0.6	0.5965	0.615
33	*Xgwm257*	2B	12	0.15	0.8857	0.895
34	*Xgwm410.2*	2B	15	0.2	0.9038	0.91
35	*Xgwm429*	2B	4	0.4	0.6654	0.715
36	*Xgwm539*	2D	13	0.15	0.8917	0.9
37	*Xgwm261*	2D	11	0.3	0.8364	0.85
38	*Xgwm102*	2D	14	0.15	0.9032	0.91
39	*Xwmc11*	3A	13	0.25	0.8702	0.88
40	*Xgwm369*	3A	14	0.15	0.9087	0.915
41	*Xgwm674*	3A	13	0.15	0.903	0.91
42	*Xgwm494*	3A	1	1	0	0
43	*Xgwm162*	3A	11	0.35	0.7974	0.815
44	*Xgwm391*	3A	8	0.55	0.6445	0.665
45	*Xwmc291*	3B	14	0.15	0.9087	0.915
46	*Xwmc326*	3B	12	0.2	0.8804	0.89
47	*Xcfa2170*	3B	1	1	0	0
48	*Xbarc84*	3B	13	0.25	0.876	0.885
49	*Xbarc206*	3B	15	0.15	0.909	0.915
50	*Xwmc687*	3B	11	0.35	0.8165	0.83
51	*Xgwm108*	3B	13	0.3	0.8551	0.865
52	*Xgwm340*	3B	14	0.15	0.9032	0.91
53	*Xgwm285*	3B	12	0.2	0.8745	0.885
54	*Xgwm547*	3B	15	0.2	0.9038	0.91
55	*Xgwm341*	3D	17	0.15	0.9258	0.93
56	*Xgwm664*	3D	11	0.25	0.8518	0.865
57	*Xgwm114*	3D	13	0.35	0.8296	0.84
58	*Xgwm3*	3D	15	0.15	0.9145	0.92
59	*Xgwm314*	3D	17	0.1	0.9312	0.935

Table 2: continued

	Primer	Ch.	Allele No	Major Allele. Frquency	PIC	Gene Diversity
60	*Xgwm161*	3D	14	0.15	0.9087	0.915
61	*Xgwm165*	4A	14	0.15	0.9032	0.91
62	*Xgwm601*	4A	13	0.2	0.8863	0.895
63	*Xgwm610*	4A	13	0.2	0.8863	0.895
64	*Xgwm637*	4A	10	0.55	0.6609	0.675
65	*Xgwm350*	4A	15	0.15	0.909	0.915
66	*Xgwm160*	4A	8	0.6	0.5965	0.615
67	*Xgwm66*	4B	14	0.2	0.8923	0.9
68	*Xgwm251*	4B	7	0.6	0.5877	0.61
69	*Xgwm368*	4B	11	0.4	0.7859	0.8
70	*Xgwm495*	4B	15	0.15	0.9145	0.92
71	*Xgwm113*	4B	9	0.3	0.8234	0.84
72	*Xgwm107*	4B	18	0.1	0.9367	0.94
73	*Xgwm165*	4D	15	0.3	0.8676	0.875
74	*Xgwm194*	4D	17	0.15	0.9258	0.93
75	*Xgwm609*	4D	13	0.2	0.8863	0.895
76	*Xgwm6*	4D	14	0.25	0.8821	0.89
77	*Xbarc48.4*	4D	13	0.15	0.8973	0.905
78	*Xgpw345*	4D	10	0.25	0.8456	0.86
79	*Xgwm624*	4D	10	0.2	0.8507	0.865
80	*Xgwm186*	5A	12	0.2	0.8691	0.88
81	*Xgwm639*	5A	18	0.1	0.9367	0.94
82	*Xgwm595*	5A	11	0.5	0.7119	0.725
83	*Xgwm410.2*	5A	15	0.15	0.9145	0.92
84	*Xgwm443*	5A	16	0.2	0.9096	0.915
85	*Xgwm415*	5A	4	0.45	0.6398	0.69
86	*Xgwm205*	5A	8	0.35	0.7898	0.81
87	*Xgwm335*	5B	11	0.3	0.8364	0.85
88	*Xgwm554*	5B	15	0.1	0.9199	0.925
89	*Xgwm271*	5B	13	0.2	0.8751	0.885

Table 2: continued

	Primer	Ch.	Allele No	Major Allele. Frquency	PIC	Gene Diversity
90	*Xgwm604*	5B	18	0.1	0.9367	0.94
91	*Xgwm190*	5D	10	0.25	0.8398	0.855
92	*Xgwm174*	5D	11	0.35	0.8165	0.83
93	*Xgwm212*	5D	19	0.1	0.9422	0.945
94	*Xgwm654*	5D	11	0.25	0.8289	0.845
95	*Xgwm121*	5D	5	0.35	0.7352	0.77
96	*Xgwm565*	5D	14	0.3	0.8614	0.87
97	*Xgwm169*	6A	16	0.1	0.9256	0.93
98	*Xgwm427*	6A	13	0.35	0.8296	0.84
99	*Xgwm613*	6B	15	0.3	0.8676	0.875
100	*Xgwm644*	6B	15	0.2	0.9038	0.91
101	*Xgwm70*	6B	13	0.15	0.903	0.91
102	*Xgwm219*	6B	11	0.35	0.8165	0.83
103	*Xgwm518*	6B	10	0.4	0.7722	0.79
104	*Xbarc196*	6D	9	0.25	0.8334	0.85
105	*Xwmc416*	6D	7	0.5	0.6732	0.7
106	*Xgwm469*	6D	16	0.15	0.9202	0.925
107	*Xgwm325*	6D	13	0.25	0.876	0.885
108	*Xgwm635*	7A	10	0.3	0.8299	0.845
109	*Xgwm332*	7A	8	0.35	0.7898	0.81
110	*Xgwm282*	7A	11	0.4	0.7859	0.8
111	*Xgwm260*	7A	6	0.35	0.7626	0.79
112	*Xgwm569*	7B	12	0.25	0.87	0.88
113	*Xgwm400*	7B	11	0.4	0.7859	0.8
114	*Xgwm297*	7B	15	0.1	0.9199	0.925
115	*Xgwm333*	7B	11	0.2	0.8743	0.885
116	*Xgwm43*	7B	15	0.15	0.909	0.915
117	*Xgwm274*	7B	15	0.25	0.8881	0.895
118	*Xgwm611*	7B	13	0.25	0.8702	0.88
119	*Xgwm146*	7B	15	0.25	0.8881	0.895

Table 2: continued

	Primer	Ch.	Allele No	Major Allele. Frquency	PIC	Gene Diversity
120	*Xgwm577*	7B	12	0.4	0.7927	0.805
121	*Xgwm46*	7B	17	0.1	0.9312	0.935
122	*Xgwm295*	7D	13	0.25	0.8702	0.88
123	*Xgwm44*	7D	12	0.35	0.823	0.835
124	*Xgwm437*	7D	16	0.25	0.894	0.9
125	*Xgwm37*	7D	8	0.3	0.798	0.82
126	*Xgwm121*	7D	4	0.35	0.6804	0.73
	Average		12.35	0.2666	106.57	105.15
	Total		1557	33.75	0.8466	0.8354

Mohammadi et al. (2009) reported the high values of SSR-based gene diversity and polymorphic information content (PIC) of 0.7 and 0.66 for 27 Iranian local commercials and adapted wheat cultivars. The monomorphism SSR markers Xgwm494 and Xcfa2170 were at chromosome location 3A and 3B respectivelty. The highest number of microsatellite loci in the existing microsatellite coverage of wheat is on the A genome and the lowest is on the D genome (Table 3).

In order to distinguish the best clustering and similarity coefficient calculation methods, the cophenetic correlation, a measure of the correlation between the similarity represented on the dendrograms and the actual degree of similarity, was calculated for each method combination. Among different methods, the highest value (r=0.70) was observed for UPGMA clustering method based on Dice (Nie & Li) similarity coefficient (Table 4). Therefore, the dendrogram constructed based on this method was used for depicting genetic diversity of genotypes (Fig. 1). Cluster analysis (Fig. 1) divided the 20 genotypes into Six groups.

The first included the wheat genotyping Roshan, Boolani, Shoele and Shotordandan. The second cluster was divided into two sub accessions. Group 2 contains wheat genotyping Arta, Mahooti, Arg, 140-PR-87, Gaspard, Kharchia. In group 2 the highest similarity value was observed between 140-PR-87 and Gaspard genotypes. Most of wheat genotypes were placed in group 3 and 4. Group 3 contains 6 genotypes as N-83-3, MV-17, KRL-4, Sistan, 107-PR-87, 139-PR-87. Group 4 contains 2 Genotypes as Moghan-3 and S-78-11. In groups 5 and 6 were only two genotypes SNH-9 and Sorkhtokhm, respectively. The genetic distances between studied genotypes were represented in Table 6. The highest genetic distance was recorded between Sorkhtokhm and Mahooti with the highest similarity index (0.960). On the other hand, the two most distantly related cultivars were Gaspard and 140-PR-87 with low similarity index (0.811) (Table 5). Since only a wide genetic base gives the opportunity to select genotypes with a trait of interest, it is essential to understand the extent and distribution of genetic variation. This type of information is particularly important for wheat as an important crop grown in the world and especially in Iran and as a result of a wide range of genetic diversity observed among all genotypes. The results have shown that it is possible both to classify the genetic diversity of elite genotypes and select genotypes or cultivars for the highest genetic diversity using SSRs, as indicated by cluster analysis. Several authors reported a narrow genetic diversity in wheat when assessed with RAPD and DNA amplification fingerprinting (DAF) (Abdollahi Mandoulakani et al., 2010), AFLPs (Khalighi et al., 2008; Shoaib and Arabi, 2006).

Table 3. Means of polymorphic information contents (PIC) for SSR markers located on each chromosome.

Chromosome	No of markers on each chromosome	No of polymorphic alleles	PIC
1A	2	20	0.8534
2A	22	293	0.8528
3A	5	60	0.8247
4A	6	73	0.8070
5A	7	84	0.8244
6A	2	29	0.8775
7A	4	35	0.7920
Average A Genomes	48	594	0.8361
1B	6	70	0.8058
2B	3	31	0.8183
3B	9	120	0.8807
4B	6	74	0.8234
5B	4	57	0.8920
6B	5	64	0.8525
7B	10	136	0.8729
Average B Genomes	43	552	0.8538
1D	2	26	0.8783
2D	3	38	0.8770
3D	6	87	0.8935
4D	7	92	08793
5D	6	70	0.8373
6D	4	45	0.8257
7D	5	53	0.8131
Average D Genomes	33	411	0.8574

Table 4. Comparison of different methods for constructing dendrogram.

Analysis Method	Cophenetic coefficient (r)		
	Simple Matching	Jaccard	Dice (Nie & Li)
UPGMA	0.58	0.68	0.70[*]
Complete Linkage	0.51	0.66	0.59

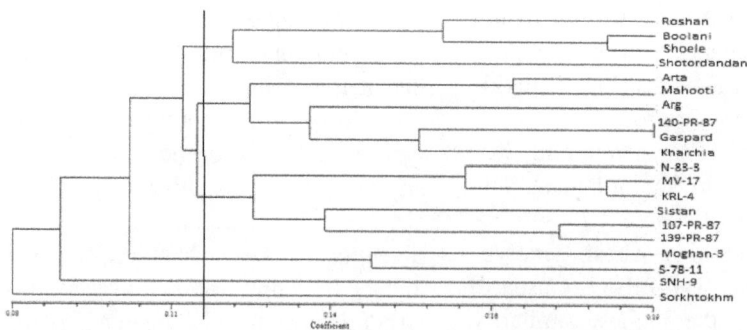

Fig. 1. A dendrogram based on genetic similarities discriminated all the wheat genotyping used in this study.

Table 5: Similarity matrix for the 20 wheat genotypes based on their microsatellite markers.

G:Genotypes , 1.Roshan, 2.Arta, 3.Moghan-3, 4.S-78-11, 5.N-83-3, 6.MV-17, 7.KRL-4, 8.Arg, 9.Shotordandan, 10.Boolani, 11.Shoele, 12.Sorkhtokhm, 13.SNH-9, 14.Sistan, 15.107-PR-87, 16.139-PR-87, 17.140-PR-87, 18.Kharchia, 19.Mahooti, 20.Gaspard.

G	1	2	3	4	5	6	7	8	9	10	11	12	13	14	15	16	17	18	19	20
1	......																			
2	0.889	......																		
3	0.937	0.897	......																	
4	0.897	0.874	0.858	......																
5	0.889	0.874	0.921	0.874	......															
6	0.929	0.889	0.905	0.921	0.834	......														
7	0.889	0.921	0.9291	0.858	0.850	0.818	......													
8	0.897	0.889	0.913	0.881	0.881	0.874	0.866	......												
9	0.937	0.881	0.921	0.921	0.889	0.897	0.881	0.897	......											
10	0.866	0.881	0.874	0.905	0.874	0.897	0.881	0.905	0.874	......										
11	0.826	0.913	0.905	0.913	0.913	0.913	0.889	0.881	0.834	0.818	......									
12	0.897	0.937	0.937	0.905	0.929	0.929	0.929	0.913	0.944	0.897	0.913	......								
13	0.944	0.897	0.921	0.921	0.905	0.921	0.905	0.897	0.905	0.905	0.929	0.921	......							
14	0.897	0.889	0.897	0.881	0.889	0.850	0.889	0.889	0.889	0.881	0.842	0.913	0.905	......						
15	0.866	0.897	0.897	0.929	0.881	0.897	0.881	0.897	0.905	0.897	0.889	0.913	0.921	0.842	......					
16	0.889	0.897	0.905	0.921	0.881	0.881	0.850	0.897	0.905	0.905	0.897	0.905	0.921	0.889	0.826	......				
17	0.850	0.866	0.881	0.889	0.889	0.897	0.889	0.874	0.889	0.866	0.897	0.897	0.905	0.866	0.874	0.818	......			
18	0.897	0.897	0.905	0.897	0.905	0.874	0.889	0.905	0.881	0.905	0.897	0.905	0.913	0.905	0.897	0.897	0.834	......		
19	0.881	0.834	0.858	0.905	0.897	0.874	0.929	0.921	0.913	0.881	0.913	0.960	0.889	0.889	0.921	0.937	0.874	0.881	......	
20	0.889	0.858	0.866	0.889	0.889	0.88	0.850	0.826	0.874	0.866	0.866	0.913	0.889	0.874	0.874	0.866	0.811	0.866	0.842	......

The knowledge about the genetic relationships of genotypes also provides useful information to address breeding programs and germplasm resource management. This type of investigation on genetic diversity is helpful for developing appropriate science based strategies for wheat breeding (Landjeva et al.,2006) and it can be a good tool of selecting genotypes in breeding programs. In conclusion, this study confirms the usefulness of SSR markers to study wheat genetic diversity. Only 36% of all primer pairs flanking wheat.

References

1. Abdollahi, M. B., Shahnejat-Bushehri, A.A., Sayed Tabatabaei, B.E., Torabi, S. and Mohammadi, H. A. 2010. Genetic diversity among wheat cultivars using molecular markers. J Crop Improv, 24:299-309.

2. Annual statistical yearbook, 2007. Croatian central bureau of statistics in RH; http://www.dzs.hr//default_e.htm.

3. Bassam, B.J. and Caetano-Anolles, G. 1993. Silver Staining of DNA Polyacrilamide geles. Appl . Biochem ., Biotechnolol, 42:181-188.

4. Botstein, D., White, R. L., Skolnick, M. and Davis, R. W. 1980. Construction of a genetic linkage map in man using restriction fragment length polymorphisms. Am. J. Hum. Genet, 32:314-331.

5. Carvalho, A., Guedes-Pinto, H., Martin- Lopes,T. and Lima-Brito, P. 2010. Genetic variability of old Portuguese bread wheat cultivar assayed by IRAP and REMAP markers. Ann Appl Biol, 156:337-345.

6. Chen, H.B., Martin, J.M., Lavin, M. and Talbert, V. 1994. Genetic diversity in hard red spring wheat based on sequence tagged-site PCR markers. Crop Sci, 34: 1629-1632.

7. Dellaporta, S.L., Wood, J. and Hicks, J.B. 1983. A plant DNA minipreparation Version II, Plant Mol Biol, 4:19-21.

8. Devos, K.M., Bryan, G.J., Collins, A.J. and Gale, M.D. 1995. Application of two microsatellite sequences in wheat storage proteins as molecular markers. Theor Appl Genet, 90: 247-252.

9. Dice, L.R. 1945. Measures of the amount of ecologic association between species. Ecology, 26: 297–302.

10. Hou, Y.C., Yan, Z.H.,Wei, Y.M. and Zheng, Y.L. 2005. Genetic diversity in barley from west China based on RAPD and ISSR analysis. barley gen Newsletter, 35:9-22.

11. Khalighi, M., Arzani, A.and Poursiahbidi, M.M. 2008. Assessment of genetic diversity in Triticum spp. and Aegilops spp. using AFLP markers. Afr J Biotech, 7:546-552.

12. Landjeva, S., Korzun, V.and Ganeva, G. 2006. Evaluation of genetic diversity among Bulgarian winter wheat (Triticum aestivum L.) varieties during the period 1925–2003 using microsatellites. Genet Resour Crop Evol, 53: 1605–1614.

13. Mohammadi, S.A., Khodarahmi, M., Jamalirad, S. and Jalalkamali, M.R. 2009. Genetic diversity in a collection of old and new bread wheat cultivars from Iran as revealed by simple sequence repeat-based analysis. Ann Appl Biol, 154:67-76.

14. Nei, M. Li, W.H. 1979. Mathematical model for studying genetic variation in terms of restriction endonucleases. Proc Nat Acad Sci, 76: 5269–5273.

15. Plaschke, J., Ganaland, M.W. and Röder, M.S. 1995. Detection of genetic diversity in closely related bread wheat using microsatellite markers. Theor Appl Genet, 91: 1001-1007.

16. Powell, W., Morgante, M., Andre, C., Hanafey, M., Vogel, J., Tingey, S. and Rafalski, A. 1996. The comparison of RFLP, RAPD, AFLP and SSR (Microsatellite) markers for germplasm analysis. Mol Breed, 2: 225–238.

17. Röder, M.S., Korzun, V., Wendehake, K., Plaschke, J., Tixier, M.H., Leroy, P. and Ganal, M.W. 1998. A Microsatellite map of wheat. Genetics, 149: 2007-2023.

18. Rohlf, F.J. 1998. NTSYS-PC: Numerical taxonomy and multivariate analysis system. Version 2.0. Applied Biostatistics. New York.

19. Senturk, A. and Ahu, A. U. 2013. Genetic Diversity of Winter Wheat (TriticumaestivumL.) Revealed by SSR Markers. Biochem Genet, 51:223–229

20. Shoaib, A. and Arabi, M.I.E.2006. Genetic diversity among Syrian cultivated and landraces wheat revealed by AFLP markers. Genet Resour Crop Evol, 53:901-906.

21. Tautz, D. 1989. Hyper Markers variability of Simple Sequences as a General Source of Polymorphic Markers. Nucl Acids Res, 17:6463-6471.

22. Taut, Z.D. and Renz, M. 1984. simple sequence are ubiquitious repetitive components of eukaryotic. Nucl Acides Res,25:4127-4138.

23. Terzopoulos, P.J. and Bebeli, P.J. 2008. Genetic diversity analysis of Mediterranean faba bean (Vicia faba L.) with ISSR markers. Field Crops Res, 108:39-44.

17

The effect of salinity stress on Na⁺, K⁺ concentration, Na⁺/K⁺ ratio, electrolyte leakage and HKT expression profile in roots of *Aeluropus littoralis*

Masoud Fakhrfeshani[1,2], Farajollah Shahriari-Ahmadi[1], Ali Niazi[3*], Nasrin Moshtaghi[1], Mohammad Zare-Mehrjerdi [4]

1. Department of Plant Breeding and Biotechnology, Ferdowsi University of Mashhad, Iran
2. Department of Plant Biotechnology, University of Jahrom, Iran
3. Institute of Biotechnology, Collage of Agriculture, Shiraz University, Iran
4. Shirvan Higher Education Complex, Iran

*Corresponding author (✉): niazi@shirazu.ac.ir

Abstract

Among abiotic stresses, salinity has been increasing over the time for many reasons like using chemical fertilizers, global warming and rising sea levels. Under salinity stress, the loss of water availability, toxicity of Na⁺ and ion imbalance directly reduces carbon fixation and biomass production in plants. K⁺ is a major agent that can counteract Na⁺ stresses, thus the potential of plants to tolerate salinity is strongly dependent on their potassium nutrition. HKTs (High-affinity K⁺ Transporters) are a family of transporters that mediate Na⁺-specific or Na⁺-K⁺ transport and play a key role in the regulation of ion homeostasis. In this study, we intended to focus on Electrolyte Leakage, ratio of K⁺/Na⁺, transcriptomic responses of a subclass two HKT in the roots of Aeluropus littoralis under salt stress. We investigated a noticeably different expression pattern over studied time points and found a snappy increase of AlHKT and rebalance of K⁺ concentration. It can be suggested that the early and high response of a Na⁺-K⁺ coupled transporter acted as a part of A. littoralis salt tolerance.

Keywords: Aeluropus littoralis, Flame photometry, Membranous HKT, Potassium, Sodium, Real-time PCR.

Introduction

Human population is growing rapidly yearly and it is estimated to reach 9 billion by 2050 (33). The current average increase of agricultural products is 32 million tons (Mt) per year but if we want to meet demand by 2050, the rate of annual increase must be about 44 Mt yearly (33). It means the production should be increased on an unprecedented scale while facing changing climate and dealing with modern nonfood requirements like green fuel. Abiotic stresses are the most important factors affecting plant growth and their yields (33). Theses stresses are even getting more important because of the climate change and reduction in water quality and availability (33). Among abiotic stresses, Salinity has been increasing over the time for many reasons like using chemical fertilizers, global warming, rising sea levels (16) and lack of drainage system (24). There was only 323 million hectare (MHa) saline land in 1980 but it is estimated to exceed 400 MHa by 2025 (16).

Physiological, biochemical and genetical studies have shown that salinity stress in plants is multifactorial, including osmotic stress (33) and cellular ion toxicity, which inhibits vital enzymes and metabolic processes (19, 26). Photosynthetic processes are among the most sensitive process to salinity; therefore, salinity stress directly reduces carbon fixation and biomass production in plants (18). Many policies can be adopted to confront this problem. Improving plant toward salinity tolerance is one of the most reliable methods that has been studying from many aspects. To improve the ability of crops to grow on saline soil, understanding traits and mechanisms that contribute to salt tolerance in wild and tolerant cultivars are necessary (23). However, mechanisms of plant adaptation to abiotic stresses, particularly for drought (33) and salinity are complex but understanding the mechanisms of salt tolerance by studying traits that contribute to tolerance separately and identifying natural occurring variation in varieties or wild relatives are scientifically applied methods to confront the complexity and to improve further salt tolerance in crops (20, 33).

Under salinity stress, the loss of water availability, toxicity of Na^+ and ion imbalance cause growth limitation so plants adopt divert mechanisms to tolerate salinity. It is repeatedly reported that K^+ deficiency and Na^+ toxicity are major restrictions of crop production worldwide (23, 24, 26, 28, 30, 36). K^+ can counteract Na^+ stresses (18) thus the potential of plants to tolerate salinity is strongly dependent on their potassium nutrition (26). K^+ composes about 10% of the plants dry weight and is the most abundant mineral cation in the plants (30). It is involved in many of functions related to enzyme activation, respiration and starch and protein synthesis (6, 30). For instance, about 50mM of K^+ is required for normal starch synthesis and 10-50mM is needed for activation of K^+ dependent enzymes (6). The optimum concentration of K^+ narrows by increasing the amount of Na^+ (26) for many reasons such as similarity of Na^+ and K^+ in their physicochemical structure. This similarity leads to competition of Na^+ and K^+ at transporters or enzymes binding sites that can result in K^+ deficiency and inhibition of biochemical processes that are dependent on K^+ (26). So the capacity of a cell to maintain a high K^+/Na^+ ratio is assumed to be a critical strategy in salt tolerance (26). For instance, it is reported that animal cells maintain the K^+/Na^+ ratio around 20 by regulating the K^+ and Na^+ concentration around 100mM and 5mM respectively (26). In plants, the optimum con-

centration of K^+ is reported to be about 100-150mM and the minimum value of K^+/Na^+ is about one (26). In contrast, the soil K^+ concentration ranges from 1 to 0.1mM (30, 35) and in some cases, the K^+/Na^+ ratio is less than 0.02 (26). So an efficient and controllable potassium supply system should be available for plants (35).

HKTs (High-affinity K^+ Transporters) are a family of transporters that mediate Na^+-specific transport (subclass one) or Na^+-K^+ transport (subclass two) and play a key role in regulation of ion homeostasis and contribute to osmotic adjustment and Na^+ detoxification in plants (20). They have been identified in all studied plants (6). The members of this family were identified in 1994 and were called HKT because their mutation in Xenopus laevis oocytes had led to defective K^+ uptake (6). According to the studies focused on isolated HKTs and their distribution in a variety of crops, it is demonstrated that they are involved in K^+ related salt tolerance responses rather than K^+ nutrition (6). A. littoralis is a wild halophyte member of Poaceae, native to coastal zones that is salt and drought tolerant (4, 35). Although A. littoralis is usually exposed to high saline conditions, it grows normally without any toxic symptoms and can maintain a high K^+/Na^+ ratio under a high salt environment (35).

Retention of a high K^+/Na^+ ratio is defined as a determinative trait in salt tolerance (26). In addition to this, wild relative of crops are important as a naturally selected gene sources. Based on this facts, this study was performed to focus on concentration of Na^+, K^+ and their ratio along with the transcriptomic expression pattern of a HKT2 of A. littoralis's roots at a 15-days span of salinity stress.

Materials and Methods

Plant material

The seeds of A. littoralis were purchased from ICRASN (Isfahan Center for Research of Agricultural Science and Natural Resources). The seeds were planted in sand pots in growth chamber (12/12h day/night, temp. 18°C/22°C). After 21 days, the seedlings were transferred to continuously aerated hydroponic systems containing Yoshida nutrition medium (10, 15) [pH= 5.5, 16/8h day/night, temp. 25°C]. The pH was checked and rebalanced every day and the medium was refreshed weekly. The salinity stress started after 21-days establishment period in hydroponic medium. The salinity stress experiment was conducted based on a completely randomized design with 12 sampling concentration/time points and three biological replications. In fact, our study took place in two phases. A pre-experiment was carried out to evaluate some physiological responses of *A. littoralis* to salinity and finding the best concentration/time points for main physiological and transcriptomic studies. For this purpose, we conducted the salinity stress and sampling at 2, 4, 6, 8, 12, 24 and 48 hours of 100, 200, 300, 400, 500 and 600 mM NaCl and simultaneously with their equivalent controls, all with three replications (7×6×2×3 samples). At the end of pre- experiment, the fresh and dry weights (data not shown), electrolyte leakage (Figure 1), Na$^+$ (Figure 2) and K$^+$ concentration (Figure 3) of them was measured. According to these data, the time points with the significant change were chosen for sampling in the main experiment.

Finally, the sampling of roots were done at 6h/100mM, 6h/200mM, 6h/300mM, 48h/300mM, 72h/300mM and 264h/300mM time/concentration points. The sampling of control plants was also done at the same time / concentration points concurrently with treated ones. Collected root tissues were snap-frozen using liquid nitrogen and were stored at -80°C immediately.

qRT PCR

Real time RT-PCR was performed by isolating RNA from root tissues using Denazist Column RNA Isolation Kit (#S-1020, Iran) followed by DNase digestion (Thermo Scientific DNase I #EN0525, USA). Synthesis of first strand cDNAs was carried out by using Thermo Scientific RevertAid First Strand cDNA Synthesis Kit (#K1622). For quantitative real time PCR 5μL of diluted, (1 to16) cDNA were used as templates and the reaction of cDNA synthesis was done using RealQ Plus Master Mix (AMPLIQON #A314402, Denmark) in a Bioer thermo cycler (Applied Bioer, LineGeneK,

Hangzhou, China). Due to the absence of *A. littoralis HKT's* gene sequence in databases, primers for the amplification of membranous *AlHKT* were designed based on the membranous *HKT*2s (subfamily 2) of *Poaceae*, specially CDs of *Zea maize* and *Setaria italica* using Bioedit® v7.0.9 to align sequences and Primer Premier© v5 to design primers. Finally, the following primers were used for amplification of membranous *AlHKT* (product size 369bp):

AlHKT　5' GAAACCGAGCAACCCTGAC 3'
　　　　 5' AATCCTAAGTATCTAACGCTC 3'

In order to normalize the qPCR data, primers of *βActin* (product size 113bp) and *elongation factor* (product size 90bp) genes were designed in the following sequences based on the above mentioned method and examined.

β Actin　5' TTGCTGGCCGAGACCTTAC 3'
　　　　 5' GGCGAGCTTTTCCTTGATG 3'

elfactor　5' ACCTTCTCTGAATACCCTCCTCTG 3'
　　　　 5' CTTCTCCACACTCTTGATGACTCC 3'

The PCR condition for *AlHKT* was 94°C/10sec, 63°C/20sec and 72°C/20sec and for *βActin* it was 94°C/10sec, 55°C/20sec and 72°C/20sec. The calculation of relative gene expression was done based on methods that explain expression ratio equal to $2^{-\Delta\Delta Ct}$ (31).

Sodium and Potassium Concentration

The measurement of Na$^+$, K$^+$ concentration was performed using flame photometry (#pfp7, Jenway. UK). Because tissue mass of some samples were very low, the preparation of the samples was done using 15mg of root tissues according to a wet ashing method (11). In this method, instead of direct heat of oven, heat and concentrated nitric acid are used to degrade organic material of tissues.

Statistical analysis

As mentioned above, our experiment was performed based on a completely randomized design with three biological replications in the samplings. To increase the reliability of gene expression analysis, real time PCR experiments were done with two identical technical replications. The statistically analyzes of the data were done using Tukey's range test (α = 0.05; MINITAB v16.).

Results and Discussion

The PCR production results of all primers were coincided with bioinformatically predicted lengths. After examination of housekeeping genes, *ßActin* selected as the reference gene for normalizing the qPCR data because it showed less influence under salinity stress. The expression pattern of *AlHKT* showed that it differs noticeably over concentration/time points; in fact its expression reduced to ⅓ at the initiation of stress (6h** – Figure 4) but increased snappily after 54h (Figure 4). The highest expression ratio of *AlHKT* was observed at 6h/200mM NaCl equal to 54h from initiation of stress (Figure 4). The highest expression at this time point was more than eleven (11.57) times higher than controls but the expression was reduced and remained equal to untreated samples after that (102h** - Figure 4). The report of Horie et al. (19) on rice also showed similar result that low-K+ concentrations (less than 3 mM) induced the expression of all *OsHKT* genes in roots, but mRNA accumulation was inhibited by the presence of 30 mM Na+. This drop in gene expression can be attributed to this assumption that the studied *HKT* genes do not carry regulation elements related to rapid and continuous expression. It can be assumed that any mechanisms other than continues *HKT* expression helped them to overcome salinity stress. In our study, it was assumed that because Yoshida medium's K+ content (2 mM) is lower than other common hydroponic mediums like Hoagland (4-5 mM), so this high accumulation of *HKT* seems to be necessary to obviate K+ deficiency.

It is thought that Na+ and K+ homeostasis are crucial for salt-tolerance in plants (19; 8). Most plant species are sensitive to high concentrations of sodium (Na+), which produce combined Na+ toxicity and osmotic stress (18). It is also successively reported that K+ is the most efficient monovalent cation in both biophysical and biochemical processes (6). Thus, plants designed and gathered efficient acquisition, redistribution and homeostasis systems for Na+ sequestration (5, 8, 14) or increasing cytoplasmic potassium (K+) levels relative to Na+ that would lead to increased salt tolerance in plants (18). Up to now, plant membranous potassium transporters are categorized into three families (14) but according to the reports of Horie *et al* (2009), the major salt-tolerance trait in monocot crops is based on some of the *HKT*-mediated mechanisms.

In line with the changes of *HKT* expression, we also investigated the ratio of sodium and potassium concentration in samples. The Na+ concentration in treated samples increased constantly from 10 to 35 mg g⁻¹DW over the

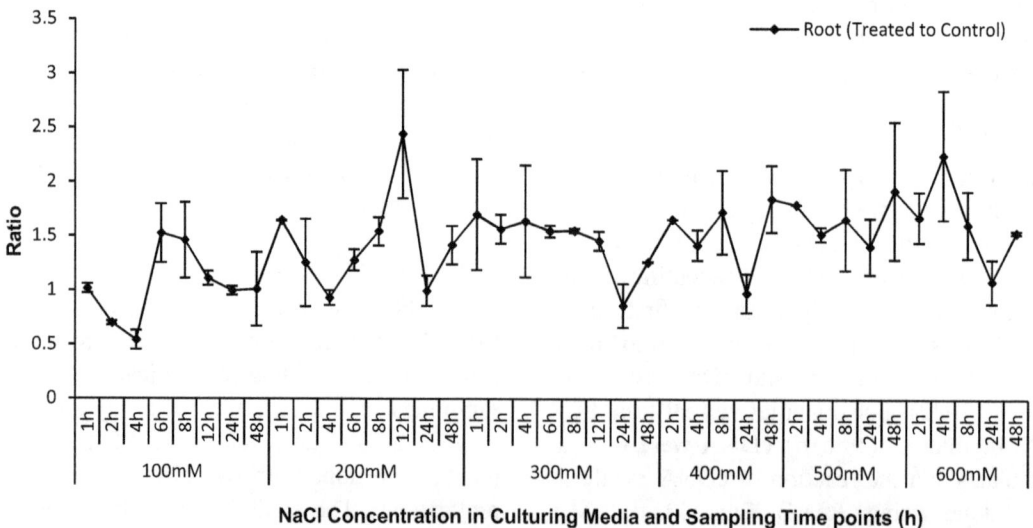

Figure 1. Treated to Control Electrolyte leakage ratio in roots of *Aeluropus littoralis* under salinity stress of pre-test.

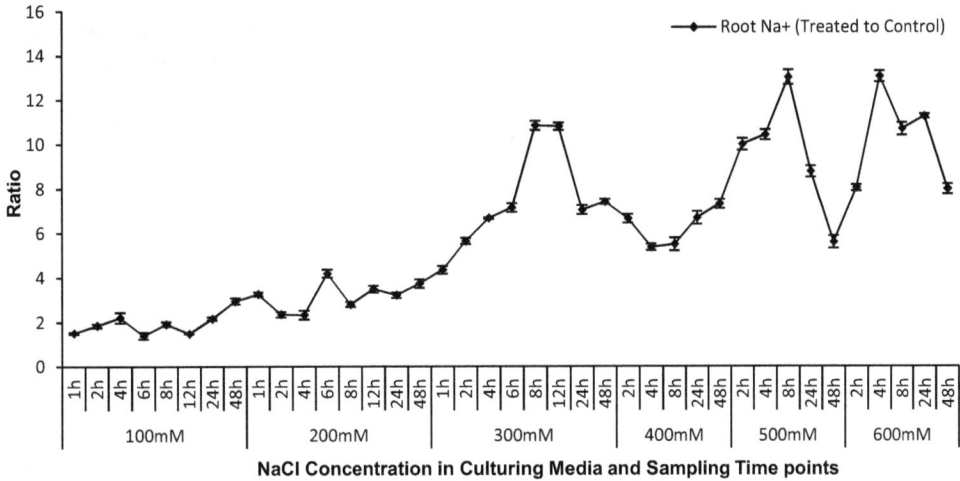

Figure 2. Treated to Control Na$^+$ Concentration ratio in roots of *Aeluropus littoralis* under salinity stress of pre-test.

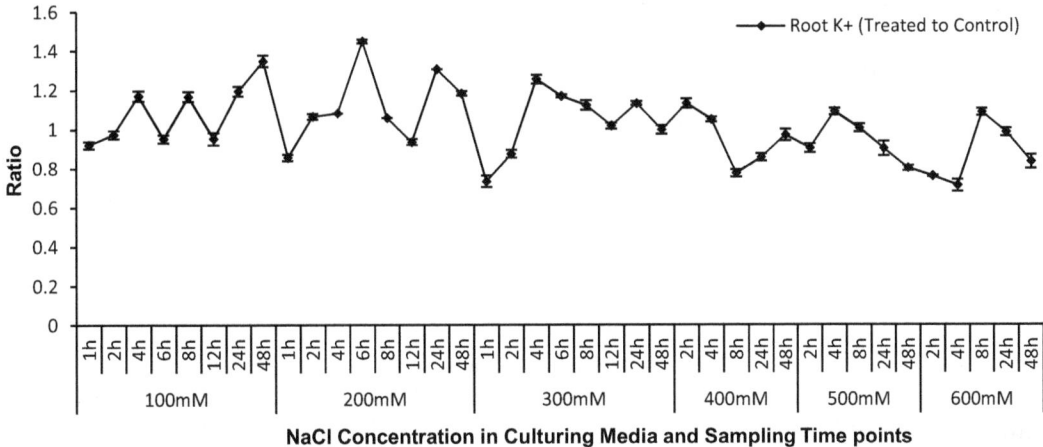

Figure 3. Treated to Control K$^+$ Concentration ratio in roots of *Aeluropus littoralis* under salinity stress of pre-test.

stress span (Figure 5). However, it seemed that the concentration of Na$^+$ had been constantly maintained about 5-10mg g^{-1}DW in untreated samples. In contrast, as it is illustrated in figure 6, the concentration of the K$^+$ in root tissue of control samples increased and maintained about 45mg g^{-1}DW, however it decreased gradually after 48 hours from initiation of stress that could be because of the increasing of tissue mass (in vegetative growth). It is also reported that Aeluropus root biomass increases under salt stress (4). The concentration of potassium ion in treated samples decreased in contrast to control samples. This decrease in concentration has continued until 54 hours since initiation of salinity stress and then increased and continued

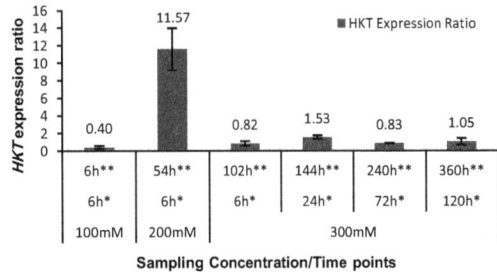

Figure 4. The treated to control expression pattern of HKT in *Aeluropus littoralis* root under salinity stress.

* Shows hours from the initiation of mentioned concentration.

** Shows hours from the initiation of stress.

in a fluctuation like control samples but in a statistically non-significant lower level. It seems that the reduction of accessible K$^+$ or lower

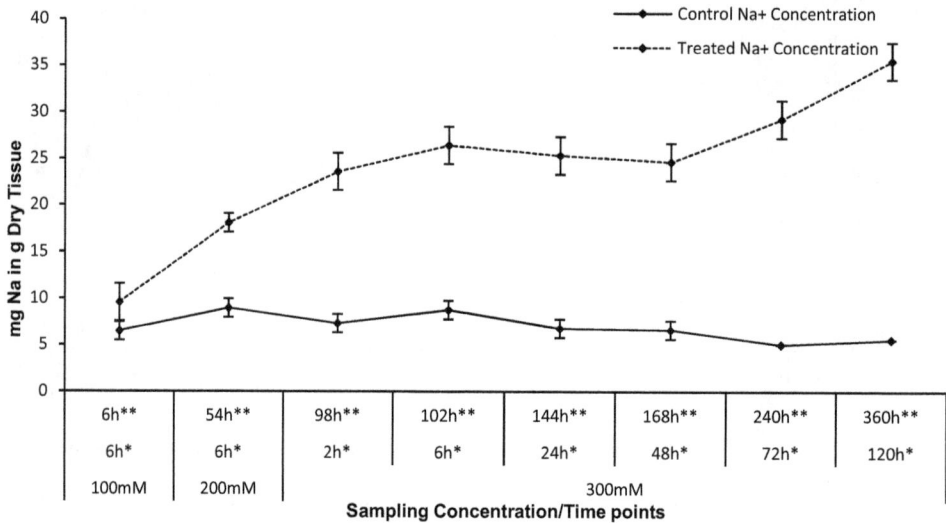

Figure 5. The fluctuation pattern of Na$^+$ concentration in *Aeluropus littoralis* root of control and under salinity stress samples.
* Shows hours from the initiation of mentioned concentration.
** Shows hours from the initiation of stress.

expression of HKT (Figure 1) led to the early reduction of K$^+$ in treated samples; as mentioned, the expression of membranous HKT increased dramatically (about 11.5 times) in 54 hours after the initiation of stress (= 6h*/200mM, Figure 5).

The increase of HKT proteins as a result of the increases in transcriptomic level may lead to accumulation of K$^+$ in roots after 72 hours in stressed samples (= 24h/300mM). The deferment between the transcriptomic increase of HKT and accumulation of K$^+$ could be attributed to the time for translation and activation of HKT proteins.

The concentration of Na$^+$ maintained about 5-10mg g^{-1}DW in control samples but increased significantly in stressed samples and led to the decrease of K$^+$/Na$^+$ ratio in treated samples (Figure 7).

This increase of Na$^+$ can be attributed to multiple reasons. The first reason can be the competition of Na$^+$ and K$^+$ to influx into the root cells; The electrophysiological similarity of Na$^+$ and K$^+$ (6), in addition to the higher accessibility of Na$^+$, may lead to its higher seepage into root cells. It is also reported that in the absence of potassium, certain potassium channels can conduct sodium (25). Secondly, it could be assumed that Na$^+$/ K$^+$ simporters have been

activated to increase the concentration of K$^+$ in response to the decline of K$^+$. This may have happened along with compartmentization of Na$^+$ in organelles like vacuoles (however it needs further studies to be approved). *A. littoralis* is a member of the four genera (*Spartina, Aeluropus, Distichlis* and *Chloris*) of Poacea family that are able to excrete salts and *A. littoralis* do it with a high selectivity in favor of sodium opposing K$^+$ and Ca$^+$ (4). So the Na$^+$ may be transported to shoot tissues to be excreted via specific glands. Third, because of electrophysiological similarity of sodium and potassium (6), in potassium deficiency, it can be replaced with Na$^+$ for some biological processes with lower specificity to potassium (12). Cramer et al. (9) and Mäser et al. (28) also reported that K$^+$ counteracts Na$^+$ stress, while Na$^+$, in turn, can to a certain degree, alleviate K$^+$ deficiency. So, the gradual increase in Na$^+$ concentration and HKT expression observed in our study could be assumed as a strategy of *A. littoralis* using Na$^+$ to maintain the biological process (33). Fourth, as it is reported, high- affinity potassium transporters (HKTs) are a large super family of transporters in plants, bacteria, and fungi (2, 14, 28, 32) It has been suggested that these transporters play crucial roles in salinity tolerant via removal of Na$^+$

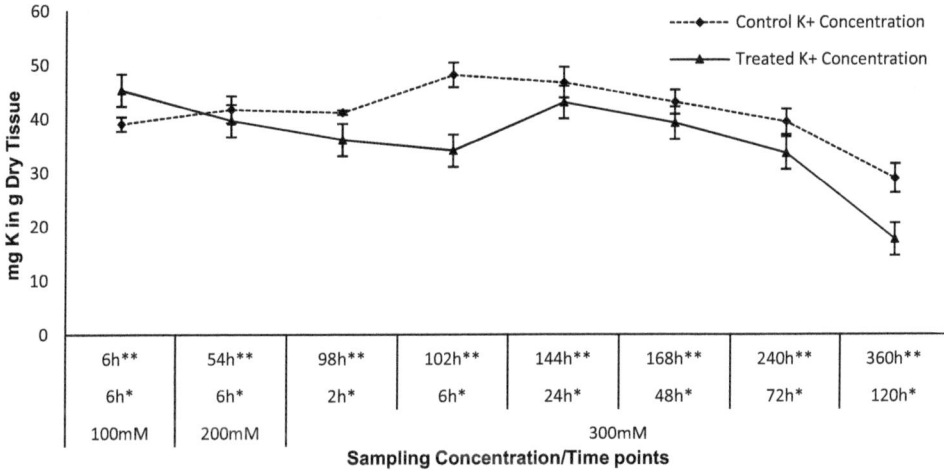

Figure 6. The fluctuation pattern of K$^+$ concentration in *Aeluropus littoralis* roots of control and under salinity stress samples.
* Shows hours from the initiation of mentioned concentration.
** Shows hours from the initiation of stress.

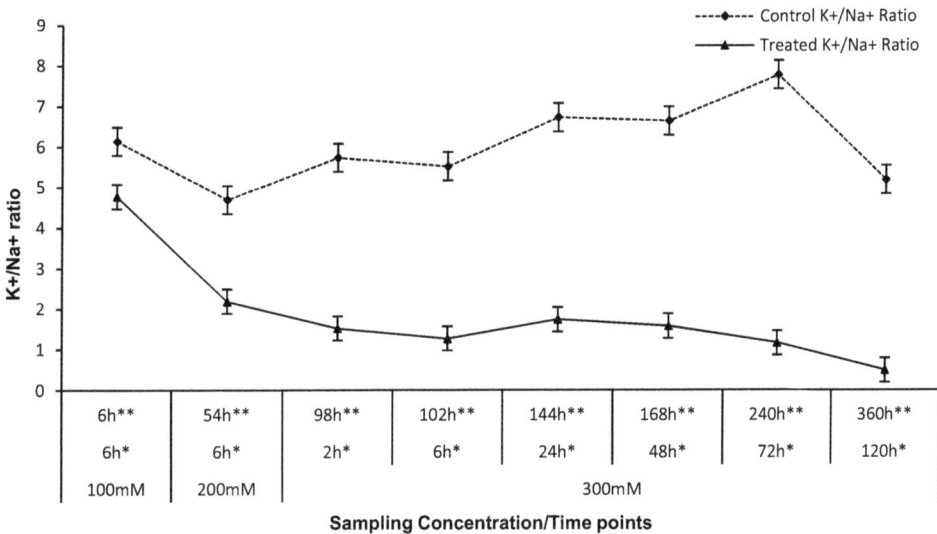

Figure 7. The fluctuation pattern of K$^+$/Na$^+$ ratio in *Aeluropus littoralis* root under salinity stress.
* Shows hours from the initiation of mentioned concentration.
** Shows hours from the initiation of stress.

from the xylem during salinity stress (21, 22). Based on the amino acid sequences, *HKTs* are categorized into two subfamilies called 1 and 2 (32). The members of two subfamilies are different in their cation (Na$^+$ or K$^+$) selection (6). Members of subfamily 1 are all Na$^+$ specific transporters (6, 20) but the other subfamily members are Na$^+$/K$^+$ uniporters (20). It is demonstrated that they are more permeable to K$^+$ compared with the subfamily 1 of *HKTs* and

the transcript level of them has increased in K$^+$ starvation stress in wheat (*Triticum aestivum*), rice (*O. sativa*) and barley (*Hurdeum vulgare*) (19). However, according to the reports of Bañuelos et al. (3), the uniport and symport activities of HKTs are also dependent on their protein density in the membrane. For instance, high level expression of barley's HKT1 (*H. vulgaris*) in yeast led to K$^+$ or Na$^+$ uniport activity, but when protein expression decreased significantly, the transporter acted in the K^+/Na^+ symport mode (6). So, the other reason for Na^+ accumulation in *A. littoralis* during salinity stress in this study on can be attributed to K^+ or Na^+ uniport activity of HKTs. However, physiological data indicate that plants have a higher capacity for K^+ uptake and distribution than for Na^+ (1, 37). The fifth reason that may explain the accumulation of sodium ion in our study is that photosynthesis is vulnerable to salinity (18) so to protect this vital process via maintaining high K^+/Na^+ ratio in photosynthetically active tissues. *A. littoralis,* alike many other species (7, 17) adopted a strategy to transport Na^+ via the phloem from shoot tissues to roots. As it is approved, Na^+/K^+ homeostasis is a key parameter in plant salinity tolerance (16, 36, 37) and the HKT family members are the most important transporters related to this balance (29) but little is known about their behavior in wild plants (2). In conclusion, our study has provided an expression analysis of *AlHKT* transporter for six concentration/time points of *A. littoralis* under salt stress by real-time PCR (Figure 4). Because it showed the highest level of expression at early point, and the increase of K^+ concentration occurred after that, we predict that K^+ inwardly flux in this species is one of the main salinity tolerance mechanism too and its HKT proteins were able to uptake K^+ efficiently in the presence of high concentration of Na^+ (Figure 6). However, the relationship between the primary structure of this transporter and Na^+ insensitivity needs to be further investigated but different Na^+ sensitivity of high-affinity K^+ uptake transporters among species is approved and attributed to their amino acid sequence and structure (35). For instance, when the *Capsicum annum* HAK1 was functionally characterized in yeast, it showed sensitivity to millimolar concentrations of Na^+ (27). HAK of *Cymodocea nodosa, (CnHAK1)* was insensitive to Na^+ (13) and the expression of HAKs from *Mesembryanthemum crystallinum* (*McHAK1* and *McHAK4*) led to growth of yeast in 150 mM NaCl (34). The early and highly expression of HKT gene studied in our experiment and its permanency and lasting activity in high concentration of Na^+ make its sequence and regulatory elements as a prospective candidate for future empirical studies as well as transforming it and improving salinity tolerance in many Poaceae crops.

Acknowledgements

The authors are grateful to the Biotechnology Institute of Shiraz University for technical support of this research. Many thanks to Khalil Malekzadeh for critically reading the manuscript.

References

1. Ahmad, I. and Maathuis, F. J. 2014. Cellular and tissue distribution of potassium: physiological relevance, mechanisms and regulation. Journal of plant physiology 171(9): 708-714.
2. Babgohari, M. Z., Niazi, A., Moghadam, A. A., Deihimi, T. and Ebrahimie, E. 2013. Genome-wide analysis of key salinity-tolerance transporter (HKT1; 5) in wheat and wild wheat relatives (A and D genomes). In Vitro Cellular & Developmental Biology-Plant 49(2): 97-106.
3. Bañuelos, M. A., Haro, R., Fraile-Escanciano, A. and Rodríguez-Navarro, A. 2008. Effects of polylinker uATGs on the function of grass HKT1 transporters expressed in yeast cells. Plant and cell physiology 49(7): 1128-1132.
4. Barhoumi, Z., Djebali, W., Smaoui, A., Chaïbi, W. and Abdelly, C. 2007. Contribution of NaCl excretion to salt resistance of Aeluropus littoralis (Willd) Parl. Journal of plant physiology 164(7): 842-850.
5. Barkla, B. J. and Blumwald, E. 1991. Identification of a 170-kDa protein associated with the vacuolar Na+/H+ antiport of Beta vulgaris. Proceedings of the National Academy of Sciences 88(24): 11177-11181.
6. Benito, B., Haro, R., Amtmann, A., Cuin, T. A. and Dreyer, I. 2014. The twins K+ and Na+ in plants. Journal of plant physiology 171(9): 723-731.
7. Berthomieu, P., Conéjéro, G., Nublat, A., Brackenbury, W. J., Lambert, C., Savio, C., Uozumi, N., Oiki, S., Yamada, K. and Cellier, F. 2003. Functional analysis of AtHKT1 in Arabidopsis shows that Na+ recirculation by the phloem is crucial for salt tolerance. The EMBO Journal 22(9): 2004- 2014.
8. Chen, Z., Zhou, M., Newman, I. A., Mendham, N. J., Zhang, G. and Shabala, S. 2007. Potassium and sodium relations in salinised barley tissues as a basis of differential salt tolerance. Functional Plant Biology 34(2): 150-162.
9. Cramer, G. R., Lynch, J., Läuchli, A. and Epstein, E. 1987. Influx of Na+, K+, and Ca2+ into roots of salt-stressed cotton seedlings effects of supplemental Ca2+. PlantPhysiology 83(3): 510-516.

10. Datta, K. and Datta, S. K. (2006). Indica rice (Oryza sativa, BR29 and IR64). Agrobacterium Protocols, Springer: 201- 212.

11. Enders, A. and Lehmann, J. 2012. Comparison of wet- digestion and dry-ashing methods for total elemental analysis of biochar. Communications in soil science and plant analysis 43(7): 1042-1052.

12. Evans, H. J. and Sorger, G. J. 1966. Role of mineral elements with emphasis on the univalent cations. Annual review of plant physiology 17(1): 47-76.

13. Garciadeblas, B., Benito, B. and Rodríguez-Navarro, A. 2002. Molecular cloning and functional expression in bacteria of the potassium transporters CnHAK1 and CnHAK2 of the seagrass Cymodocea nodosa. Plant molecular biology 50(4-5): 623-633.

14. Gierth, M. and Mäser, P. 2007. Potassium transporters in plants–involvement in K+ acquisition, redistribution and homeostasis. FEBS letters 581(12): 2348-2356.

15. Gregorio, G. B., Senadhira, D. and Mendoza, R. D. 1997. Screening rice for salinity tolerance. International Rice Research Institute discussion paper series (22).

16. Hakim, M., Juraimi, A. S., Hanafi, M., Ali, E., Ismail, M. R., Selamat, A. and Karim, S. R. 2014. Effect of salt stress on morpho-physiology, vegetative growth and yield of rice. Journal of Environmental Biology 35: 317-326.

17. Hauser, F. and Horie, T. 2010. A conserved primary salt tolerance mechanism mediated by HKT transporters: a mechanism for sodium exclusion and maintenance of high K+/Na+ ratio in leaves during salinity stress. Plant, cell & environment 33(4): 552-565.

18. Horie, T., Hauser, F. and Schroeder, J. I. 2009. HKT transporter-mediated salinity resistance mechanisms in Arabidopsis and monocot crop plants. Trends in plant science 14(12): 660-668.

19. Horie, T., Yoshida, K., Nakayama, H., Yamada, K., Oiki, S. and Shinmyo, A. 2001. Two types of HKT transporters with different properties of Na+ and K+ transport in Oryza sativa. The Plant Journal 27(2): 129-138.

20. Huang, S., Spielmeyer, W., Lagudah, E. S. and Munns, R. 2008. Comparative mapping of HKT genes in wheat, barley, and rice, key determinants of Na+ transport, and salt tolerance. Journal of experimental botany 59(4): 927-937.

21. James, R. A., Blake, C., Byrt, C. S. and Munns, R. 2011. Major genes for Na+ exclusion, Nax1 and Nax2 (wheat HKT1; 4 and HKT1; 5), decrease Na+ accumulation in bread wheat leaves under saline and waterlogged conditions. Journal of Experimental Botany 62(8): 2939- 2947.

22. James, R. A., Davenport, R. J. and Munns, R. 2006.

23. Kao, W.-Y. 2011. Na, K and Ca Contents in Roots and Leaves of Three Glycine Species Differing in Response to NaCl Treatments. Taiwania 56(1): 17-22.

24. Khan, M., Shirazi, M., Khan, M. A., Mujtaba, S., Islam, E., Mumtaz, S., Shereen, A., Ansari, R. and Ashraf, M. Y. 2009. Role Of Proline, K/Na Ratio And Chlorophyll Content In Salt Tolerance Of Wheat (Triticum aestivum L.). Pak. J. Bot 41(2): 633-638.

25. Korn, S. J. and Ikeda, S. R. 1995. Permeation selectivity by competition in a delayed rectifier potassium channel. Science 269(5222): 410.

26. Maathuis, F. J. and Amtmann, A. 1999. K+ nutrition and Na+ toxicity: the basis of cellular K+/Na+ ratios. Annals of Botany 84(2): 123-133.

27. Martínez-Cordero, M. A., Martínez, V. and Rubio, F. 2004. Cloning and functional characterization of the high-affinity K+ transporter HAK1 of pepper. Plant molecular biology 56(3): 413-421.

28. Mäser, P., Gierth, M. and Schroeder, J. I. 2002. Molecular mechanisms of potassium and sodium uptake in plants. Plant and soil 247(1): 43-54.

29. Mian, A., Oomen, R. J. F. J., Isayenkov, S., Sentenac, H., Maathuis, F. J. M. and Véry, A.-A. 2011. Over-expression of an Na+- and K+-permeable HKT transporter in barley improves salt tolerance. The Plant Journal 68(3): 468-479.

30. Nieves-Cordones, M., Alemán, F., Martínez, V. and Rubio, F. 2014. K+ uptake in plant roots. The systems involved, their regulation and parallels in other organisms. Journal of plant physiology 171(9): 688-695.

31. Pfaffl, M. W. 2001. A new mathematical model for relative quantification in real-time RT–PCR. Nucleic acids research 29(9): e45-e45.

32. Platten, J. D., Cotsaftis, O., Berthomieu, P., Bohnert, H., Davenport, R. J., Fairbairn, D. J., Horie, T., Leigh, R. A., Lin, H.-X. and Luan, S. 2006. Nomenclature for HKT transporters, key determinants of plant salinity tolerance. Trends in plant science 11(8): 372-374.

33. Roy, S. J., Tucker, E. J. and Tester, M. 2011. Genetic analysis of abiotic stress tolerance in crops. Current opinion in plant biology 14(3): 232-239.

34. Su, H., Golldack, D., Zhao, C. and Bohnert, H. J. 2002. The expression of HAK-type K+ transporters is regulated in response to salinity stress in common ice plant. Plant Physiology 129(4): 1482-1493.

35. Su, Q., Feng, S., An, L. and Zhang, G. 2007. Cloning and functional expression in Saccharomyces cereviae of a K+ transporter, AlHAK, from the graminaceous halophyte, Aeluropus littoralis. Biotechnology letters 29(12): 1959-1963.

36. Véry, A.-A., Nieves-Cordones, M., Daly, M., Khan, I., Fizames, C. and Sentenac, H. 2014. Molecular biology of K+ transport across the plant cell membrane: what do we learn from comparison between plant species? Journal of plant physiology 171(9): 748-769.

37. Wigoda, N., Moshelion, M. and Moran, N. 2014. Is the leaf bundle sheath a "smart flux valve" for K+ nutrition? Journal of plant physiology 171(9): 715-722.

18

The responses of *L-gulonolactone oxidase* and HKT2;1 genes in *Aeluropus littoralis'* shoots under high concentration of sodium chloride

Khalil Malekzadeh[1], Ali Niazi[2], Farajollah Shahriari-Ahmadi[1*],
Amin Mirshamsi-Kakhaki[1], Mohammad Zare-Mehrjerdi[3]

1 Department of Biotechnology and Plant Breeding, Faculty of Agriculture, Ferdowsi University of Mashhad, Iran
2 Biotechnology Institute, Faculty of Agriculture, Shiraz University, Iran
3 Shirvan Higher Education Complex, Iran
   *Corresponding Author (✉): shahriari@um.ac.ir

Abstract
Among abiotic stresses, salinity has been increasing over the time for many reasons like using chemical Salinity is one of the most important abiotic stresses that limit crop growth and production. Salt stress influences plants in two ways: by affecting ion toxicity and increasing osmotic stress. Ion homeostasis, the excretion of Na$^+$ and using antioxidant systems are the major strategies of salt tolerance in plants. Na$^+$ and K$^+$ transporters with enzymes that are involved in detoxification of reactive oxygen species play key roles in salt tolerance in plants. The aim of this study was to investigate the responses of high affinity K$^+$ transporter2;1 gene (*HKT2;1*) which is involved in regulation of ion homeostasis and *L-gulonolactone oxidase* (*GLOase*) which is involved in the ascorbic acid biosynthesis pathway, under different concentrations of NaCl over different time points in *Aeluropus littoralis* shoots. Results from Real Time PCR data showed that expressions of both genes were influenced by external and internal concentrations of Na+ and the internal K+ content. *AlHKT2;1* was significantly upregulated by increasing Na+ concentration at all time points. Furthermore, its highest expression level in shoots occurred after 6 days in 300mM NaCl in shoots which was 25folds more than untreated shoots. *AlGLOase* expression levels increased 54 h after initiation of salt stress. These results indicate that *AlHKT2;1* and *AlGLOase* respond to different salinity conditions and probably are part of the mechanisms involved in tolerance to high salt concentrations in A. littoralis
Keywords: *Aeluropus littoralis*, Ascorbic Acid, Gene Expression, HKT Genes, Salt Stress.

Introduction

The main aim of plant breeding is to increase crops yields and generate new cultivars that tolerate different growth conditions. World population is growing rapidly and it is needed to produce more food annually (12, 25).
Abiotic stresses have significant effects on crops and limit their growth and yield (31). Among abiotic stresses, salinity is a major stress that affects more than 20 percent of irrigated agricultural land worldwide (6). In arid areas, due to improper use of natural resources and applying unsuitable technologies, especially irrigation in crop production, considerable parts of cultivated lands have faced salinity

(6, 17). Unfortunately, because of co-occurrence of several types of stresses and a variety of mechanisms that occur simultaneously, in many cases, the mechanisms that crops use tolerate abiotic stresses and maintain their yield, are poorly understood (25). Not only abiotic stresses have complicated signals, but also they have complex responses in plants (12, 21, 25). In addition, the variation in sensitivity of crops to abiotic stresses in different growth stages makes difficulties to apply a reverse genetics approach to study the tolerance of crops (8, 19). Therefore, identifying natural variation of abiotic stress tolerance in varieties, landraces and wild relatives of crops, and studying the traits that are involved in tolerance are highly prior approaches (25). Salinity is a soil condition characterized by a high concentration of soluble salts. Sodium chloride is highly soluble and the most abundant salt in saline soils (21) which causes plants to expose Na^+ toxicity (20). High concentration of salt also makes water uptake harder and leads to water deficiency. Osmotic stress of water shortage leads to the generation of reactive oxygen species (ROS) (23). When ROS accumulates in a large amount, cell membranes will damage and chlorophyll degradation will occur that will lead to the decline of net photosynthesis (5, 23). Plants have different strategies to reduce pernicious influences of high concentrated sodium, such as limiting the entry of sodium ion into root cells, Na^+ exclusion from leaf blades, cellular compartmentalization of excessive Na^+ into the vacuole and increasing K^+ uptake to rebalance K^+/Na^+ ratio (6, 21, 30) and improving ROS detoxification systems including enzymatic and non-enzymatic antioxidants, such as ascorbic acid (1, 6, 10, 22). Ascorbic acid (AsA) is the most abundant antioxidant that plays a main role in minimizing the damage caused by ROS in plants. It is found in all plant tissues and its highest concentration is in green matured leaves (10).

The potential of plants to tolerate salt stress is also significantly dependent on K^+ availability (20). Potassium plays an essential role in a wide range of both biophysical and biochemical processes such as enzyme activation, protein synthesis, photosynthesis, osmoregulation, stomatal movement, energy transfer, phloem transport, cation-anion balance and stress tolerance (4, 30). K^+ deficiency can be usually observed under salinity stress. Although most plants are adaptable to different concentrations of external K^+, but the range of physiological concentrations of K^+ is strictly limited in the presence of increasing amounts of sodium (20). The increasing potassium in saline culture solution not only takes up K^+ concentrations in plant tissue, but also decreases Na^+ content. Moreover it was shown that it led to growth increase and salt tolerance. Most reports show that it is not only pure internal concentration of Na^+ that affects on salt tolerance, but the K^+/Na^+ ratio is also more important to determine plant

salt tolerance (30). The maintenance of K^+/Na^+ ratio depends on coordinated activities of ion transport systems located at plasma and vacuolar membranes (20). Transporters, belonging to the HKT (high affinity potassium transporter) family play a key role in salt tolerance by regulating transportation of Na^+ and K^+ (4, 12, 30). Two classes of HKT Transporters have been identified with different properties and cell membrane location to regulate Na^+ and K^+ homeostasis (14, 15). Studies on crops HKT transporters indicated that they are more effective in potassium related salt tolerant responses than potassium nutrition (4). This study is carried out due to the role of ascorbic acid and potassium in salt tolerance and the importance of *A. littoralis* as a salt tolerant wild member of *Poaceae*. In the present study, the responses of *L-gulonolactone oxidase* (*GLOase*) a gene in ascrobate synthesis pathway and a gene of HKT family (subclass II) are evaluated under salt stress condition. Moreover some physiological traits related to K^+ balance in shoots of *A. littoralis* are investigated

Materials And Methods

Plant material

The *A. littoralis* seeds were received from ICRASN (Isfahan Center for Research of Agricultural Science and Natural Resources). Seeds were grown in pots containing sand and remained in a growth chamber (12/12 day/night, Temp.18/22). Twenty-day old seedlings were moved to a hydroponic system containing Yoshida solution (11) (pH: 5-6, 16/8 day/night, Temp. 25°C). The pH of the Yoshida solution in the hydroponic system was checked every day and the medium was refreshed every eight days. The salinity stress was induced on two-month old seedlings by adding NaCl in the solution gradually at a rate of 100 mM per 48 h until reaching 300 mM. Half of these plants were grown as control plants on a non-stress condition. The salinity stress experiment was conducted based on a completely randomized design with three biological replicates. The sampling of shoots was done at 6h/100mM, 6h/200mM, 6h/300mM, 48h/300mM, 6d/300mM and 11d/300mM. The sampling of control plants was also done simultaneously with treated ones.

Sodium and Potassium Concentrations

To determine the concentrations of Na^+ and K^+ in the shoot tissues, the sampling was done simultaneously with qRT-PCR sampling with three biological replicates. After sampling, each sample was weighed and then placed in a 72°C oven for 48 hours. The measurement of Na^+ and K^+ concentrations was performed according to a wet ashing method. The protocol was adjusted for 15 mg of tissues (7).

RNA extraction and qRT-PCR

RNA was extracted from shoots using Denazist Column RNA Isolation Kit (#S-1020, Iran) followed by DNase digestion (Thermo Scientific DNase I #EN0525, USA). The synthesis of first strand cDNAs was performed in a 20 µL reaction containing 1.5 µg total RNA by using Thermo Scientific RevertAid First Strand cDNA Synthesis Kit (#K1622).

GLOase primers were designed according to an EST sequence related to L-gulonolactone oxidase (gb: EE594874.1) and AlHKT primers were designed based on the membranous HKT2;1 of A lagopoides (gb:KP081769.1). In order to normalize the qRT-PCR data, primers for two housekeeping genes: 6Actin and Elongation factor (Elf) were designed based on A. littoralis gene sequence and EST databases (Table 1). qPCR reactions were carried out using RealQ Plus Master Mix (AMPLIQON #A314402). Moreover, 5µL of 1/16 diluted cDNA of each samples were used as a template in 20 µL final volume. qRT-PCR was performed on a Bioer thermo cycler machine (Applied Bioer, LineGeneK) with PCR condition of 94°C/10 min, 40 cycles of 94°C/10sec, Ta (Table 1) /20sec and 72°C/25sec. Normalization of qRT-PCR data was done by geometric averaging of multiple internal control genes (28) and the relative gene expression was calculated according to the comparative C_T method (26).

Data analysis

All experiments were performed based on a completely randomized statistical design with three biological replicates. The gene expression analysis was done with two identical technical replicates to increase the reliability of data. Statistical tests of significance were performed using the ANOVA in MINITAB v16 following the Tukey multiple comparison procedure ($\alpha = 0.05$) to to separate the means.

Table 1. Gene specific and reference primers used for qRT-PCR in A. littoralis

Primer	Primer sequence 5'→3'	Ta (°C)	Product (bp)
GLOase	GCCAGGGTCTGCCGCTC TCAGTTATTGCCGCTGCTTG	61	136
AlHKT2;1	GAAACCGAGCAACCCTGAC ATCCTAAGTATCTAACGCTC	63	369
βActin	TGCTGGCCGAGACCTTAC GGCGAGCTTTTCCTTGATG	59	113
Elf	ACCTTCTCTGAATACCCTCCTCTG CTTCTCCACACTCTTGATGACTCC	65	90

Results

The Na^+ concentration in treated plants increased from 4.4 to 25 mg g^{-1}DW shoot tissues and then it maintained constantly after ten days of inducing salt stress. The increasing of sodium concentration was not continuous and was constant between 6h/200mM and 6h/300mM NaCl (Figure 1). But, it appeared that the concentration of Na^+ was consistent for about 2.5 mg g^{-1}DW in the shoot of control Aeluropus plants at all time points. Also, as it is shown in Figure 2, the K^+ concentration in the shoot tissue of control plants did not have significant changes and maintained about 38 mg g^{-1}DW at all time points. In contrast, concentration of K^+ in treated shoots decreased gradually and reduced from 40 to 17 mg g^{-1}DW.

The increasing sodium and decreasing K^+ concentration in the shoot of treated plants led to re-

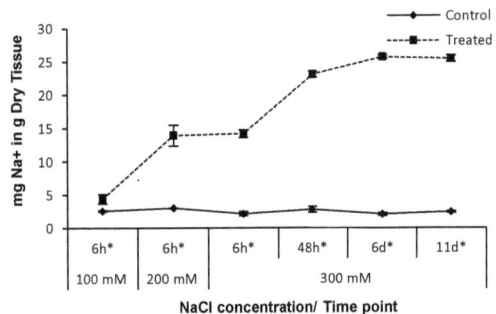

Figure 1. The fluctuation pattern of Na+ concentration in A. littoralis shoot in the control and salt treated samples with three biological replicates.

*Shows times from the initiation of the concentration.

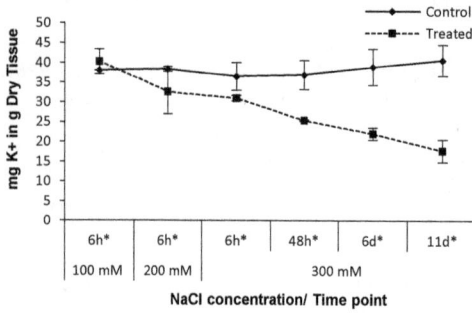

Figure 2. The fluctuation pattern of K+ concentration in *A. littoralis'* shoot in the control and salt treated samples with three biological replicates.

*Shows time from the initiation of the concentration.

duction in K^+/Na^+ ratio significantly (Figure 3). As it is illustrated, this ratio decreased rapidly for 6h/200mM NaCl, and after that, it decreased slowly. The shoot dry/fresh weight (Dw/Fw) ratio was calculated for both control and salt treated plants. As it is shown in Figure 4, Dw/Fw ratio had been constantly maintained for approximately 0.15 in control plants, however in treated plants it increased from 0.15 to 0.18 for 6h/200mM NaCl and then remained constant.

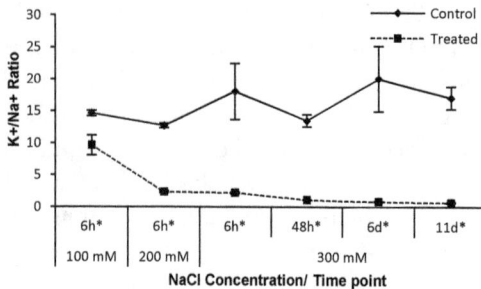

Figure 3. K^+/Na^+ ratio in shoot of *A. littoralis* in the control and salt treated samples with three biological replicates.

*Shows time from the initiation of the concentration.

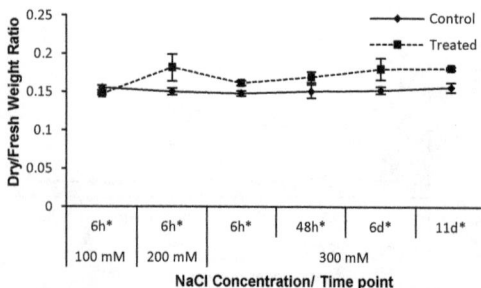

Figure 4. Dry/Fresh weight Ratio in *A. littoralis'* shoot in the control and salt treated samples with three biological replicates.

*Shows time from the initiation of the concentration.

The expression level of L-gulonolactone oxidase gene (*GLOase*) in shoots was affected by NaCl treatment. The results showed *GLOase* relative expression was constant for 6 h in 100mM NaCl, but then it increased for 48h/300mM NaCl, and it was about 4folds higher in treated plant shoots than untreated ones (Figure 5).

Subsequently, the level of *GLOase* transcript decreased and it reached a third of the level in the controls. Also transcript levels of *AlHKT2* showed a significant increase at all time points in shoots of salt treated plants compared to the controls (Figure 6).

The increase of *AlHKT2* expression in treated shoots was about 2 times more than the controls for 6h/100mM and 48h/300mM NaCl. 6 days after reaching NaCl concentration to 300mM, the mRNA abundance of *AlHKT2* in the shoots of salt treated plants showed the highest level and it was 25.34 times more than in shoots of the controls.

Discussion

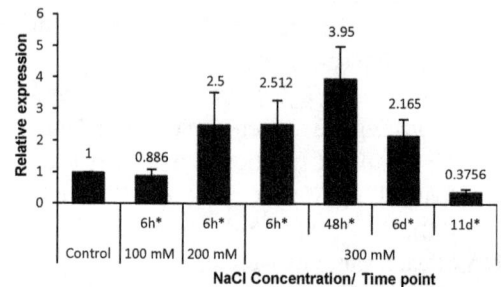

Figure 5. Relative expression pattern of *L-gulonolactone oxidase* in *A. littoralis'* shoot under salinity stress.

*Shows time from the initiation of the concentration.

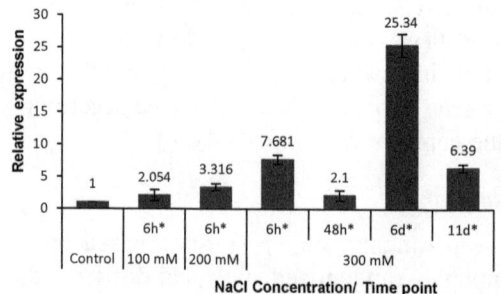

Figure 6. Relative expression pattern of *ALHKT2.1* in *A. littoralis'* shoot under salinity stress.

*Shows time from the initiation of the concentration.

In most plants, the redelivery of Na^+ from shoots to roots is done in a small proportion of the Na^+ that is sent from the roots to the shoots, so a large proportion of delivered Na^+ to shoot remains in it, and the concentration of Na^+ will reach the toxic level and cause cell death (21). *A. littoralis* uses the salt excretion strategy to avoid increase of Na^+ concentration in shoot tissues. There are lots of trichomes and salt glands in the surface of *Aeluropus'* leaves that specialized in excreting NaCl (2, 3).

In our study, the initiation of salt excretion was observed five days after the addition of salt in the medium. As the investigated results implicated the concentration of Na^+ was constant during 300mM NaCl treatment which was 25 mg g^{-1}DW. The accumulation of Na^+ in shoots increased by adding NaCl to culture media which is in agreement with the results of Barhoumi *et al.* 2007. As it is illustrated in Figure 1, the constancy of Na^+ content in shoots between 6h/200mM and 6h/300mM NaCl is probably due to beginning salt excretion form leaves and the increasing of Na^+ content after that time points may be due to adding more salt to culture media.

In contrast, it is reported that the concentration of potassium in salt stressed *Aeluropus* shoots decreased by increasing NaCl concentration and this decline was more severe in shoots than roots (3). Similar results were shown in our study. Several studies indicated K^+ starvation affects regulating *HKT* genes (9, 14).

All members of HKT2 are expected to be K^+ transporters (29). In rice, two members of the subclass 2, OsHKT2;1 and OsHKT2;2 act as Na^+-K^+ symporters (16, 29). As previously mentioned, by increasing Na^+ concentration and reducing inner K^+ content, the expression of *AlHKT2;1* showed a significant increase in salt treated shoots. Jabnoune *et al.* used in situ hybridization and indicated HKT2;1 was expressed in rice in vascular tissues, phloem, xylem and mesophyll cells and expression patterns of HKT2;1 in the leaves were not modified by different growth conditions (16). Our study showed that increasing expression levels of *AlHKT2;1* were variable in different salt concentrations and time points. The role of HKT2;1 is not well understood in shoot tissues, so, enhancing in transcription levels of *HKT2;1* by this plant probably serves to get rid of Na^+ in cells and move to outside of shoot tissues or catalyze K^+ influx to shoots. In some monocot plants like rice and barley, the expression of *HKT2;1* in roots is increased based on low external K^+ availability. *HKT2;1* is also expressed in leaves, but at a lower level than in roots (29). It seems *HKT2;1* has a role in osmoregulatory processes in leaf tissues and bulliform cells (29).

Another reaction to salt stress is osmotic stress which causes disruption in water absorption of plants (6, 21) so, plants adjust their water potential to more negative levels such as salinity increase (24). It was found that the water content was about 13% less in treated plants compared with controls. This decline in water content probably resulted in the decrease of growth rate in vegetative growth of salt treated plants. Water deficiency caused by salt stress, as well as ion toxicity and K^+ starvation strongly affects photosynthesis and causes oxidative stress (6). Most of salt tolerant genotypes apply antioxidative defense systems for the detoxification of ROS (6, 12, 21). Among plant antioxidants, the ascorbic acid acts as an agent in reducing harmful effects of ROS (1, 10, 13). Moreover, AsA plays a vital role in multiple physiological processes including photosynthesis, photoprotection, the control of cell cycles and cell elongation, gene regulation, and senescence (18). L- gulonolactone oxidase (GLOase) is one of the key enzymes as well as the final enzyme that catalyzes the transformation of L-Gulonolactone to ascorbic acid (13, 27). It is reported that in the salt stressed wheat, AsA leaf contents decreased however, adding AsA in culture media enhanced AsA leaf contents and improved growth (1). As it is illustrated in Figure 5, *GLOase* relative expression showed a significant increase at 6h/200mM NaCl and continued until 6d/300mM NaCl in treated shoots. This enhancing may be correlated with reduction in K^+ and water content, and increase in Na^+ concentration. These parameters that were previously mentioned induce oxidative stress that results in increasing ROS. It seems *littoralis* plants under these conditions, upregulated *GLOase* to produce more AsA and detoxified ROS and helped plants to tolerate salt stress. As it is reported over expression of *GLOase* in transgenic potato which was done by increasing AsA content, improved salt tolerance (13). Also, engineered *Arabidopsis* for over expression of *GLOase* showed the same results (18). But, we observed a significant reduction in *GLOase* expression level for 11d/300mM NaCl, compared with controls and other time points in salt treated shoots. It is probable because of the recovery pathway of ascorbic acid that produces AsA from Monodehydroascorbate (Monodehydroascorbate is generated by reaction of ASA with ROS) that provides AsA in suitable levels that are needed and it probably has a negative effect on *GLOase* expression (27).

In conclusion, according to the fact that tolerance to salinity is a polygenic trait, and our investigation shows that the expression of *AlHKT2;1* and *AlGLOase* genes are a body of complex reactions of *A. littoralis* to salinity, So, introducing of these genes and regulatory elements of them can be assumed as reliable element to improve other relative crops toward salinity stress.

Acknowledgements

The authors are grateful to the Biotechnology Institute of Shiraz University for the technical support of this research. Many thanks to Masoud Fakhrfeshani and Hawwa Gabier for critically reading and editing the manuscript.

References

1. Athar, H. R., Khan, A. and Ashraf, M. 2008. Exogenously applied ascorbic acid alleviates salt-induced oxidative stress in wheat. Environ Exp Bot, 63: 224-231.
2. Barhoumi, Z., Djebali, W., Abdelly, C., Chaïbi, W. and Smaoui, A. 2008. Ultrastructure of *Aeluropus littoralis* leaf salt glands under NaCl stress. Protoplasma, 233:195-202.
3. Barhoumi, Z., Djebali, W., Smaoui, A., Chaïbi, W. and Abdelly, C. 2007. Contribution of NaCl excretion to salt resistance of *Aeluropus littoralis* (Willd) Parl. J Plant Physiol, 164: 842-850.
4. Benito, B., Haro, R., Amtmann, A., Cuin, T. A. and Dreyer, 2014. The twins K^+ and Na^+ in plants. J Plant Physiol, 171: 723-731.
5. Cakmak, I. 2005. The role of potassium in alleviating detrimental effects of abiotic stresses in plants. J Plant Nutr Soil Sci, 168: 521-530.
6. Chinnusamy, V., Jagendorf, A. and Zhu, J. K. 2005. Understanding and improving salt tolerance in plants. Crop Sci, 45: 437-448.
7. Enders, A. and Lehmann, J. 2012. Comparison of wet- digestion and dry-ashing methods for total elemental analysis of biochar. Commun Soil Sci Plant Anal, 43: 1042- 1052.
8. Flowers, T. J. and Yeo, A. R. 1981. Variability in the resistance of sodium chloride salinity within rice (Oryza sativa L.) varieties. New Phytol, 88:363-373.
9. Garciadeblás, B., Senn M. E., Bañuelos, M. A., and Rodríguez-Navarro, A. 2003. Sodium transport and HKT transporters: the rice model. Plant J, 34:788-801.
10. Gill, S. S. and Tuteja, N. 2010. Reactive oxygen species and antioxidant machinery in abiotic stress tolerance in crop plants. Plant Physiol Biochem, 48: 909-930.
11. Gregorio, G. B., Senadhira, D. and Mendoza, R. D. 1997. Screening rice for salinity tolerance. International Rice Research Institute discussion paper series (22).
12. Gupta, B. and Huang, B. 2014. Mechanism of salinity tolerance in plants: physiological, biochemical, and molecular characterization. Int J Genomics, 1-18.
13. Hemavathi, Upadhyaya, C. P., Akula, N., Young, K. E., Chun, S. C., Kim, D. H. and Park, S. W. 2010. Enhanced ascorbic acid accumulation in transgenic potato confers tolerance to various abiotic stresses. Biotechnol Lett, 32:321-330.
14. Horie, T., Hauser, F. and Schroeder, J. I. 2009. HKT transporter-mediated salinity resistance mechanisms in Arabidopsis and monocot crop plants. Trends Plant Sci, 14: 660-668.
15. Horie, T., Yoshida, K., Nakayama, H., Yamada, K., Oiki, S. and Shinmyo, A. 2001. Two types of HKT transporters with different properties of Na^+ and K^+ transport in *Oryza sativa*. The Plant J, 27: 129-138.
16. Jabnoune, M., Espeout, S., Mieulet, D., Fizames, C., Verdeil, J. L., Conéjéro, G., Rodríguez-Navarro, A., Sentenac, H., Guiderdoni, E., Abdelly, C. and Véry, A. A. 2009. Diversity in expression patterns and functional properties in the rice HKT transporter family. Plant Physiol, 150:1955-1971.
17. Koocheki, A. and M. N. Mohalati. 1994. Feed value of some halophytic range plants of arid regions of Iran. In Squires, V. R. and Ayoub, A. T. (eds.). Halophytes as a resource for livestock and for rehabilitation of degraded lands. Kluwer Academic Publishers, Printed in the Netherlands, pp 249-253.
18. Lisko, K. A., Torres, R., Harris, R. S., Belisle, M., Vaughan, M. M., Jullian, B., Chevone, B. I.,

Mendes, P., Nessler, C. L. and Lorence, A. 2013. Elevating vitamin C content via overexpression of myo-inositol oxygenase and l-gulono-1,4-lactone oxidase in *Arabidopsis* leads to enhanced biomass and tolerance to abiotic stresses. In Vitro Cell Dev Biol Plant, 49: 643-655.

19. Lutts, S., Kinet, J. M. and Bouharmont, J. 1995. Changes in plant response to NaCl during development of rice (*Oryza sativa* L.) varieties differing in salinity resistance. J Exp Bot, 46:1843-1852.

20. Maathuis, F. J. M. and Amtmann, A. 1999. K nutrition and Na^+ toxicity: the basis of cellular K^+/Na^+ ratios. Ann Bot, 84: 123-133.

21. Munns, R. and Tester, M. 2008. Mechanisms of salinity tolerance. Annu Rev Plant Biol, 59: 651-81.

22. Naliwajski, M. R. and Sklodowska, M. 2014. The oxidative stress and antioxidant systems in cucumber cells during acclimation to salinity. Biol Plantarum, 58: 47-54.

23. Parida, A. K. and Das, A. B. 2005. Salt tolerance and salinity effects on plants: a review. Ecotoxicol Environ Saf, 60: 324-349.

24. Qin, J., Dong, W. Y., He, K. N., Yu, Y., Tan, G. D., Han, L., Dong, M., Zhang, Y. Y., Zhang, D., Li, A. Z. and Wang, Z. L. 2010. NaCl salinity-induced changes in water status, ion contents and photosynthetic properties of *Shepherdia argentea* (Pursh) Nutt. Seedlings. Plant Soil Environ, 56: 325-332.

25. Roy, S. J., Tucker, E. J. and Tester, M. 2011. Genetic analysis of abiotic stress tolerance in crops. Curr Opin Plant Biol, 14: 232-239.

26. Schmittgen, T. D. and Livak, K. J. 2008. Analyzing real- time PCR data by the comparative C_T method. Nat Protoc, 3:1101-1108.

27. Smirnoff, N. 2005. Ascorbate, tocopherol and carotenoids: metabolism, pathway engineering and functions. In Smirnoff, N.(eds). Antioxidants and reactive oxygen species in plants. Blackwell Publishing Ltd, pp 53-86.

28. Vandesompele, J., De Preter, K., Pattyn, F., Poppe, B., Van Roy, N., De Paepe, A. and Speleman, F. 2002. Accurate normalization of real-time quantitative RT-PCR data by geometric averaging of multiple internal control genes. Genome Biol, 3: 0034.1–0034.11.

29. Véry, A.-A., Nieves-Cordones, M., Daly, M., Khan, I., Fizames, C. and Sentenac, H. 2014. Molecular biology of K^+ transport across the plant cell membrane: what do we learn from comparison between plant species? J plant physiol, 171: 748-769.

30. Wang, M., Zheng, Q., Shen, Q. and Guo, S. 2013. The critical role of potassium in plant stress response. Int J Mol Sci, 14: 7370-7390.

31. Witcombe, J. R., Hollington, P. A., Howarth, C. J., Reader, S. and Steele, K. A. 2008. Breeding for abiotic stresses for sustainable agriculture. Philos Trans R Soc Lond B Biol Sci, 363:703-716.

19

Proline accumulation and osmotic stress: an overview of *P5CS* gene in plants

Sahand Amini[1], Cyrus Ghobadi[2] and Ahad Yamchi[3*]

[1] Department of Agricultural Biotechnology, College of Agriculture Isfahan University of Technology, Isfahan, Iran

[2.] Department of Horticultural Sciences, College of Agriculture Isfahan University of Technology, Isfahan, Iran

[3.] Department of Plant Breeding and Biotechnology, College of Plant Production Gorgan University of Agriculture Science and Natural Recourses, Gorgan, Iran

*Corresponding Author (✉): yamchi@gau.ac.ir

Abstract

Under osmotic stresses, proline accumulation is an important response of plants to these conditions. Proline is a compatible osmolyte which affects many cellular and molecular aspects of a plant in both normal and stressful situations. Proline is shown to be involved in plant development in normal conditions and in conferring resistance to a plant under biotic and abiotic stresses. Therefore, many surveys have already been conducted to unveil its mechanisms and signaling pathways, so that it might be considered as an insight into resolving growing challenges of agriculture, drought and soil salinity. Δ^1-pyrroline-5-carboxylate synthetase (P5CS), one of the two main enzymes in the proline biosynthesis pathway of the glutamate precursor, has been demonstrated to play a significant role in proline accumulation in plants under water stresses. Regarding the role of P5CS under the osmotic stress, there are controversial observations in various plants which casts doubts regarding whether P5CS is a rate-limiting enzyme in the pathway or not. Obviously, transgene *P5CS* is proved to give higher resistance to transgenic plants under drought and salinity, by elevating proline content. In this review of literature, proline and its identified various functions in plants, characteristics of P5CS enzyme, signals, inducers and inhibitors of *P5CS* gene and the expression pattern of *P5CS* under differential conditions in studied plant species are discussed. Finally, some of the important features of the transgenic plants overexpressing *P5CS* have been summarized.

Keywords: Abiotic Stress, Overexpressing, Proline, P5CS, Transgenic Plants

Introduction

It is not impenetrable today that the most influential obstacles to achieve high-yielding crops are osmotic stresses, particularly drought and soil salinity whereby dehydration causes loss of millions of tons of crops every year and half of the arable lands has already become arid due to soil salinization (66) inasmuch as they are growing at the worrisome rate of 3 hectare per minute (39). These stresses affect the plant with lowering the amount of water available for it, as well as osmotic potential of cells. These conditions cause deficiency in normal development and growth of the plant, reduction in its fertility and even death of the plant in

severe and prolonged stresses (8). Therefore, investigation on improving plants resistance to osmotic stresses has drawn much attention among researchers. Proline, the most accumulated osmolyte, which accumulates to high levels in many plants in various stressful conditions such as drought, salinity, high and low temperature. Photo-damage, heavy metals and even pathogens, have been proven to play a significant role in adapting the plants to water stresses (30, 66, 69). Proline which is conspicuously more than just an osmoprotectant, has many functions under normal and stressful conditions in plants. Proline, as a cyclic amino acid, is an important part of many proteins involved in osmotic regulation, the plant cell wall and membrane. So, it is essential for their stability (62). Proline deficiency in plants causes defect in growth and development of flowers and seeds of *Arabidopsis thaliana* (33). Also, proline metabolism and catabolism, which help maintaining redox balance of the cell is required for efficient flowering and seedling of plants (18). It probably affects flowers and seeds by transporting carbon, nitrogen and reducing agents to them (66). Moreover, Proline deficiency is reported to postpone the flowering time (62). Accordingly, on long days, it was shown that one of the main proline biosynthesis pathway genes, *P5CS*, is the target gene for CONSTANS transcription factor, which mediates flowering in long days (49). Inhibition of proline biosynthesis has resulted in morphological abnormalities in leaves, inflorescence, epidermal and mesophyll cells and the vascular system of *Arabidopsis thaliana* (41). Proline is also involved in cell division (18) and embryogenesis (62) in plants. Despite its numerous roles in plant normal development, proline is most renowned for its functions under stressed conditions. Highly soluble in water, proline is a compatible solute, which confers resistance to many plants from algae and aquatic plants to higher plants like *Arabidopsis*, halophytes and various crops, under osmotic tensions. In water-deficit conditions, it retains osmotic potential (11, 29, 37) and redox balance of cells (21, 68, 72), scavenges free radicals and ROS as an antioxidant (24, 50, 54), protects macromolecules from denaturation as a chemical chaperone (45) and regulates cytosolic pH (54). Finally, under stressful conditions, proline is a nitrogen and carbon provider after rehydration (3), source of energy (66), metal chelator (34) and signal molecule (70). In plants, proline is synthesized from two precursors, L- Glutamate (Glu) and Arginine/Ornithine (Orn) (25). Although Glu pathway is believed to be dominant in many stressful and normal conditions, except for the case of excessive nitrogen (9), Orn pathway has been reported to play a crucial role in adult *Arabidopsis* under osmotic stress, while in young plants both pathways cooperate to accumulate proline (48). Recently, it has been shown that after the removal of

severe stress, Orn cooperates with the Glu in the biosynthesis of proline (3). However, Glu pathway seems to have a central role in synthesizing proline under water stresses. In Glu pathway, glutamic-γ-semialdehyde (GSA) is synthesized by L-Glutamic acid including both phosphorylation and reduction activities of Δ^1-pyrroline-5-carboxylate synthetase (P5CS). In *E. coli*, GSA is formed by two distinct enzymes, γ-glutamyl Kinase (GK) and GSA dehydrogenase or γ-glutamyl phosphate reductase (GPR). GSA, then turns to Δ^1-pyrroline-5-carboxylate (P5C) in spontaneous cyclization reaction. Finally, P5C Reductase (P5CR) reduces P5C to L-Proline (25).

Proline degradation pathway regulation is also important for its accumulation under abiotic stresses and plant life. Proline accumulation in plants is the result of both increased proline biosynthesis and decreased proline degradation (76). Furthermore, the activation of proline degradation after rehydration (31, 42) is crucial to provide reducing potential for mitochondria and so as to continue the respiratory cycle of cells and plant life (62). Besides, inhibition of proline degradation is reported to be highly toxic in *Arabidopsis* (35, 40). Proline is degraded by two oxidation reactions in which, it is first oxidized by proline dehydrogenase/oxidase (ProDH) to P5C and after converting to GSA, spontaneously, P5C dehydrogenase (P5CDH) oxidizes GSA to Glu (20).

Generally, proline biosynthesis and a degradation cycle are required for balanced redox potential under water deficit pressure, flowering and seedling in normal development of plant (18, 19, 66). Proline distribution, intracellular and intercellular, also plays a critical role in rendering osmotic-resistance to various tissues under stress (66). Proline biosynthesis of Glu is mostly occurring in cytosol of plant cells and in chloroplast, when faced with water deficit (33, 63). Under stressful conditions, it is accumulated in cytosol to induce water diffusion into cells (36), while in the absence of stress, proline is transported to organelles, particularly vacuole and plastid (33). Vacuole, distributes proline to cytosol, in lack of water (76). By rehydration, proline is transported to mitochondria, where it is degraded to Glu (33).

Transportation of proline between tissues and its accumulation is not the same in reported surveys. It seems that they are greatly dependent on the condition and species, but what is similar in all related studies is that proline can travel long distances as far as the height of the plant from roots to flowers, through both xylem and/or phloem. Proline is largely synthesized and accumulated in roots and leaves of plants under water stress (3, 12) and then transported mainly to meristems, dividing cells of root apex and sexual organs (33, 67). Under normal conditions, proline accumulates chiefly in pollen, seeds and fruits (51, 56, 66).

In metabolism and catabolism of proline under water stress, P5CS and ProDH enzymes are believed to be rate- limiting and play a significant role in the regulation of the proline level in plants (76). P5CS, having two genes encoding it in some plants including *Arabidopsis* (57), is the only rate-limiting enzyme in the Glu-based pathway. In spite of an anomalous report in which no relation between the level of P5CS and proline was observed (56), there are many reports that show when the expression of *P5CS* is increased, proline accumulation reaches higher levels (11, 25, 51, 75, 79), which does not occur with the other enzyme of the pathway, P5CR, at least in that amount. For instance, overexpression of *P5CR* gene in tobacco led to 200 times more expression of *P5CR*, but no noticeable increase in proline level in transgenic tobacco was observed (63). On the other hand, *p5cs* mutants of *Arabidopsis* gathered less amount of proline and demonstrated no resistance to the hypo- osmotic condition (41). We believe that reconsidering different aspects of P5CS in this review, will shed light on known and unknown territories of the proline biosynthesis pathway which can be considered as an insight regarding the challenge of osmotic stresses in agriculture. Therefore, we discuss characteristics of P5CS enzyme and gene in various plants, *P5CS* expression pattern and transgenic plants over-expressing *P5CS* gene, in this review of literature.

P5CS Enzyme

As mentioned above, P5CS in plants, lo-calized in cytosol and plastid in cells (62), consists of two domains functioning as kinase and dehydrogenase enzymes. Each domain has leucine zipper sequence, which is involved in preservation of tertiary structure of the enzyme, protein-protein interaction and probably the contribution of two domains (25). GK domain of P5CS, responsible for phosphorylation of Glu, depends on ATP (11), while the GPR domain requires NADPH as reducing agent (19). Accordingly, the presence of both ATP-binding and NAD(P)H-binding motifs in P5CS has been proposed (51, 60). The Leucine-rich region has also been reported to exist in sorghum P5CS (60). The P5CS1 protein of *Arabidopsis*, the second isolated *P5CS* gene among plants after mothbean (*Vigna aconitifolia*), is a poly peptide with 717 amino acids, which is estimated to weigh about 77.7 kDa (51, 57). While the native P5CS enzyme was shown to be approximately a 450 kDa protein, it was deducible that the P5CS functions as a hexamer with six similar subunits (79). The same had also been reported with two distinct enzymes in *E. coli* (55). Leucine zipper may have a role in formation of this quaternary structure, as it mediates protein-protein dimerization (1). Later, the approximate length and molecular weight were estimated for P5CS in grapevine (56), common bean (7) and sorghum (60). The activation of P5CS enzyme is inhibited by specific amount of proline, depending on some variables especially plant species (25, 56, 62). The inactivation of P5CS enzyme in grapevine mature fruits was between 33% and 50%, depending on the proline concentration (56). After removing the proline-binding residue in *E. coli* GK enzyme, the residues responsible for this feedback inhibition were also found in *V. aconitifolia* and recently in sorghum (60, 79). Mutation of Asp 126 or Phe 129 in *Vigna*, removed this inhibition and consequently more proline accumulated in transgenic plants (79). P5CSF129A, in which Phe 129 is replaced by Ala, was later used broadly to generate plants with higher levels of proline (discussed later). These two amino acids are also conserved in *Arabidopsis* (57). However, the same mutations didn't work in sorghum, but instead, Phe 128 and

141 were recognized to be the target residues of proline in sorghum (60). It is proposed that the accumulation of proline to higher levels under stressful conditions, might be due to the inactivation of feedback inhibition in these situations (11), but this inactivation is not complete (24).

P5CS gene inducers and inhibitors

In addition to regulation in protein level, expression of P5CS is regulated in transcription and probably post- transcription levels as well. Nevertheless, induction and inhibition signaling of P5CS has not been completely discovered so far. P5CS expression is shown to be induced, as recognized till now, mainly by various environmental factors and plant hormones. In an early survey, it was reported that drought and salt stress amplify AtP5CS transcription (75). To investigate transcription of P5CS, GUS enzyme was expressed under AtP5CS1 promoter in Arabidopsis and tobacco (77). Transgenic plants displayed increased expression of GUS in dehydration and to lower contents in salinity, while low temperature (4°C) had no effect on the amount of GUS. In rice, salinity, dehydration, low temperature induced OsP5CS (26). NaCl, also mediated the expression of P5CS in tomato (11). These studies plus other similar reports (2, 4, 25, 43, 53, 60, 80) prove that P5CS gene transcription in plants although varies by species, is promoted by osmotic stresses including dehydration, salinity, high and low temperature. Recently, it has been indicated that osmotic stresses increase DNA methylation modification of P5CS gene in rice (78). DNA methylation in plants is involved in response to biotic and abiotic stresses (47). Signaling pathway for the induction of P5CS expression is however not clear yet, but different expression patterns in various stresses, suggest distinct pathways for different stresses such as cold and water stresses (21). As proline accumulates in light (15), some groups studied the effect of light on P5CS transcription in Arabidopsis. They showed that the level of AtP5CS1 mRNA was much higher in light than in dark (2, 22). However, in the saline condition, it was high in both light and darkness (22). In (22), the authors proposed that light and darkness might play an indirect role to regulate P5CS expression, probably by affecting water potential of leaves, in the meanwhile, light has a negative effect on ProDH expression. In light, photosynthesis is activated and so is sugar synthesis. Consequently, leaves face decreased osmotic potential. This stress, might be the reason for P5CS transcription promotion in light and the reverse in dark. Accordingly, no light-responsive element was found in the upstream of SbP5CS gene in sorghum (60). In freezing-resistant eskimo1 mutant Arabidopsis, the 8-fold higher expression of AtP5CS1 was observed (71). Also, in these plants abscisic acid (ABA)-dependent RAB18 gene showed constitutive expression, while RD29A, mainly regulated by ABA- independent pathway, had no considerable change in transcription level. Although freezing has not been reported to stimulate P5CS up-regulation, the cold- response signaling pathway is involved in regulating P5CS and proline amount in Arabidopsis. Furthermore, proline might have a role in giving resistance to freezing in plants.

Predicting cis-acting elements of SbP5CS promoter revealed a MeJA-responsive motif (TGACG-motif) upstream of the gene (60). MeJA is a plant hormone, produced in response to abiotic and biotic stresses. Expectedly, MeJA treatment of sorghum seedlings mediated SbP5CS expression (60). Phenolic plant hormone, salicylic acid(SA), which is mostly recognized by its role in plant growth, development, photosynthesis and defense against pathogens, was reported to affect AtP5CS2 expression positively in pathogenic condition (10). Later, finding a SA-responsive element, TCA- element, in promoter region of SbP5CS (60), increased the impact of SA on P5CS up-regulation. Likewise, a gibberellin (GA)-responsive element, GARE, was predicted as the upstream of SbP5CS gene, but no survey has been done yet to clarify involvement of GA in P5CS stimulation. Another plant stress-responsive hormone, ABA, has also been shown to induce P5CS1 and P5CS2 expression in Arabidopsis (57, 75) and rice (58). An inconsistent report which showed that exogenous ABA had no effect on expression of *GUS* under *AtP5CS1* promoter (77) led to the hypothesis that ABA might regulate *P5CS* in post-transcription level (77). But further investigations signify the positive role of ABA in *P5CS* transcription. For instance, exogenous ABA treatment induced *OsP5CS* in rice (26) and even stronger ABA-responsive element (ABRE) has been found to exist in promoter region of *AtP5CS2* and *SbP5CS* (60, 77). Besides, another ABA-responsive cis-acting element exists in the upstream of *AtP5CS1*, *AtP5CS2* and *SbP5CS* genes which is the binding element for MYB transcription factor (21, 60). Calcium signals are also believed to activate the MYBs, which promote *P5CS* transcription (74). The abiotic response of plants is assorted into two main pathways, ABA-dependent signaling pathway and ABA-independent one. While many genes are controlled by either pathways in some cases, like *RD29A*, both pathways contribute in the regulation of the gene. Interestingly, expression pattern of *AtP5CS* is more like that of *Rd29A*, rather than ABA- dependently controlled genes such as *RAB18* (64). Considering all the observations, scientists suggest that *P5CS* is among the genes which are controlled by both pathways (21, 76).

To unveil upstream signaling the pathway of *P5CS*, a research was done on it in *Arabidopsis*. It showed that phospholipase D, which is involved in water stress responses, mediates ABA signal transduction (17). However, it inhibited proline accumulation by under- regulating *P5CS1* under normal and stressful conditions (64). Calcium accumulation in cytosol is one of the first responses of plant cells to water stress. It was reported that calcium played a significant role in proline accumulation under the saline condition, but it was not sufficient for up-regulation of *P5CS* expression; while, simultaneous treatment of *Arabidopsis* with $CaCl_2$ and Phospholipase D inhibitor resulted in higher *P5CS1* mRNA level (64). Calcium, was also suggested to regulate Phospholipase D, as a downstream signal messenger (64). Despite some efforts, little is yet known in this respect and further surveys are needed to understand the *P5CS* and proline signaling pathways.

P5CS Expression under normal and osmotic stress conditions in non-transgenic plants

The expression pattern of *P5CS* gene, as an indicator of the way it affects proline accumulation in plants, was studied in various studies. In *Arabidopsis*, under the normal condition, *AtP5CS1* and *AtP5CS2* showed different expressions in various tissues. While no *P5CS* mRNA was notably expressed in roots, *AtP5CS1* transcript level was very high in Leaves, stems and flowers, with the highest level in leaves and there was no detectable amount in callus and cell suspension cultures. Conversely, *AtP5CS2* was expressed highly in dividing tissues especially in callus of *Arabidopsis* (57). In another report, the highest *AtP5CS1* mRNA was detectable in flowers, even though it was transcribed in any tissue (51). 10-day-old seedlings under 170 mM NaCl salinity, demonstrated to accumulate *P5CS* transcript just 4 hours after the treatment and reached the highest level after 8 hours. Then, this level started to decrease during 24 hours (51). Dehydration led to severe and immediate increase in *AtP5CS1* expression, reaching the maximum level in 6 hours and much lower effect on *P5CS2*. 25 mM NaCl, as was observed in previous study, promoted *P5CS* expression, but slower and in lower amount in comparison with dehydration. While *AtP5CS1* mRNA accumulated after 6 hours and persisted for 24 hours in roots, *AtP5CS2* showed a little increase in transcription just for 24 hours (57). In another work, dehydration promoted *AtP5CS1* expression an hour after treatment and its level reached the summit in 5 hours. Cold had slight impact on *P5CS* expression, too (75). In these studies, the transcript level of *AtP5CS1* was about 7 to 8 times higher in water and salt stress conditions in comparison with normal conditions. Also, proline accumulation was consistent at *P5CS* mRNA level.

In rice, *OsP5CS*, having about 75% similarity with *P5CS* in *Arabidopsis* and *Vigna*, up-regulated for 10 hours after the treatment of 250 mM NaCl and kept the trend for 24 hours. It subsided gradually to the normal level after 72 hours. Dehydration caused promoted expression of *OsP5CS* in 5 hours, which reached the maximum level in 10 hours and returned to the control amount after 24 hours. 4°C treatment of 10-day-old rice plants promoted *OsP5CS* mRNA level after 1 or 2 hours. Proline accumulation level showed consistency with *P5CS* mRNA level (26). In a recent study, activity of OsP5CS enzyme was reported to increase by about 19%, subjected to 425mM NaCl and led to proline accumulation (4).

Tomato *tomP5CS1*, interestingly consists of two distinct ORFs, just like that of *E. coli*, while *tomP5CS2* has characteristics like other plant *P5CS* genes. Though the transcript level of *tomP5CS* under 100 and 200 mM NaCl stress was about two times in comparison to control plants, rather lower than *Arabidopsis* and rice, proline accumulated in tomato much higher, up to 80 folds higher than control plants (11).

In alfalfa, *MsP5CS1* and *MsP5CS2* cDNAs were isolated and their expression under 90 mM NaCl was studied in roots of 6-day-old alfalfa seedlings (13). In this condition, *MsP5CS1* transcript showed an increase after 48 hours and it was still growing after 72 hours, but *MsP5CS2* expression got promoted at early hours, in 6 hours and it kept ascending for 72 hours after NaCl treatment. Induction level of *MsP5CS2* was obviously higher than that of *MsP5CS1*. Surprisingly, the increase in proline content of roots was not considerable during stress, while *P5CS* mRNA was accumulated up to 4 times higher than control plants. Finding two P5CS coding regions also in bean (7), shows that duplication of *P5CS* gene is common in plants. Isolated *P5CS1* and *P5CS2* genes in mentioned plants, positioned on nuclear genomes, have shown 65 to 80% similarity between two isoforms.

In grapevine, one coding region for VvP5CS was identified and its cDNA was isolated. Surprisingly, the accumulation of proline in mature fruits, up to 80 folds higher than that in leaves and roots was independent of *P5CS* transcript and the enzyme level (56).

In cactus pear (*Opuntia streptacantha*), an aquatic macrophyte, the expression pattern of isolated *P5CS* cDNA and the activity of the relative enzyme was studied. *OsP5CS* showed an increased expression under 75 to 350 mM NaCl stress after 6 to 9 days (53). This study, consistent with two other investigations (48, 69), deduced that P5CS might not be the rate-limiting enzyme in the Glu-based pathway, in salinity condition. These studies reported that while *P5CS* expression was promoted by NaCl, no sign of elevated activity of P5CS enzyme was observed, meanwhile, proline reached high amounts in cactus pear and wheat. In sorghum, *SbP5CS1* and *SbP5CS2* genes were isolated and their characteristics were evaluated in salinity and drought (60). Both mRNAs accumulated in leaves and roots, under dehydration and 250 mM NaCl stresses. However, *SbP5CS1* transcript level was obviously higher than *SbP5CS2*. Under drought conditions, the up-regulation of genes started in 3 days, while salinity provoked their expression much earlier, but in a different manner. *SbP5C1* became stimulated in 4 hours and it reached the highest level after 12 hours in leaves, but after 24 hours in roots. *SbP5CS2* reached its lower highest level in 8 hours. Generally, the transcript levels were a bit higher in roots. Proline accumulation under drought was 60 folds higher than that of the control plants in 6 days. These numbers for salinity were at most 8 folds after 48 hours. Proline accumulation was highly consistent with *P5CS1* mRNA accumulation pattern under both stresses. Observing much lower *SbP5CS2* mRNA than *SbP5CS1* under stress, the authors assumed the gene to be a house- keeping gene, which is only involved in the proline metabolism (60).

Recently, *P5CS* cDNA was isolated from a drought- and salinity-resistant halophyte, *Nitraria tangutorum*, and its expression was characterized under various osmotic conditions (80). *NtP5CS* mRNA was shown to be up- regulated under 200 mM NaCl, 10% polyethylene glycol, 50°C and 4°C stresses, with the highest amount under heat, followed by salinity. Also, the proline level was in accordance with *NtP5CS* expression. However, the lack of data about the relative enzyme activity, makes any deduction impossible.

Table 1. Transgenic plants with over-expressed *P5CS* gene.

Gene	Transformed plant species	Promoter	Proline Normal [a]	Proline Drought [b]	Proline Salinity [c]	Effects	Reference (s)
Mothbean *P5CS*	Tobacco	*CaMV 35S*	8 to 14	2	-	Later wilting under drought stress. Higher biomass level, longer roots and more number of seeds under salinity	(29)
Mothbean *P5CS*	Rice	*AIPC*	1.5 to 2.5	-	-	Increased shoot length, shoot weight and root weight under salinity. Later wilting and increased shoot weight under drought	(81)
Mothbean *P5CS*	Tobacco	*CaMV 35S*	2.5 to 3	-	1.5	Higher germination and lower free radicals under salinity	(24)
Mothbean *P5CSF129A*	Tobacco	*CaMV 35S*	5 to 6	-	3	Much higher germination and much lower free radicals under salinity	(52)
Mothbean *P5CS*	Wheat	*CaMV 35S*	12	-	2.5	Resistance to salinity and normal growth up to 200 mM NaCl	(58)
Mothbean *P5CS*	Rice	rice *Actin 1* / *AIPC*	3.2	2 to 3.2	2.2 to 3.2	Higher shoot and root weight under drought and salinity	(37)
Mothbean *P5CS*	Rice	*AIPC*	1.2 to 1.4	1.2 to 1.9	1.2 to 2	Much higher shoot and root weight under drought and salinity	(72)
Mothbean *P5CSF129A* / Arabidopsis *P5CS*	Orange	*CaMV 35S*	2	2.5	-	Higher photosynthetic activity under drought	(23)
Rice *P5CS*	Petunia	*CaMV 35S*	2 to 3	-	-	High survivor percentage after drought	(9)
Arabidopsis *P5CS1*	Potato	*CaMV 35S*	1.5 to 3.5	-	2.5 to 7	Resistance to salinity and normal growth up to 100 mM NaCl, and lower yield reduction under salinity	(46, 73)
Mothbean *P5CS*	Wheat	*AIPC*	Same as NT	2	-	Higher membrane stability, lower oxidative damages, and higher photosynthesis under drought	(5)
Arabidopsis *P5CS*	Tobacco	*CaMV 35S*	7.5 to 35	-	3.3 to 11	Control plants germinated in the presence of NaCl concentration up to 50 mM and tolerated 100 mM NaCl during growth phase while transgenic plants were able to germinate in 200 mM NaCl and tolerated up to 250 mM NaCl during growth phase	(32)
Mothbean *P5CSF129A*	Chickpea	*CaMV 35S*	2 to 6	1.3 to 2.9	-	High resistance and decreased free radicals under drought	(12)
Mothbean *P5CSF129A*	Rice	*CaMV 35S*	2.5 to 4	-	4.5 to 5.5	Higher plant height and weight, lower oxidative damages under salinity	(44)
Mothbean *P5CS*	Chickpea	*CaMV 35S*	10	-	-	lower membrane damage, and higher survival percentage under salinity	(27)
Mothbean *P5CSF129A*	Tobacco	*CaMV 35S*	8	1.8	-	No significant effect was observed	(6)
Mothbean *P5CS*	Rice	*CaMV 35S*	2.5 to 5	-	3.5 to 5	higher plant height and weight under salinity	(61)
Bean *P5CS1*	Arabidopsis	*CaMV 35S*	2.8	-	3.2 to 3.7	higher resistance under salinity	(6)
Bean *P5CS2*	Arabidopsis	*CaMV 35S*	2.7	-	7	much higher resistance under salinity	(6)
Mothbean *P5CSF129A*	Pigeonpea	*CaMV 35S*	3.5 to 5	-	4 to 4.5	Higher plant height, water and Chlorophyll content, lower oxidative damages under salinity	(61)

[a] Proline content(fold) in comparison to non-transgenic (NT) plant in normal condition

[b] Proline content(fold) in comparison to NT plant under drought stress

[c] Proline content(fold) in comparison to NT plant under salt stress

Transformation of both *NtP5CS* and *AtP5CS* in *E. coli*, demonstrated that although both transgenic strains had improved growth under drought, salinity, heat and cold, a halophyte P5CS worked more efficiently than *Arabidopsis* P5CS in conferring osmotic resistance to *E. coli* (80).

Generally speaking, while there is no explicit similarity between expression patterns of *P5CS* in studied plants, it is deducible that *P5CS* responds to dehydration more quickly than to salinity, but many variables such as plant species, stress severity and plant organs are seemingly influential which should be mentioned. Considering the role of P5CS in proline biosynthesis as a rate-limiting enzyme, it seems that some researches' doubtfulness and inconsistent reports cannot lead to an accurate conclusion on the matter and it needs to be investigated more. These studies should be comprehensive, meaning that both Glu and Orn pathways as well as both P5CS and P5CR along with their transcripts and enzyme activity levels should be taken into account. The effect of elements like plant species, genotypes, a plant organ and its developmental stage, type of stress, severity and its duration transcriptional and post-transcriptional regulation of P5CS, with an unknown signaling pathway for proline and P5CS accumulation have complicated the role of P5CS and it is still unidentified.

P5CS overexpression in plants

Despite lack of sufficient knowledge about P5CS in plants, it is proved that except for one report, all transgenic plants overexpressing *P5CS* gene, are resistant to osmotic stresses. These pieces of evidence are summarized in Table 1. This resistance is the result of accumulating proline in higher levels than control plants (Table 1). The overexpression of *P5CS* gene has resulted in higher survival rate, improved tolerance and higher yield under osmotic stresses in important crops such as wheat, rice and potato. For example, by ectopic expression of Mothbean *P5CS* in wheat, the transgenic lines could tolerate the salinity up to concentration of 200 mM, which is a great success. Also, the overexpression of Mothbean *P5CSF129A* in Tobacco, increased

the proline level twice as did *P5CS* of Mothbean, and consequently, the transgenic lines with *P5CSF129A*, showed much higher germination and much lower free radicles. Except for three works, in which *P5CS* gene has been transformed under inducible AIPC promoter (58, 81), all groups have overexpressed P5CS under a constitutive promoter. AIPC is a stress- inducible heterologous promoter which responds to ABA (59). As ABA accumulation is a plant response to osmotic stresses (14), an ABA-responsive promoter is accordingly induced in stressful conditions. It is believed that overexpressing genes under inducible promoters reduces undesirable side effects in transgenic plants in normal conditions (16, 65, 81). However, some of these studies reported no or an insignificant defect on plant development and growth in spite of using a constitutive promoter for *P5CS* gene transformation (5, 12, 23, 72). Owing to the inconspicuous role of P5CS in some studies, discussed above, simultaneous overexpression of both P5CS and P5CR might result in more resistant plants, at least under the salt stress. Moreover, overexpressing these genes under osmotic-responsive promoters such as *rd29A* and recently characterized *Oshox24*, which results in higher resistance under osmotic stresses, with lowest phenotypic abnormality in normal conditions (28, 38), is suggested to be studied in further investigations.

References

1. Abel, T. and Maniatis, T. 1989. Gene regulation. Action of leucine zippers. *Nature*, 341: 24-25.
2. Abrahám, E., Rigó, G., Székely, G., Nagy, R., Koncz, C. and Szabados, L. 2003. Light-dependent induction of proline biosynthesis by abscisic acid and salt stress is inhibited by brassinosteroid in Arabidopsis. *Plant Mol. Biol.*, 51:363-372.
3. An, Y., Zhang, M., Liu, G., Han, R. and Liang, Z. 2013. Proline Accumulation in Leaves of Periploca sepium via Both Biosynthesis Up-Regulation and Transport during Recovery from Severe Drought. *PLOS ONE*, 8:e69942.
4. Bagdi, D.L. and Shaw, B.P. 2013. Analysis of proline metabolic enzymes in *Oryza sativa* under NaCl stress. *J Env. Biol./Aca. Env. Biol. India*, 34:677-681.
5. Bhatnagar-Mathur, P., Vadez, V., Devi, M.J., Lavanya, M., Vani, G. and Sharma, K.K. 2009. Genetic engineering of chickpea (*Cicer arietinum*

L.) with the *P5CSF129A* gene for osmoregulation with implications on drought tolerance. *Mol. Breeding*, 23:591-606.

6. Chen, J.B., Yang, J.W., Zhang, Z.Y., Feng, X.F. and Wang, S.M. 2013. Two *P5CS* genes from common bean exhibiting different tolerance to salt stress in transgenic Arabidopsis. *J. Genet.*, 92:461-469.

7. Chen, J., Zhang, X., Jing, R., Blair, M. W., Mao, X. and Wang, S. 2010. Cloning and genetic diversity analysis of a new P5CS gene from common bean (*Phaseolus vulgaris L.*). *Theor. Appl. Genet.*, 120:1393-1404.

8. Colmenero-Flores, J.M., Campos, F., Garciarrubio, A. and Covarrubias, A.A. 1997. Characterization of Phaseolus vulgaris cDNA clones responsive to water deficit: identification of a novel late embryogenesis abundant-like protein. *Plant Mol. Biol.*, 35:393-405.

9. Delauney, A.J., Hu, C.A., Kishor, P.B. and Verma, D.P. 1993. Cloning of ornithine-e-amino transferase cDNA from Vigna aconitifolia by trans-complementation in Escherichia coliand regulation of proline biosynthes. *J. Biol. Chem.*, 268:18673-18678.

10. Fabro, G., Kovács, I., Pavet, V., Szabados, L. and Alvarez, M.E. 2004. Proline Accumulation and AtP5CS2 Gene Activation Are Induced by Plant-Pathogen Incompatible Interactions in Arabidopsis. *Mol. Plant-Microbe Interact.*, 17:343-350.

11. Fujita, T., Maggio, A., Garcia-Rios, M., Bressan, R.A. and Csonka, L.N. 1998. Comparative Analysis of the Regulation of Expression and Structures of Two Evolutionarily Divergent Genes for Δ^1-Pyrroline-5- Carboxylate Synthetase from Tomato. *Plant Physiol.*, 118:661-674.

12. Ghanti, S.K., Sujata, K.G., Kumar, B.V., Karba, N.N., Janardhan Reddy, K., Rao, M.S. and Kishor, P.K. 2011. Heterologous expression of *P5CS* gene in chickpea enhances salt tolerance without affecting yield. *Biologia Plantrum*, 55:634-640.

13. Ginzberg, I., Stein, H., Kapulnik, Y., Szabados, L., Strizhov, N., Schell, J. and Zilberstein, A. 1998. Isolation and characterization of two different cDNAs of Δ^1- pyrroline-5-carboxylate synthase in alfalfa, transcriptionally induced upon salt stress. *Plant Mol. Biol.*, 38:755-764.

14. Giraudat, J., Parcy, F., Bertauche, N., Gosti, F., Leung, J., Morris, P.C. and Vartanian, N. 1994. Current advances in abscisic acid action and signalling. *Plant Mol. Biol.*, 26:1557-1577.

15. Goas, G., Goas, M. and Larher, F. 1982. Accumulation of free proline and glycine betaine in Aster tripolium subjected to a saline shock: a kinetic study related to light period. *Physiol. Plant.*, 55:383-388.

16. Grover, A., Kapoor, A., Satya Lakshmi, O., Agarwal, S., Sahi, C., Katiyar-Agarwal, S. and Himanshu, D. 2001. Understanding molecular alphabets of the plant abiotic stress responses. *Curr. Sci.*, 80:206-216.

17. Hallouin, M., Ghelis, T., Brault, M., Bardat, F., Cornel, D., Miginiac, E. and Jeannette, E. 2002. Plasmalemma Abscisic Acid Perception Leads to RAB18 Expression via Phospholipase D Activation in *Arabidopsis* Suspension Cells. *Plant Physiol.*, 130:265-272.

18. Hare, P.D. and Cress, W.A. 1996. Tissue-specific accumulation of transcript encoding Δ^1-pyrroline-5- carboxylate reductase in Arabidopsis thaliana. *Plant Growth Regul.*, 19:249-256.

19. Hare, P.D. and Cress, W.A. 1997. Metabolic implications of stress-induced proline accumulation in plants. *Plant Growth Regul.*, 21:79-102.

20. Hare, P.D., Cress, W.A. and Van Staden, J. 1998. Dissecting the roles of osmolyte accumulation in plants. *Plant Cell Environ.*, 21:535-553.

21. Hare, P.D., Cress, W.A. and Van Staden, J. 1999. Proline synthesis and degradation: a model system for elucidating stress-related signal transduction. *J. Exp. Bot.*, 50:413- 434.

22. Hayashi, F., Ichino, T., Osanai, M. and Wada, K. 2000. Oscillation and Regulation of Proline Content by *P5CS* and *ProDH* Gene Expressions in the Light/Dark Cycles in Arabidopsis thaliana L. *Plant Cell Physiol.*, 41:1096-1101.

23. Hmida-Sayari, A., Gargouri-Bouzid, R., Bidani, A., Jaoua, L., Savouré, A. and Jaoua, S. 2005. Overexpression of Δ^1- pyrroline-5-carboxylate synthetase increases proline production and confers salt tolerance in transgenic potato plants. *Plant Sci.*, 169:746-752.

24. Hong, Z., Lakkineni, K., Zhang, Z. and Verma, D.S. 2000. Removal of Feedback Inhibition of Δ^1-Pyrroline-5- Carboxylate Synthetase Results in Increased Proline Accumulation and Protection of Plants from Osmotic Stress. *Plant Physiol.*, 122:1129-1136.

25. Hu, C.A., Delauney, A.J. and Verma, D.P. 1992. A bifunctional enzyme (Δ^1-pyrroline-5-carboxylate synthetase) catalyzes the first two steps in proline biosynthesis in plants. *Proc. Natl. Acad. Sci.*, 89:9354- 9358.

26. Igarashi, Y., Yoshiba, Y., Sanada, Y., Yamaguchi-Shinozaki, K., Wada, K. and Shinozaki, K. 1997. Characterization of the gene for Δ^1-pyrroline-5-carboxylate synthetase and correlation between the expression of the gene and salt tolerance in *Oryza sativa L. Plant Mol. Biol.*, 33:857-865.

27. Karthikeyan, A., Pandian, S.K. and Ramesh, M. 2011. Transgenic indica rice cv. ADT 43 expressing a Δ^1- pyrroline-5-carboxylate synthetase (*P5CS*) gene fromVigna aconitifoliademonstrates salt tolerance. *Plant Cell Tiss Organ Cult (PCTOC)*, 107:383-395.

28. Kasuga, M., Liu, Q., Miura, S., Yamaguchi-Shinozaki, K. and Shinozaki, K. 1999. Improving
plant drought, salt and freezing tolerance by gene
transfer of a single stress- inducible transcription
factor. *Nat. Biotech.*, 17:287-291.
29. Kishor, P.K., Hong, Z., Miao, G. H., Hu, C.-A.A.
and Verma, D.S. 1995. Overexpression of del-
ta1-Pyrroline-5- Carboxylate Synthetase Increases
Proline Production and Confers Osmotolerance in
Transgenic Plants. *Plant Physiol.*, 108:1387-1394.
30. Kishor, P.K., Sangam, S., Amrutha, R.N., Sri Lax-
mi, P., Naidu, K.R., Rao, K.S. and Sreenivasulu,
N. 2005. Regulation of proline biosynthesis, degradation, uptake and transport in higher plants:
Its implications in plant growth and abiotic stress
tolerance. *Curr. Sci.*, 88:424- 438.
31. Kiyosue, T., Yoshiba, Y., Yamaguchi-Shinozaki,
K. and Shinozaki, K. 1996. A nuclear gene encoding mitochondrial proline dehydrogenase,an enzyme involved in proline metabolism, is upregulated by proline but downregulated by dehydration
in *Arabidopsis. Plant Cell Online*, 8:1323-1335.
32. Kumar, V., Shriram, V., Kishor, P.K., Jawali, N.
and Shitole, M.G. 2010. Enhanced proline accumulation and salt stress tolerance of trans-
genicindica rice by over- expressing *P5CSF129A*
gene. *Plant Biotech. Rep.*, 4:37- 48.
33. Lehmann, S., Funck, D., Szabados, L. and Rent-
sch, D. 2010. Proline metabolism and transport in
plant development. *Amino Acids*, 39:949-962.
34. Liang, X., Zhang, L., Natarajan, S.K. and Becker,
D.F. 2013. Proline Mechanisms of Stress Survival.
Antioxidants Redox Signal., 19:998-1011.
35. Mani, S., Van de Cotte, B., Van Montagu, M. and
Verbruggen, N. 2002. Altered levels of proline dehydrogenase cause hypersensitivity to proline and
its analogs in Arabidopsis. *Plant Physiol.*, 12:73-
83.
36. Matoh, T., Watanabe, J. and Takahashi, E. 1987.
Sodium, potassium, chloride and betaine concentrations in isolated vacuoles from salt-grown Atri-
plex gmelini leaves. *Plant Physiol.*, 84:173-177.
37. Molinari, H. C., Marur, C. J., Filho, J. B., Kobayashi, A. K., Pileggi, M., Júnior, R. L. and Vie-
ira, L. E. 2004. Osmotic adjustment in transgenic
citrus rootstock Carrizo citrange (*Citrus sinensis*
Osb. x *Poncirus trifoliata* L. Raf.) overproducing
proline. *Plant Sci.*, 167:1375-1381.
38. Nakashima, K., Jan, A., Todaka, D., Maruyama,
K., Goto, S., Shinozaki, K. and Yamaguchi-Shinozaki, K. 2014. Comparative functional analysis
of six drought-responsive promoters in transgenic
rice. *Planta*, 239:47-60.
39. Naliwajski, M. R. and Sklodowska, M. 2014. Proline and its metabolism enzymes in cucumber cell
cultures during acclimation to salinity. *Protoplasma*, 251:201-209.
40. Nanjo, T., Fujita, M., Seki, M., Kato, T., Tabata,
S. and Shinozaki, K. 2003. Toxicity of free proline
revealed in an Arabidopsis T-DNA-tagged mutant
deficient in proline dehydrogenase. *Plant Cell
Physiol.*, 44:541-548.
41. Nanjo, T., Kobayashi, M., Yoshiba, Y., Sanada, Y.,
Wada, K., Tsukaya, H. and Shinozaki, K. 1999.
Biological functions of proline in morphogenesis
and osmotolerance revealed in antisense transgenic *Arabidopsis thaliana. The Plant J*, 18:185-193.
42. Peng, Z., Lu, Q. and Verma, D. S. 1996. Reciprocal regulation of Δ^1-pyrroline-5-carboxylate synthetase and proline dehydrogenase genes controls
proline levels during and after osmotic stress in
plants. *Mol. General Genet.*, 253:334-341.
43. Porcel, R., Azcón, R. and Ruiz-Lozano, J. M.
2004. Evaluation of the role of genes encoding for
Δ^1-pyrroline- 5-carboxylate synthetase (*P5CS*)
during drought stress in arbuscular mycorrhizal
Glycine max and Lactuca sativa plants. *Physiol.
Mol. Plant Pathol.*, 65:211-221.
44. Pospisilova, J., Haisel, D. and Vankova, R. 2011.
Responses of Transgenic Tobacco Plants with Increased Proline Content to Drought and/or Heat
Stress. *Am. J. Plant Sci.* 2:318-324.
45. Rajendrakumar, C.S., Reddy, B.V. and Reddy,
A.R. 1994. Proline–protein interactions: Protection of structural and functional integrity of M4
lactate dehydrogenase. *Biochem. Biophys. Res.
Commun.*, 201:957-963.
46. Rastgar J. F., Yamchi A., Hajirezaei M. and
Karkhane A.A. 2011. Analysis of Growth and
Germination Stage of T2 Generation of *P5CS*
Over- expressing Tobacco Plant *Nicotiana tabac-
cum* cv. Xanthi Exposed to Osmotic Stress. *African J. Biotech.* 10:8539-8552
47. Richards, E. J. 2006. Inherited epigenetic varia-
tion- revisiting soft inheritance. *Nat. Rev. Genet.*,
7:395-401.
48. Rout, N. P. and Shaw, B. P. 1998. Salinity tolerance in aquatic macrophytes: probable role of proline, the enzymes involved in its synthesis and C4
metabolism. *Plant Sci.*, 136:121-130.
49. Samach, A., Onouchi, H., Gold, S.E., Ditta, G.S.,
Schwarz-Sommer, Z., Yanofsky, M.F. and Coup-
land, G. 2000. Distinct Roles of CONSTANS Target Genes in Reproductive Development of Arabidopsis. *Science*, 288:1613-1616.
50. Saradhi, P.P., AliaArora, S. and Prasad, K.S. 1995.
Proline accumulates in plants exposed to UV radiation and protects them against UV induced
peroxidation. *Biochem. Biophys. Res. Commun.*,
209:1-5.
51. Savouré, A., Jaoua, S., Hua, X.-J., Ardiles, W.,
Montagu, M.V. and Verbruggen, N. 1995. Isolation, characterization and chromosomal location
of a gene encoding the Δ^1- pyrroline-5-carboxyl-
ate synthetase in Arabidopsis thaliana. *FEBS Letters*, 372:13-19.

52. Sawahel, W.A. and Hassan, A.H. 2002. Generation of transgenic wheat plants producing high levels of the osmoprotectant proline. *Biotechnol. Letters,* 24:721-725.

53. Silva-Ortega, C.O., Ochoa-Alfaro, A.E., Reyes-Agüero, J.A., Aguado-Santacruz, G.A. and Jiménez-Bremont, J.F. 2008. Salt stress increases the expression of *P5CS* gene and induces proline accumulation in cactus pear. *Plant Physiol. Biochem.,* 46:82-92.

54. Sivakumar, P., Sharmila, P. and Pardha Saradhi, P. 2000. Proline alleviates salt-stress induced enhancement in ribulose-1,5-bisphosphate oxygenase activity. *Biochem. Biophys. Res. Commun.,* 279:512-515.

55. Smith, C.J., Deutch, A.H. and Rushlow, K.E. 1984. Purification and characteristics of a gamma-glutamyl kinase involved in Escherichia coli proline biosynthesis. *J. Bacteriol.,* 157:545-551.

56. Stines, A.P., Naylor, D.J., Hoj, P.B. and van Heeswijck, R. 1999. Proline Accumulation in Developing Grapevine Fruit Occurs Independently of Changes in the Levels of Δ^1-Pyrroline-5-Carboxylate Synthetase mRNA or Protein. *Plant Physiol.,* 120:923-931.

57. Strizhov, N., Abraham, E., Ökrész, L., Blickling, S., Zilberstein, A., Schell, J. and Szabados, L. 1997. Differential expression of two *P5CS* genes controlling proline accumulation during salt-stress requires ABA and is regulated by ABA1, ABI1 and AXR2 in Arabidopsis. *The Plant J.,* 12:557-569.

58. Su, J. and Wu, R. 2004. Stress-inducible synthesis of proline in transgenic rice confers faster growth under stress conditions than that with constitutive synthesis. *Plant Science,* 166:941-948.

59. Su, J., Shen, Q., Ho, T. D. and Wu, R. 1998. Dehydration- stress-regulated transgene expression in stably transformed rice plants. *Plant Physiol.,* 117:913-922.

60. Su, M., Li, X.F., Ma, X.Y., Peng, X.J., Zhao, A.G., Cheng, L.Q. and Liu, G.S. 2011. Cloning two *P5CS* genes from bioenergy sorghum and their expression profiles under abiotic stresses and MeJA treatment. *Plant Sci.,* 181:652-659.

61. Surekha, C., Kumari, K.N., Aruna, L.V., Suneetha, G., Arundhati, A. and Kishor, P.K. 2014. Expression of the Vigna aconitifolia *P5CSF129A* gene in transgenic pigeonpea enhances proline accumulation and salt tolerance. *Plant Cell Tiss. Organ Cult. (PCTOC),* 116:27- 36.

62. Szabados, L. and Savouré, A. 2009. Proline: a multifunctional amino acid. *Trends Plant Sci.,* 15:89-97.

63. Szoke, A., Miao, G.H., Hong, Z. and Verma, D.S. 1992. Subcellular location of Δ^1-pyrroline-5-carboxylate reductase in root/nodule and leaf of soybean. *Plant Physiol,* 99:1642-1649.

64. Thiery, L., Leprince, A.-S., Lefebvre, D., Ghars, M., Debarbieux, E. and Savouré, A. 2004. Phospholipase D Is a Negative Regulator of Proline Biosynthesis in Arabidopsis thaliana. *J. Biol. Chem.,* 279:14812-14818.

65. Vaucheret, H., Béclin, C. and Fagard, M. 2001. Post- transcriptional gene silencing in plants. *J. Cell Sci.,* 114:3083-3091.

66. Verbruggen, N. and Hermans, C. 2008. Proline accumulation in plants: a review. *Amino Acids,* 35:753- 759.

67. Verslues, P. E. and Sharp, R. E. 1999. Proline accumulation in maize (*Zea mays L.*) primary roots at low water potentials. II. Metabolic source of increased proline deposition in the elongation zone. *Plant Physiol.,* 119:1349-1360.

68. Verslues, P.E., Lasky, J.R., Juenger, T.E., Liu, T.W. and Kumar, M.N. 2014. Genome-Wide Association Mapping Combined with Reverse Genetics Identifies New Effectors of Low Water Potential-Induced Proline Accumulation in Arabidopsis. *Plant Physiol.,* 164:144- 159.

69. Wang, Z.Q., Yuan, Y.Z., Ou, J.Q., Lin, Q.H. and Zhang, C.F. 2007. Glutamine synthetase and glutamate dehydrogenase contribute differentially to proline accumulation in leaves of wheat (*Triticum aestivum*) seedlings exposed to different salinity. *J. Plant Physiol.,* 164:695-701.

70. Werner, J.E. and Finkelstein, R.R. 1995. Arabidopsis mutants with reduced response to NaCl and osmotic stress. *Physiol. Plant,* 93:659-666.

71. Xin, Z. 1998. eskimo1 mutants of Arabidopsisare constitutively freezing tolerant. *PNAS,* 95:7799-7804.

72. Yamada, m., Morishita, H., Urano, K., Shinozaki, N., Yamaguchi-Shinozaki, K., Shinozaki, K. and Yoshiba, Y. 2005. Effects of free proline accumulation in petunias under drought stress. *J. Exp. Bot.,* 56:1975-1981.

73. Yamchi A., Rastgar J. F., Mousavi A., Karkhanei A.A. and Renu. 2007 Proline Accumulation in Transgenic Tobacco as a Result of Expression of Arabidopsis Δ^1- Pyrroline-5-carboxylate synthetase (*P5CS*) During Osmotic Stress. *J. Plant Biochem. Biotech.* 16:9-15

74. Yoo, J.H., Park, C.Y., Kim, J.C., Heo, W.D., Cheong, M.S., Park, H.C. and Kim, M.C. 2005. Direct interaction of a divergent CaM isoform and the transcription factor, MYB2, enhances salt tolerance in Arabidopsis. *J. Biol. Chem.,* 280:3697-3706.

75. Yoshiba, Y., Kiyosue, T., Katagiri, T., Ueda, H., Mizoguch, T., Yamaguchi-Shinozaki, K. and Shinozaki, K. 1995. Correlation between the induction of a gene for Δ^1-pyrroline-5-carboxylate synthetase and the accumulation of proline in Arabidopsis thalianaunder osmotic stress. *Plant J.,* 7:751-760.

76. Yoshiba, Y., Kiyosue, T., Nakashima, K., Yamaguchi- Shinozaki, K. and Shinozaki, K. 1997. Regulation of Levels of Proline as an Osmolyte in Plants under Water Stress. *Plant Cell Physiol.,* 38:1095-1102.

77. Zhang, C.S., Lu, Q. and Verma, D.S. 1997. Characterization of Δ^1-pyrroline-5-carboxylate synthetase gene promoter in transgenic *Arabidopsis thaliana* subjected to water stress. *Plant Sci.,* 129:81-89.

78. Zhang, C.Y., Wang, N.N., Zhang, Y.H., Feng, Q.Z., Yang, C.W. and Liu, B. 2013. DNA methylation involved in proline accumulation in response to osmotic stress in rice (*Oryza sativa*). *Genet. Mol. Res.,* 12:1269-1277.

79. Zhang, C.S., Lu, Q. and Verma, D.S. 1995. Removal of Feedback Inhibition of Δ^1-pyrroline-5-carboxylate synthetase, a Bifunctional Enzyme Catalyzing the First Two Steps of Proline Biosynthesis in Plants. *J. Biol. Chem.,* 270:20491-20496.

80. Zheng, L., Dang, Z., Li, H., Zhang, H., Wu, S. and Wang, Y. 2014. Isolation and characterization of Δ^1-pyrroline-5- carboxylate synthetase (*NtP5CS*) from *Nitraria tangutorum* Bobr. and functional comparison with its *Arabidopsis* homologue. *Mol. Biol. Rep.,* 41:563-572.

81. Zhu, B., Su, J., Chang, M., Verma, D. S., Fan, Y.-L. and Wu, R. 1998. Overexpression of a Δ^1-pyrroline-5- carboxylate synthetase gene and analysis of tolerance to water-and salt-stress in transgenic rice. *Plant Sci.,* 139:41- 48.

PERMISSIONS

All chapters in this book were first published in JPMB, by Genetics and Agricultural Biotechnology Institute of Tabarestan; hereby published with permission under the Creative Commons Attribution License or equivalent. Every chapter published in this book has been scrutinized by our experts. Their significance has been extensively debated. The topics covered herein carry significant findings which will fuel the growth of the discipline. They may even be implemented as practical applications or may be referred to as a beginning point for another development.

The contributors of this book come from diverse backgrounds, making this book a truly international effort. This book will bring forth new frontiers with its revolutionizing research information and detailed analysis of the nascent developments around the world.

We would like to thank all the contributing authors for lending their expertise to make the book truly unique. They have played a crucial role in the development of this book. Without their invaluable contributions this book wouldn't have been possible. They have made vital efforts to compile up to date information on the varied aspects of this subject to make this book a valuable addition to the collection of many professionals and students.

This book was conceptualized with the vision of imparting up-to-date information and advanced data in this field. To ensure the same, a matchless editorial board was set up. Every individual on the board went through rigorous rounds of assessment to prove their worth. After which they invested a large part of their time researching and compiling the most relevant data for our readers.

The editorial board has been involved in producing this book since its inception. They have spent rigorous hours researching and exploring the diverse topics which have resulted in the successful publishing of this book. They have passed on their knowledge of decades through this book. To expedite this challenging task, the publisher supported the team at every step. A small team of assistant editors was also appointed to further simplify the editing procedure and attain best results for the readers.

Apart from the editorial board, the designing team has also invested a significant amount of their time in understanding the subject and creating the most relevant covers. They scrutinized every image to scout for the most suitable representation of the subject and create an appropriate cover for the book.

The publishing team has been an ardent support to the editorial, designing and production team. Their endless efforts to recruit the best for this project, has resulted in the accomplishment of this book. They are a veteran in the field of academics and their pool of knowledge is as vast as their experience in printing. Their expertise and guidance has proved useful at every step. Their uncompromising quality standards have made this book an exceptional effort. Their encouragement from time to time has been an inspiration for everyone.

The publisher and the editorial board hope that this book will prove to be a valuable piece of knowledge for researchers, students, practitioners and scholars across the globe.

LIST OF CONTRIBUTORS

B. Dolatabadi
Department of Agronomy and Plant Breeding, Sari University of Agricultural Sciences and Natural Resources, Sari, Iran
Genetics and Agricultural Biotechnology Institute of Tabarestan, Sari University of Agricultural Sciences and Natural Resources, Sari, Iran

Gh. Ranjbar
Department of Agronomy and Plant Breeding, Sari University of Agricultural Sciences and Natural Resources, Sari, Iran

M. Tohidfar
Agricultural Biotechnology Research Institute of Iran, Karaj, Iran

A. Dehestani
Genetics and Agricultural Biotechnology Institute of Tabarestan, Sari University of Agricultural Sciences and Natural Resources, Sari, Iran

Farnaz Goodarzi and Abbas Hassani
Department of Horticulture, Urmia University, Urmia, Iran

Reza Darvishzadeh
Department of Plant Breeding and Biotechnology, Urmia University, Urmia, Iran
Institute of Biotechnology, Urmia University, Urmia, Iran

M. Alizadeh and H. Askari
Department of Biotechnology, Faculty of New Technologies Engineering, Shahid Beheshti University, G. C., Evin, Tehran, Iran

S. Awasthi and J. P. Lal
Department of Genetics and Plant Breeding, Institute of Agricultural Sciences, Banaras Hindu University, Varanasi, India

N. Eyvaznejad
Department of Plant Breeding and Biotechnology, Urmia University, Urmia, Iran

Gholamhossein Hosseini
Cotton Research Institute of Iran

Hajar Abedinpour, Nad Ali Babaeian Jelodar and Gholam Ali Ranjbar
Department of Plant Breeding, Sari Agricultural Sciences and Natural Resources University, Sari, Iran

Behrouz Golein
Iran Citrus Research Institute, Ramsar, Iran

S. H. Hashemi-Petroudi and Gh. A. Nematzadeh
Genetic and Agricultural Biotechnology Institute of Tabarestan, Sari University of Agricultural Sciences and Natural Resources, Iran

H. Askari
Department of Biotechnology, Faculty of New Technologies and Energy Engineering, Shahid Beheshti University, G. C., Tehran, Iran

S. Ghahary
Department of Agronomy, Tarbiat Modares University, Iran

Mahboubeh Davoudi Pahnekolayi, Ali Tehranifar and Mahmoud Shoor
Horticultural Department, College of Agriculture, Ferdowsi University of Mashhad, Mashhad, Iran

Leila Samiei
Research Center for Plant Sciences, Ferdowsi University of Mashhad, Mashhad, Iran

M. Habibi, S. Malekzadeh-Shafaroudi, H. Marashi, N. Moshtaghi and M. Nassiri
Department of Biotechnology and plant breeding, Faculty of Agriculture, Ferdowsi University of Mashhad, Mashhad, Iran

S. Zibaee
Razi Vaccine and Serum Research Institute, Mashhad, Iran

M. Arefrad, Gh. Nematzadeh and M. Karimi
Genetics and Agricultural Biotechnology Institute of Tabarestan (GABIT), Sari, Mazandaran, Iran

N. Babaian Jelodar and S. K. Kazemitabar
Department of Biotechnology, Sari Agricultural Sciences and Natural Resources University (SANRU). Sari, Mazandaran, Iran

Behzad Shahin Kaleybar, Sara Kabirnattaj and Ghorban Ali Nematzadeh
Genetics and Agricultural Biotechnology Institute of Tabarestan (GABIT), Sari Agricultural Sciences and Natural Resources University, Sari, Iran

Seyyed Kamal Kazemitabar
Department of Agronomy and Plant Breeding, Sari Agricultural Sciences and Natural Resources University, Sari, Iran

Seyede Mona Salim Bahrami
Department of Biology, University of Mazandaran, Babolsar, Iran

E. Mehrazar and A. Izadi-Darbandi
Department of Agronomy and Plant Breeding Sciences, College of Aburaihan, University of Tehran, Tehran, Iran

M. Mohammadi and G. Najafian
Seed and Plant Improvement Institute, Cereals Research Department, Karaj, Iran

N. Moradi
Agricultural Biotechnology, Genetics and Agricultural Biotechnology Institute of Tabarestan. (GABIT), Sari, Iran

H. Badakhshan and Gh. Mirzaghaderi
Department of Biotechnology, Faculty of Agriculture, University of Kurdistan, Sanadaj, Iran

H. Mohammadzadeh and M. R Zakeri
Master Degree in Agricultural Biotechnology

Zeinab Mohammadi, R. Mollaheydari Bafghi and Atefeh Sabouri
Department of Agronomy & Plant Breeding, University of Guilan, Rasht, Iran

Sedigheh Mousanejad and A. Baghizadeh
Department of Plant Protection, University of Guilan, Rasht, Iran
Department of Biotechnology, Institute of Sciences and High Technology and Environmental Sciences, Graduate University of Advanced Technology, Kerman-Iran

Gh. Mohammadi-Nejad
Department of Agronomy and plant Breeding, College of Agriculture and Center of Excellence for Abiotic Stress in Cereal Crop., Shahid Bahonar University of Kerman-Iran

B. Nakhoda
Department of System Biology, Agricultural Biotechnology Research Institute of Iran, Karaj–Iran

Masoud Fakhrfeshani
Department of Plant Breeding and Biotechnology, Ferdowsi University of Mashhad, Iran
Department of Plant Biotechnology, University of Jahrom, Iran

Farajollah Shahriari-Ahmadi and Nasrin Moshtaghi
Department of Plant Breeding and Biotechnology, Ferdowsi University of Mashhad, Iran

Ali Niazi
Institute of Biotechnology, Collage of Agriculture, Shiraz University, Iran

Mohammad Zare-Mehrjerdi
Shirvan Higher Education Complex, Iran

Khalil Malekzadeh and Amin Mirshamsi-Kakhaki
Department of Biotechnology and Plant Breeding, Faculty of Agriculture, Ferdowsi University of Mashhad, Iran

Sahand Amini
Department of Agricultural Biotechnology, College of Agriculture Isfahan University of Technology, Isfahan, Iran

Cyrus Ghobadi
Department of Horticultural Sciences, College of Agriculture Isfahan University of Technology, Isfahan, Iran

Ahad Yamchi
Department of Plant Breeding and Biotechnology, College of Plant Production Gorgan University of Agriculture Science and Natural Recourses, Gorgan, Iran

Index

www.ingramcontent.com/pod-product-compliance
Lightning Source LLC
Chambersburg PA
CBHW062002190326
41458CB00009B/2945